AF166402

Communications
in Computer and Information Science 2205

Series Editors

Gang Li , *School of Information Technology, Deakin University, Burwood, VIC, Australia*
Joaquim Filipe , *Polytechnic Institute of Setúbal, Setúbal, Portugal*
Zhiwei Xu, *Chinese Academy of Sciences, Beijing, China*

Rationale

The CCIS series is devoted to the publication of proceedings of computer science conferences. Its aim is to efficiently disseminate original research results in informatics in printed and electronic form. While the focus is on publication of peer-reviewed full papers presenting mature work, inclusion of reviewed short papers reporting on work in progress is welcome, too. Besides globally relevant meetings with internationally representative program committees guaranteeing a strict peer-reviewing and paper selection process, conferences run by societies or of high regional or national relevance are also considered for publication.

Topics

The topical scope of CCIS spans the entire spectrum of informatics ranging from foundational topics in the theory of computing to information and communications science and technology and a broad variety of interdisciplinary application fields.

Information for Volume Editors and Authors

Publication in CCIS is free of charge. No royalties are paid, however, we offer registered conference participants temporary free access to the online version of the conference proceedings on SpringerLink (http://link.springer.com) by means of an http referrer from the conference website and/or a number of complimentary printed copies, as specified in the official acceptance email of the event.

CCIS proceedings can be published in time for distribution at conferences or as post-proceedings, and delivered in the form of printed books and/or electronically as USBs and/or e-content licenses for accessing proceedings at SpringerLink. Furthermore, CCIS proceedings are included in the CCIS electronic book series hosted in the SpringerLink digital library at http://link.springer.com/bookseries/7899. Conferences publishing in CCIS are allowed to use Online Conference Service (OCS) for managing the whole proceedings lifecycle (from submission and reviewing to preparing for publication) free of charge.

Publication process

The language of publication is exclusively English. Authors publishing in CCIS have to sign the Springer CCIS copyright transfer form, however, they are free to use their material published in CCIS for substantially changed, more elaborate subsequent publications elsewhere. For the preparation of the camera-ready papers/files, authors have to strictly adhere to the Springer CCIS Authors' Instructions and are strongly encouraged to use the CCIS LaTeX style files or templates.

Abstracting/Indexing

CCIS is abstracted/indexed in DBLP, Google Scholar, EI-Compendex, Mathematical Reviews, SCImago, Scopus. CCIS volumes are also submitted for the inclusion in ISI Proceedings.

How to start

To start the evaluation of your proposal for inclusion in the CCIS series, please send an e-mail to ccis@springer.com.

A. Mirzazadeh · Zohreh Molamohamadi ·
Efran Babaee Tirkolaee ·
Gerhard-Wilhelm Weber · Janny Leung
Editors

Optimization and Data Science in Industrial Engineering

First International Conference, ODSIE 2023
Istanbul, Turkey, November 16–17, 2023
Proceedings, Part II

 Springer

Editors
A. Mirzazadeh ⓘ
Kharazmi University
Tehran, Iran

Zohreh Molamohamadi ⓘ
Kharazmi University
Tehran, Iran

Efran Babaee Tirkolaee ⓘ
Istinye University
Istanbul, Türkiye

Gerhard-Wilhelm Weber ⓘ
Poznań University of Technology
Poznań, Poland

Janny Leung ⓘ
University of Macau
Macao, China

ISSN 1865-0929   ISSN 1865-0937 (electronic)
Communications in Computer and Information Science
ISBN 978-3-031-81457-0   ISBN 978-3-031-81458-7 (eBook)
https://doi.org/10.1007/978-3-031-81458-7

This Springer imprint is published by the registered company Springer Nature Switzerland AG
The registered company address is: Gewerbestrasse 11, 6330 Cham, Switzerland

If disposing of this product, please recycle the paper.

Preface

The International Conference on Optimization and Data Science in Industrial Engineering (ODSIE 2023) was held virtually in Istanbul, Turkey on Nov. 16–17, 2023 by Istinye University. It provided an energetic knowledge-transferring atmosphere for participants (as several comments revealed).

ODSIE 2023 attracted the attention of students and professionals internationally. The subjects covered included, but were not limited to, "Industry 4.0, IoT and smart manufacturing", "Digital twin and virtual commissioning", "Sustainable and smart cities", "Artificial intelligence and expert systems", "Metaheuristic algorithms with applications in IE", "Machine learning algorithms", "Big data analytics and Data mining", "Robotic process automation", "Decision support systems", "E-Government, E-Commerce and E-Learning", "Supply chain design and logistics", "Optimization- and Data Science-based case studies of manufacturing/service industries", "Quantitative finance and risk modelling", and "Other fields of study related to Optimization and Data Science with applications in IE".

ODSIE 2023 was honored to be enriched by outstanding keynote speakers and workshop organizers from UK, Tunisia, Saudi Arabia, India, Pakistan, and Iran. In this event, fifteen universities from USA, UK, Czech Republic, Tunisia, Malaysia, Algeria, India, Morocco, and Turkey were present as scientific sponsors.

The conference included participation of 50 countries. The geographical diversity of international scientific committee members was from 27 countries.

The conference team received 311 English and Turkish papers, which were reviewed by at least three international reviewers (single-blind reviews). Considering the reviewers' comments in the first round of review, the papers were reviewed once more with stricter criteria to select the most appropriate ones for Springer's publication. Finally, the two-round review process resulted in the selection of 35 papers (around 11%) for Springer.

The review criteria in this step were: content; originality; relevance; contribution to the professional literature; significance and potential impact of the paper; language accuracy; study validity; accuracy of methodology and analysis; paper organization, required relevant data, citations and references; adequate reference to background information; consistency of references, symbols and units throughout the paper; quality and clarity of tables and figures.

The papers in ODSIE 2023 were presented in 21 panel sessions: "Optimization Problems Related to Logistics, Transportation and Manufacturing Systems", "Operations Research in Optimization", "Artificial Intelligence and Expert Systems", "Sustainable Business", "From Data to Knowledge - Analysis for Strategies and Challenges", "Experiment and Optimization in Research", "Industry 4.0, IoT and Smart Manufacturing", "Computational Thinking and Artificial Intelligence in Industrial Engineering", "Inventory Planning, Production and Scheduling", "Decision Support Systems",

"Other Fields of Study Related to Optimization and Data Science", "Machine Learning Solutions for Engineering Problems", "Data Analytics in Intelligent Transportation Systems", "Project Management and Fintech", "Advancing Innovation and Collaboration in Science, Engineering and Information Technology", "Artificial Intelligence for Biomedicine and Healthcare", "Machine Learning Algorithms", "Advancing Operations Management through Optimization and Data Science", "E-Government, E-Commerce and E-Learning" and "Analytics for Sustainable Society".

ODSIE 2023 included five applied workshops by outstanding lecturers with good-sized audiences. The workshop subjects were very well received by the participants, and were entitled: "From Sustainability to Circularity: Transforming Supply Chains for an Eco-centric Economy"; "Unlocking the Potential of Drones: Exploring Diverse Applications"; "A Multidisciplinary Landscape of Data Sciences"; "Indispensable Arbitration of Leaders Is Truly Conducive for the Utmost Clarity of an Organization"; and "Leading the Charge: Optimizing Sustainable Finance Strategies Through Data-Driven Decision-Making in Industrial Engineering".

December 2023

A. Mirzazadeh
Zohreh Molamohamadi
Efran Babaee Tirkolaee
Gerhard-Wilhelm Weber
Janny Leung

Organization

Conference Chairs

Abolfazl Mirzazadeh	Kharazmi University, Iran
Erfan Babaee Tirkolaee	Istinye University, Istanbul, Turkey
Nadi Serhan Aydın (Co-chair)	Istinye University, Istanbul, Turkey

Istinye University

Erkan Ibiş	Rector
Hatice Gülen	Vice Rector
Mehmet Alper Tunga	Dean of the Faculty of Engineering and Natural Sciences

Conference Coordinators

Leila Chehreghani (Planning Manager)	Ulster University, UK
Zohreh Molamohamadi (Scientific Coordinator)	Kharazmi University, Iran

Program Committee Chairs

Efran Babaee Tirkolaee	Istinye University, Turkey
Janny Leung	University of Macao, China President of IFORS
Zohreh Molamohamadi	Kharazmi University, Iran
A. Mirzazadeh	Kharazmi University, Iran
Gerhard-Wilhelm Weber	Poznań University of Technology, Poland

Editorial Committee Members

Alexandre Dolgui	IMT Atlantique, France
Arpan Kumar Kar	Indian Institute of Technology Delhi, India

Safa Bhar Layeb	University of Tunis El Manar, Tunisia
Kathryn Stecke	University of Texas at Dallas, Naveen Jindal School of Management, USA
Kok Lay Teo	JIMO Editor in Chief, Sunway University, Malaysia, Curtin University, Australia
Chefi Triki	University of Kent, UK

Steering Committee Members

Erfan Babaee Tirkolaee	Istinye University, Turkey
Babek Erdebilli	Yildirim Beyazit University, Turkey
Josef Jablonsky	Prague University of Economics and Business, Czech Republic
Janny M. Y. Leung	University of Macao, China President of IFORS
Taicir Moalla Loukil	University of Sfax, Tunisia President of The Tunisian Operational Research Society
A. Mirzazadeh	Kharazmi University, Iran
Mehmet Alper Tunga	Istinye University, Turkey
Gerhard-Wilhelm Weber	Poznań University of Technology, Poland

PhD Thesis Competition Jury

Saliha Karadayı Üsta (Chair)	Istinye University, Turkey
Emre Çakmak (Co-chair)	Istinye University, Turkey
Tatiana Tchemisova (Member)	University of Aveiro, Portugal

Scientific Committee Members

Adeyinka Peter Ajayi	Redeemer's University, Nigeria
Sadia Samar Ali	King Abdulaziz University, Saudi Arabia
Tofigh AllahViranloo	Istinye University, Turkey
Bernardo Almada-Lobo	Porto University, Portugal
Sırma Zeynep Alparslan Gök	Suleyman Demirel University, Turkey
Fayçal Belkaid	Abou Bekr Belkaid University of Tlemcen, Algeria
Sanjib Biswas	Calcutta Business School, India

Marilisa Botte	University of Naples, Italy
Jaouad Boukachour	University of Le Havre-Normandy, France
Emre Çakmak	Istinye University, Turkey
Yavuz Can	Friedrich-Alexander University, Germany
Leopoldo Eduardo Cárdenas-Barrón	Tecnológico de Monterrey, Mexico
Aybike Özyüksel Çiftçioğlu	Manisa Celal Bayar University, Turkey
Dursun Delen	Oklahoma State University, USA
Elif Kılıç Delice	Ataturk University, Turkey
Souhail Dhouib	University of Sfax, Tunisia
Serap Ergun	Isparta University of Applied Sciences, Turkey
Paulina Golinska	Poznań University of Technology, Poland
Ömer Faruk Görçün	Kadir Has University, Turkey
Sylwia Gwoździewicz	Jacob of Paradies University, Poland
Mostafa Hajiaghaei-Keshteli	Tecnológico de Monterrey, Mexico
Sarfaraz Hashemkhani Zolfani	Northern Catholic University, Chile
Marwa Hasni	University of Sfax, Tunisia
Dinh Tran Ngoc Huy	International University of Japan, Japan
Saliha Karadayi Üsta	Istinye University, Turkey
Michael G. Kay	North Carolina State University, USA
Selçuk Korucuk	Giresun University, Turkey
Eloisa Macedo	University of Aveiro, Portugal
Beata Mrugalska	Poznań University of Technology, Poland
Ammar Odeh	King Hussein School for Computing Sciences, Jordan
Fabiana Lucena Oliveira	Amazonas State University (UEA), Brazil
Mustapha Oudani	International University of Rabat, Morocco
Yavuz Selim Ozdemir	Ankara Science University, Turkey
Çağrı Özgün Kibiroğlu	Halic University, Turkey
Dragan Pamučar	University of Belgrade, Serbia
Stefan Pickl	Bundeswehr University, Germany
Sankar Kumar Roy	Vidyasagar University, India
Rubén Ruiz Garcia	Polytechnic University of Valencia, Spain
İlkay Saraçoğlu	Halic University, Turkey
Yaroslav Sergeyev	University of Calabria, Italy
Vladimir Simic	University of Belgrade, Serbia
Esra Sipahi Dungul	Aksaray University, Turkey
Renata Sotirov	Tilburg University, The Netherlands
Majid Tavana	La Salle University, USA
Tatiana Tchemisova	University of Aveiro, Portugal
Hadi Rezaei Vandchali	University of Tasmania, Australia
Ibrahim Yilmaz	Yildirim Beyazit University, Turkey

Keynote Speakers

Chefi Triki	University of Kent, UK
Safa Bhar Layeb	LR-OASIS, National Engineering School of Tunis, University of Tunis El Manar, Tunisia

Workshop Organizers

Alireza Goli	University of Isfahan, Iran
Sadia Samar Ali	King Abdulaziz University, Saudi Arabia
Maria Anjum	Lahore College for Women University, Pakistan
Rudrarup Gupta	Tagore School of Rural Development and Agriculture Management, Kalyani University, India
Sameera Fernandes	Century Financial, UAE and Garden City University, India

Executive Committee Members

Kutay Alada	Istinye University, Turkey
Anderson Chukwudi Oki	Istinye University, Turkey
Gizem Karatepe	Istinye University, Turkey
Alper Koçak	Istinye University, Turkey
Muhammet Ali Şencan	Istinye University, Turkey
Sadaf Soltani	IT Expert in DevOps University, Iran
Cansu Tayar	Istinye University, Turkey

Reviewers

Otman Abdoun	University of Abdelmalek Essaadi, Morocco
Mohamed Abdelgadir	University of Technology Malaysia, Malaysia
Salma Abid	University of Rabat, Morocco
Mohamed Nezar Abourraja	Le Havre Normandy University, France
Negar Afzali Behbahani	Islamic Azad University, Iran
Maher Agi	Rennes School of Business, France
Hadi Ahmed	Cairo University, Egypt
Sara Ahmed	Concordia University, Canada
Ghada Alhudhud	King Saud University, Saudi Arabia
Aylin Alkaya	Nevsehir Haci Bektas Veli University, Turkey

Richard Allmendinger	University of Manchester, UK
Juscelino Almeida Dias	Technical University of Lisbon, Portugal
Shabnam Amirnezhad Barough	Yıldırım Beyazit University, Turkey
El-Saed Ammar	Tanta University, Egypt
Le Thi Diep Anh	National Economics University, Vietnam
Hichem Aouag	Batna University, Algeria
Mst. Anjuman Ara	Bangladesh Army University of Science and Technology, Bangladesh
Selcen Aslan Ozsahin	TOBB University of Economics and Technology, Turkey
Mamoon Atout	Dubai Electricity and Water Authority, UAE
Nadi Serhan Aydin	Istinye University, Turkey
Salih Aytar	Suleyman Demirel University, Turkey
Youssef Baddi	Chouaib Doukkali University, Morocco
Rahmi Baki	Aksaray University, Turkey
Srinivasan Balan	North Carolina State University, USA
Igor Barahona	King Fahd University of Petroleum & Minerals, Saudi Arabia
Haripriya Barman	Vidyasagar University, India
Faycal Belkaid	University of Lorraine, France
Adil Bellabdaoui	National School of Computer Science and Systems Analysis, Morocco
Mohamed Amine Ben Rabia	National School of Computer Science and Systems Analysis, Morocco
Abderaouf Benghalia	Algiers I University, Algeria
Jamal Benhra	National School of Computer Science and Systems Analysis, Morocco
Abdelaziz Berrado	Mohammed V University, Morocco
Subir Bhattacharya	National Institute of Technology Durgapur, India
Satyajit Bhunia	Midnapore City College, India
Papatya Sevgin Bicakci	Başkent University, Turkey
Shazia Bilal	COMSATS Institute of Information Technology, Pakistan
Sanjib Biswas	Calcutta Business School, India
Bonaventure Boniface	Sabah University, Malaysia
Eleonora Bottani	University of Parma, Italy
Marilisa Botte	University of Naples, Italy
Jaouad Boukachour	Le Havre Normandy University, France
Gercek Budak	Yıldırım Beyazıt University, Turkey
Victor Camargo	Federal University of São Carlos, Brazil
Patricia Cano-Olivos	Universidad Popular Autónoma del Estado de Puebla, Mexico

Nancy-Paulina Carren	Universidad Popular Autónoma del Estado de Puebla, Mexico
Michal Cerny	Prague University of Economics and Business, Czech Republic
Ying-Hua Chang	Tamkang University, Taiwan
Maxime Chassaing	University of Skövde, Sweden
Duong Cuong	Hanoi University of Science and Technology, Vietnam
Soumen Kumar Das	Vidyasagar University, India
Manoranjan De	Vidyasagar University, India
Bikash Koli Dey	Hongik University, South Korea
Oshmita Dey	Techno India University, India
Luis C. Dias	University of Coimbra, Portugal
Oleksandr Dluhopolskyi	West Ukrainian National University, Ukraine
Aman Dua	NIFTEM University, India
Mustafa Ekici	Çanakkale Onsekiz Mart University, Turkey
Abdellatif El Afia	Mohammed V University, Morocco
Karim El Bouyahyiouy	Mohammed V University, Morocco
Adiba El Bouzekri el Idrissi	Doukkali University, Morocco
Nizar El Hachemi	Mohammed VI Polytechnic University, Morocco
Mamdouh El Haj Assad	University of Sharjah, UAE
Mohamed El Merouani	Abdelmalek Essaâdi University, Morocco
Moulay Driss El Ouadghiri	Moulay Ismail University, Morocco
Ali Emrouznejad	Surrey Business School, UK
Ergun Eraslan	Yildirim Beyazit University, Turkey
Nuh Erdogan	Robert Gordon University, UK
Serap Ergun	Isparta University of Applied Sciences, Turkey
Petr Fiala	Prague University of Economics and Business, Czech Republic
Sukono Firman	Universitas Padjadjaran, Indonesia
Martin Flegl	Tecnológico de Monterrey, Mexico
Vijay Gahlawat	National Institute of Food Technology Entrepreneurship and Management, India
Askar Garad	Muhammadiyah Yogyakarta University, Indonesia
Rakesh Garg	Amity University, Noida, India
Michel Gendreau	University of Montreal, Canada
Peiman Ghasemi	University of Calgary, Canada
Shyamali Ghosh	Vidyasagar University, India
Bibhas C. Giri	Jadavpur University, India
Beata Glinkowska-Krauze	University of Łódź, Poland
Alireza Goli	University of Isfahan, Iran
Paulina Golinska	Poznań University of Technology, Poland

Kristens Gudfinnsson   University of Skövde, Sweden
Mete Gundogan   Ankara Yildirim Beyazit University, Turkey
Shunsheng Guo   Wuhan University of Technology, China
Seyyed Ali Haddad Sisakht   Iowa State University, USA
Achraf Haibi   Moulay Ismail University, Morocco
Sondes Hammami   University of Carthage, Tunisia
Sepehr Hanjani   University of Putra, Malaysia
Amit V. Hans   Guru Gobind Singh Indraprastha University, India
Khandaker Hasan   United International University, Bangladesh
Gholamreza Haseli   Monterrey Institute of Technology, Mexico
Muhammad Hoque   Management College of Southern Africa,
                                    South Africa
Tarak Housein   Libya, Libya
Olena Hrechyshkina   Polessky State University, Belarus
Chang-Ling Hsu   Ming Chuan University, Taiwan
Jianwen Huang   China Three Gorges University, China
Qimin Huang   Case Western Reserve University, USA
Dina M. Ibrahim   Qassim University, Saudi Arabia
Hamid Reza Irani   University of Tehran, Iran
M. Y. Jaber   Ryerson University, Canada
Ajay Jain   GGSIP University, India
Madhu Jain   IIT Roorkee, India
Suresh Jakhar   Indian Institute of Management Lucknow, India
Jishu Jana   Vidyasagar University, India
Nsikan John   Covenant University, Nigeria
Sudhanshu Joshi   Doon University, India
Deepshikha Kalra   Management Education and Research Institute,
                                    India
Mohamad Amin Kaviani   University of Cyprus, Cyprus
Michael Kay   North Carolina State University, USA
Mohd Khairol Anuar Ariffin   Putra University, Malaysia
Hamiden Khalifa   Cairo University, Egypt
Soheyl Khalilpourazari   Kharazmi University, Iran
Manjeet Kharub   Institute of Management Technology, Ghaziabad,
                                    India
Mazdak Khodadadi-Karimvand   University of Science and Culture, Iran
Reza Kiani Mavi   Edith Cowan University, Australia
Iryna Kramarenko   National University of Shipbuilding, Ukraine
Aalok Kumar   Jaipuria Institute of Management, India
Pavan Kumar   VIT Bhopal University, India
Martina Kuncova   Prague University of Economics and Business,
                                    Czech Republic

Tanmoy Kundu	National University of Singapore, Singapore
Flevy Lasrado	University of Wollongong in Dubai, UAE
Mohamed Lazaar	Mohammed V University, Morocco
Emily Lee	California State University, USA
Ramesh Lekurwale	Vidyasagar University, India
Viraj Lele	DHL, USA
Viacheslav Liashenko	Academy of Economic Science of Ukraine, Ukraine
Sunil Luthra	SIET Jhajjar, India
Vinayak Madiwale	Precast India Infrastructures Private Ltd., India
Gour Mahata	Sidho Kanho Birsha University, India
Hamidreza Mahboobinejad	Wollongong University, Australia
S. Maheswaram	University of Peradeniya, Sri Lanka
Arunava Majumder	Lovely Professional University, India
Dragana Makajic-Nikolic	Belgrade University, Serbia
Paulo Manrique	National University of Mexico, Mexico
Malek Masmoudi	University of Sharjah, UAE
Tafadzwa Matiza	North-West University, South Africa
Abu Md Mashud	Hajee Mohammad Danesh Science and Technology University, Bangladesh
Jan Medlock	Oregon State University, USA
Adeel Mehmood	COMSATS University, Pakistan
Fatima-Zahra Mhada	Mohammed V University, Morocco
Sudipta Midya	Vidyasagar University, India
Shashi Mishra	Banaras Hindu University, India
Froilan Mobo	Philippine Merchant Marine Academy, Philippines
Nanees Mohamed	Helwan University, Egypt
Essaaidi Mohammed	Mohammed V University, Morocco
Arijit Mondal	Vidyasagar University, India
Rahul Mor	National Institute of Food Technology Entrepreneurship and Management, India
Chanicha Moryadee	Suan Sunandha Rajabhat University, Thailand
Regaieg Mouna	University of Sfax, Tunisia
Mahabub Musa	Yusuf Maitama Sule University, Kano, Nigeria
Mahantesh M. Nadakatti	KLS Gogte Institute of Technology, India
Peter Nadeem	University of Derby, UK
Ahmed Nait Sidi Moh	Jean Monnet University, France
Mehdi Najib	International University of Rabat, Morocco
Behnam Nakhai	Millersville University of Pennsylvania, USA
Amir Hossein Nasiri	University of Putra, Malaysia
Luka Neralic	University of Zagreb, Croatia

Luan-Thanh Nguyen	HUFLIT-Ho Chi Minh City University of Foreign Languages-Information Technology, Vietnam
Lewis Njualem	Texas Tech University, USA
Mohammed Nusari	Lincoln University College, Malaysia
Mehmet Onur Olgun	Suleyman Demirel University, Turkey
Khaoula Ouaddi	National School for Computer Science, Morocco
Rachid Oucheikh	Norwegian University of Science and Technology, Norway
Mustapha Oudani	International University of Rabat, Morocco
Kenza Oufaska	International University of Rabat, Morocco
Amar Oukil	Sultan Qaboos University, Oman
Fatima Ouzayd	National School for Computer Science, Morocco
Mohamed Ouzineb	National Institute of Statistics and Applied Economics, Morocco
İsmail Ozcan	Suleyman Demirel University, Turkey
Manvinder Pahwa	Manipal University Jaipur, India
Ash Pain	UMT, Pakistan
Brojeswar Pal	University of Burdwan, India
Osman Palanci	Suleyman Demirel University, Turkey
Iztok Palcic	University of Maribor, Slovenia
Anupama Panghal	National Institute of Food Technology Entrepreneurship and Management, India
Sarla Pareek	Banasthali Vidyapith, India
Hiren Patel	Ganpat University, India
Asim Paul	Vidyasagar University, India
Sanjoy Paul	University of Technology Sydney, Australia
Dina Pereira	University of Beira Interior, Portugal
Magfura Pervin	Vidyasagar University, India
Jaroslav Pushak	Lviv State University of Internal Affairs, Ukraine
Juanjuan Qin	Tianjin University of Finance and Economics, China
Kumar Rahul	National Institute of Food Technology Entrepreneurship and Management, India
Shivani Rana	Hoshiarpur S.D. College, India
Shalendra Rao	Mohanlal Sukhadia University, India
Svetlana Rastvortseva	National Research University Higher School of Economics, Russian Federation
D. Raut	Indian Institute of Management Mumbai, India
Goncalo Reis Figueira	INESC TEC, FEUP, Portugal
Naoufal Rouky	Le Havre Normandy University, France
Darin Rungklin	Suratthani Rajabhat University, India
Sahara Sahara	Bogor Agricultural University, Indonesia
Tkatek Said	University Ibn Tofail, Morocco

Luis Salinas	New Jersey Institute of Technology, USA
Guruprasad Samanta	Indian Institute of Engineering Science and Technology, India
Sushant Samir	PEC University of Technology, India
Shib Sankar Sana	Kishore Bharati Bhagini Nivedita College, India
Thomy Saputro	Porto University, Portugal
Fanny Saruchera	University of the Witwatersrand, South Africa
Seyyed Hadi Seifi	Mississippi State University, USA
Marc Sevaux	University of South Brittany, France
Bhavin Shah	IIM Sirmaur, India
Janmejai Shah	Graphic Era University, India
Nita Shah	Gujarat University, India
Asadullah Shaikh	Najran University, Saudi Arabia
Manu Sharma	Graphic Era University, India
Ali Akbar Sheikh	University of Burdwan, India
Arun Kumar Shettigar	National Institute of Technology Karnataka, India
Om Ji Shukla	National Institute of Technology Patna, India
Parag Siddique	University of Louisville, USA
Vladimir Simic	University of Belgrade, Serbia
Avanish Singh Chauhan	Manipal University, India
Mustafa Soba	Usak University, Turkey
Roya Soltani	Khatam University, Iran
Karan Sukhija	Panjab University, India
Aneerav Sukhoo	Ministry of ITC and Operations, Mauritius
Yulin Sun	Southwestern University of Finance and Economics, China
Faustino Taderera	National University of Science and Technology, Oman
Nezih Tayyar	Usak University, Turkey
Ali Tehci	Ordu University, Turkey
Stefania Testa	University of Genoa, Italy
Vikas Thakur	Norwegian University of Science and Technology, Norway
Sunil Tiwari	National University of Singapore, Singapore
Mehdi Toloo	Technical University of Ostrava, Czech Republic
Achraf Touil	Hassan 1st University, Morocco
Rakesh Tripathi	Dr. APJ Abdul Kalam Technical University, India
Pinar Usta	Isparta University of Applied Science, Turkey
Tatapudi Vasista	Srinivas University Mangalore, India
Agnes Nalini Vincent	University of Third Age, Mauritius
Gongming Wang	Tsinghua University, China

Wenqing Wu	Southwest University of Science and Technology, China
Liangping Wu	Sichuan Normal University, China
Mehmet Yetim	Suleyman Demirel University, Turkey
Abdullah Yildizbasi	Ankara Yıldırım Beyazıt University, Turkey
Nurullah Yılmaz	Suleyman Demirel University, Turkey
Ramadan Zenedean	Cairo University, Egypt
Jian Zhang	Sichuan Normal University, China
Haining Zheng	Massachusetts Institute of Technology, USA
Karim Zkik	ESAIP School of Engineers, France
Shakib Zohrehvandi	Technical University of Kosice, Slovakia

Scientific Sponsors

International University of Rabat, Morocco

The University of Texas at Dallas, USA

Prague University of Economics and Business, Czech Republic

Kent Business School, UK

**University of Sfax, Tunisia
Faculty of Economics and Management**

University of Sfax, Tunisia

Sunway University, Malaysia

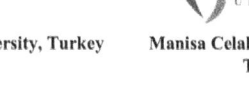

Ankara Science University, Turkey

Manisa Celal Bayar University, Turkey

Manufacturing Engineering Laboratory of Tlemcen, Algeria

Abou Bekr Belkaid Tlemcen University, Algeria

Global Academy, Turkey

Suleyman Demirel University, Turkey

Calcutta Business School, India

Halic University, Turkey

Contents – Part II

Digital Transformation of Supply Chain and Logistics Systems

Contents – Part I

Advances of Artificial Intelligence/Operational Research Tools in Healthcare

Technology, Learning and Analytics
in Intelligent Systems

Key Factors of Mobile Applications in Government Services (MG-App) Adoption in Indonesia: An Empirical Study Based on Netnography and Text Mining Approach

Nur Rahmah Syah Ramdani[(✉)] and Zulkarnain Zulkarnain

Universitas Indonesia, Depok, Indonesia
{nur.rahmah11,zulkarnain17}@ui.ac.id

Abstract. The development of e-Government has established the use of mobile applications (mG-App) for service delivery, yet studies on their adoption within developing countries, notably Indonesia, are still limited. Given user acceptance challenges, this study analyzes social media discourse to empirically investigate the key factors contributing to the success of mG-App adoption in Indonesia. Social media discourse significantly impacts government initiatives by fostering consensus or dissent. While prior studies on digital co-production often focus on formal-rational factors using the Unified Acceptance and Use of Technology (UTAUT), this research aims to bridge this gap by integrating the rational factors outlined by UTAUT with the intricate social context. Leveraging aspect-based sentiment analysis and social media netnography, this study examines citizens' perceptions and highlights mG-App adoption factors from two perspectives: government-to-citizens and citizens-to-citizens, using public reviews from X/Twitter as primary data. By innovatively integrating qualitative with quantitative methods for data processing, this study offers insights into new technology adoption, contributing to the evolving landscape of digital co-production in government services. It elucidates technical factors like performance expectancy, effort expectancy, and facilitating conditions while delving into social dimensions such as accountability, transparency, trust, privacy, and cost. Findings reveal widespread public concerns over mG-App performance and effort expectancy in Indonesia, shedding light on how the UTAUT social environment concept influences public resistance. Additionally, the study offers practical recommendations to optimize mG-App performance, secure third-party support, instill trust, and enhance efficiency, ultimately elevating user experience and driving adoption in Indonesia's public sector.

Keywords: Mobile Applications · Government Services · Netnography · UTAUT · LDA · Text Mining

1 Introduction

Electronic Government (e-Government), defined as "the use of information and communication technologies (ICTs) and its application by the government for the provision of information and public services to the citizens," [1] has witnessed significant

© The Author(s), under exclusive license to Springer Nature Switzerland AG 2025
A. Mirzazadeh et al. (Eds.): ODSIE 2023, CCIS 2205, pp. 3–19, 2025.
https://doi.org/10.1007/978-3-031-81458-7_1

transformation. The emergence of COVID-19 has compelled governments to go beyond conventional practices, prompting them to utilize various tools to enhance and transform their public services [2]. In response, there is a growing trend in the adoption of digital solutions in the government to boost the effectiveness and efficiency of public service and engage in co-production with stakeholders. A further step in this e-Government transformation is the mobile-enabled service delivery known as mobile applications in government services (mG-App) to serve its citizens better. This integration of mobile ICT significantly shapes public service delivery, integrating citizen engagement processes as a vital aspect of co-production [2]. Nevertheless, in the context of developing countries, implementing mG-App presents more challenges, such as concerns related to privacy and security, peoples' readiness, legal issues, compatibility, power limitation, limited bandwidth and slow download speeds, and the high cost of internet access via mobile [3]. This study explicitly examines Indonesia as a case from a developing country.

Today, the Indonesian government is implementing mG-App for public services, with an emphasis on addressing issues related to payment schemes. Indonesia's five most downloaded mG-Apps include Mobile JKN, SIGNAL, Digital Korlantas POLRI, M-Passport, and M-Pajak. Conventional payment methods in Indonesia have been problematic, often leading to illegal procedures and fees facilitated by officials or brokers [4]. The introduction of mG-Apps, such as SIGNAL (Samsat Digital National) for annual vehicle tax payments, aims to offer online One Stop Service, eliminating the need for physical visits to the One-stop Administration Services Office (Samsat) and ensuring transparent transactions [5]. However, considered as a solution for this issue, these digitalized services often suffer from a lack of participation. According to the developer's report in April 2022 for the SIGNAL application, the total number of transactions amounted to approximately 80,000, less than 1% of the more than 1 million downloads. This figure remains minimal compared to the total number of registered vehicles, approximately 143 million units [6]. Despite lacking competitors in the market, mG-Apps are anticipated to be a preferred choice for the services. Active participation in usage is crucial, particularly in terms of public services. Considering the contemporary societal setting, community plays a role that extends beyond mere recipients; it involves active co-production in developing public policies and services [7].

Previous studies on digital co-production have predominantly employed the Unified Acceptance and Use of Technology (UTAUT), with a primary focus on formal-rational factors [8]. Yet, there is a scarcity of scholarly insights into the adoption of new technology in such a developing country within its own socioeconomic complexity. This research addresses this gap by exploring how public perspectives influence the co-production of digital government services, taking into account the complexities of the social context and the rational factors outlined by UTAUT. To delve into these dynamics, this study leverages the function of social media in enabling discourse [9] and employs netnography to empirically extend the well-established UTAUT framework within the contemporary context of mobile application use in the public sector. Social media serves as a platform for citizens to express opinions and engage in discussions regarding mG-Apps. In contrast to related studies using UTAUT that largely rely on surveys and Structural Equation Modeling (SEM) [10–13], this research adopts a text mining approach, integrating it with netnography as an innovative method. This combined method is deemed less time-consuming, cost-effective, and capable of acquiring larger data sizes than traditional survey methods.

This study expands the research landscape on accountability in the use of mG-App in several respects. Firstly, it employs a quantitative approach (text mining) to investigate the adoption factors contributing to the success of mG-App in Indonesia, a developing country. Given the limited existing research on mG-App adoption in Indonesia, this approach could offer valuable viewpoints to assist the government in better understanding the requirements of its citizens, ultimately resulting in the successful provision of public services. Secondly, employing an innovative method that combines quantitative and qualitative aspects, it utilizes netnography to qualitatively demonstrate how public sentiment on social media can influence the co-production of mobile applications, encompassing both government-to-citizens and citizens-to-citizens perspectives. Lastly, this study deepens scholarly understanding of the social dimensions involved in the co-production of mG-App services, illuminating aspects such as accountability, transparency, trust, privacy, and cost.

In line with this, the study seeks to examine the empirical factors influencing the successful adoption of mG-App in Indonesia from the user's perspective. Accordingly, the research question posed is: What are the empirical factors contributing to the success of mG-App adoption in Indonesia? The structure of the paper is outlined as follows: an explanation of UTAUT, co-production, netnography, and text mining; a detailed presentation of the methodology, encompassing the data collection process; subsequent sections include the presentation of results and an analytical discussion. The final section of the paper addresses the study's conclusion and limitations.

2 Literature Review

2.1 The UTAUT

Whether private or public, services involve the participation of potential users, the acceptance of technology by users, and their active engagement [2]. The Unified Theory of Acceptance and Use of Technology (UTAUT) is a framework or model widely used to understand and predict factors of acceptance and use of technology. Numerous studies have utilized and extended technology acceptance models like the TAM, the Innovation Diffusion Theory, the Theory of Reasoned Action, and particularly the UTAUT [14], to study the adoption of e-Government services [10]. The UTAUT model identifies four constructs—performance expectancy, effort expectancy, social influence, and facilitating conditions—as direct influencers on user acceptance and usage behavior [14]. In a different model developed for mG-App adoption in the United Arab Emirates (UAE), research demonstrated that social influence, cost, trust, a range of services, and demographic factors, alongside TAM constructs, had a substantial impact on influencing users' behavior toward mG-App services [12]. Additional factors extended from UTAUT, such as transparency [13], and information quality [11], etc., have also been considered. These previous related studies conducted surveys to study this digital adoption. Nevertheless, this method is acknowledged as relatively time-consuming, expensive, and limited in data size.

2.2 Co-production

In this period of digital government, co-production manifests in diverse forms, encompassing crowdsourcing, open data, and social media, evolving from a rational approach to a more social one, thereby reinforcing social interaction [7]. Defined as the "voluntary or involuntary involvement of public service users in [...] design, management, delivery and evaluation of public services" [7], this multifaceted co-production paradigm becomes increasingly essential for fostering effective engagement, transparency, and accountability in the delivery of public services. The shift to a collaborative model for data-driven service creation requires redefining the roles of public and private actors, guided by the New Public Service [15] principles. This approach prioritizes citizen-centricity, service provision over control, and valuing partnerships. Collaborative innovation entails stakeholders redefining roles [16]: politicians engage in dialogue to set agendas; public managers coordinate collaboration; private entities become responsible partners; and citizens participate as co-producers rather than passive recipients. Consequently, co-produced data-driven public services highlight both data utilization and the adaptation of stakeholder roles to enhance public value creation.

2.3 Netnography

In 1995 R.V. Kozinets [17] coined the term 'netnography,' a qualitative method for examining social interaction and communities in digital communications [18]. It provides valuable insights into social phenomena by identifying themes, patterns, and propositions, enabling researchers to observe textual discourse [2]. Initially utilized in marketing research, ethnography has expanded across various domains. Netnography's success lies in its flexibility, affordability, and practicality. Setting it apart from other investigative methods, netnography stands in several key areas [19]: (1) Cultural focus, emphasizing human understanding to delve into the complexities, contexts, and meanings within phenomena, as in this case, the citizens of Indonesia; (2) Utilization of social media data, capturing interactions across diverse online platforms like Twitter, Facebook, and Instagram, accessible through various devices; (3) Active participation, netnographic researchers actively engage with the communities they study, becoming reflective learners in the process. While netnography has traditionally been centered on qualitative design, its development has significant potential to incorporate a combination of quantitative research methods [20].

2.4 Text Mining

Sentiment analysis (SA) from online texts, also called opinion or text mining, has been a highly active researched domain within natural language processing (NLP) from the beginning of the century [21, 22]. Analyzing Indonesian text mining poses particular challenges due to the complexity of the multi-ethnic language used on social media. To address this, this research utilizes assistance from the Google Translate API to translate the text into English, facilitating easier pre-processing. According to [21] discovering aspects for further analysis is often necessary, determining whether the sentiment is positive, negative, or neutral for each aspect—an approach called aspect-based sentiment analysis (ABSA). For this ABSA study, Latent Dirichlet Allocation (LDA) is chosen for topic modeling. This technique is favored as the most popular topic modeling technique, capable of overcoming limitations present in alternative techniques like Latent Semantic Analysis (LSA), Probabilistic Latent Semantic Analysis (pLSA), and TF-IDF (term frequency-inverse document frequency) [23, 24]. The LDA is an unsupervised machine-learning technique introduced by [25] to reveal latent topics within a collection of documents by employing a probability distribution across a vocabulary [24]. Determining the ideal number of topics is a crucial aspect of topic modeling using the LDA algorithm. Perplexity and average coherence values are calculated quantitatively to assess different topic numbers. Perplexity measures the effectiveness of a probability model in predicting unstructured data, with lower scores indicating better models, and coherence score shows words that co-occur more frequently, with values closer to zero as a negative log probability value [24].

3   Methodology

3.1   Research Methods Using Netnography from X/Twitter

This research utilized netnography to gather an extensive dataset for analyzing public discussions related to mG-App. The data were sourced from X/Twitter, a widely used contemporary platform recognized for its transparency and concise statements, limited to 280 characters per post [26]. The data collection spanned from January 2022 to July 2023. This study identified two discourse categories: "government-to-citizens" and "citizens-to-citizens." The netnography process involved three stages outlined in Table 1, encompassing five steps: defining research questions, selecting and identifying communities, data collection, analysis, and research reporting [17]. The topics were then interpreted and theorized, employing the analytical framework of UTAUT alongside the additional social dimensions from the social environment. The primary goal of using data-driven netnography is to equip researchers with tools to explore extensive information. This approach empowers netnographers to delve deeper into online conversations, saving valuable time that would otherwise be spent sorting through information to identify common topics, themes, and codes.

3.2   Research Methods Using Text Mining

The text mining method for this research will utilize aspect-based sentiment analysis, employing LDA for topic modeling and Flair for sentiment detection. The initial stage before data processing is text pre-processing, which converts raw human language data into a machine-readable format. The pre-processing phase encompasses various sub-stages, such as data selection, case folding, spelling normalization, filtering, stemming, and tokenization. Python programming was employed to execute these steps. Topic coherence is used to evaluate the topic modeling.

Table 1. Netnography Stages.

Step 1: Government-to-citizens	
Goal	Examination and identification of citizens' responses towards government accounts or government posts related to mG-App
Selection	Data collected from tweets, retweets, comments on official government accounts on the applications such as "@BPJSKesehatanRI" for Mobile JKN, "@SamsatDigital" for SIGNAL, "@korlantasid" for the Digital Korlantas POLRI, "@ditjen_imigration" for M-Passport and "@DitjenPajakRI" for M-Pajak
Data	1500 tweets/comments
Topic Modeling Result	Long process paying via SIM application, need help resolving errors when using the application, failure to use services in the application, need help updating personal data in the application, payment transactions process in the application
Step 2: Citizens-to-citizens	
Goal	1. Exploration and identification of citizens' discussions about mG-App 2. Exploration and identification of themes beyond the Stage 1 approach
Selection	Conduct searches with the keywords of the applications accompanied by keywords from the identification process during Phase 1, namely "government app data," "trust government applications," "security of government public data," etc., and from public discussion accounts discussing mG-App in Indonesia
Data	1500 tweets/comments
Topic Modeling Result	Fulfillment of data requirements and administrative fees, registration process and makes payments, more budget for the application, still need to come to the office even using the application, online test using application, technical errors that cause failure to use application, applications that are integrated with other systems, use of application for online service extension
Step 3: Examination, understanding, and theorizing	
Goal	Develop aggregated topics comprehension; and contextualizing themes, and creating a big picture
Selection	Multiple iterations of analysis and identify aggregated topics
Data	Dataset from step 1 and step 2
Aggregated Topic Result	'Performance Expectancy,' 'Effort Expectancy,' 'Facilitating Conditions,' 'Accountability and Transparency,' 'Trust,' 'Privacy,' and 'Cost.'

4 Results and Discussion

From 3000 tweets/comments, Fig. 1 illustrates the sentiment obtained using Flair, revealing a higher prevalence of negative sentiments toward the mG-Apps topic than positive sentiments. This suggests that the topics generated through the LDA method are likely to revolve around negative sentiments. Hence, the expected topics generated from these tweets/comments are likely to be in a negative perspective.

After classifying the sentiments, the best number of topics needs to be determined by checking the coherence score for each number of topics. The best coherence score in Step 1 (Government-to-Citizens) is achieved with five topics (see Appendix 1), followed by a decrease with 14 topics and so forth. By using LDA, the topic modeling result from the extracted words in Step 1 is shown in Table 2, wherein these extracted topics from keywords are then correlated with the UTAUT. Similarly, the best coherence score in Step 2 (Citizens-to-Citizens) is associated with six topics (see Appendix 2) but experiences another increase at eight topics before declining at 14 topics and beyond. The selection of 8 topics is on purpose as it aims to identify themes beyond the Step 1 approach. By using LDA, the topic modeling result from the extracted words in Step 2 is shown in Table 3, wherein these extracted topics from keywords are then correlated with the UTAUT.

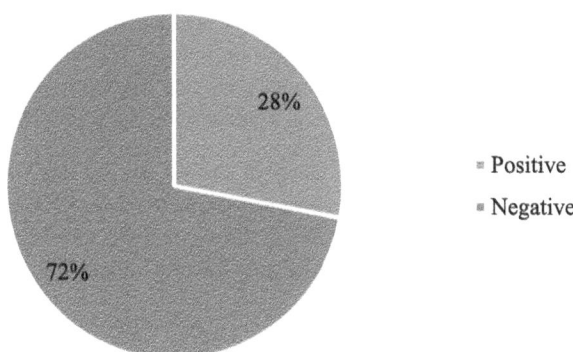

Fig. 1. Semantic Distribution Result.

The discourses from the results of both Step 1 and Step 2 constituted the foundation for further results examined in Step 3 (examination, understanding, and theorizing). The 13 initially emerging themes were condensed into six aggregated topics to strengthen theoretical framing during this phase. Three of the topics are categorized into the technical dimensions, while the other three are categorized into the social dimensions of co-production, as shown in Fig. 2.

In terms of technical dimensions, the predominant public discourse revolves around 'Performance Expectancy.' Citizens often express complaints on the X/Twitter platform when encountering errors while using mG-Apps, exemplified by instances like this one:

admin, why does it keep failing? I've checked the terms and conditions again, but it always fails. I've been trying for 3 days, but it always fails. (#1649563123349860352) as a reply to a post from @korlantasid

This is aligned with other studies that have identified performance expectancy as a key predictor influencing personal intention to use mG-App [2, 10–13]. Another technical dimension, 'Effort Expectancy,' is also a subject of discussion, primarily focusing on citizens' efforts to use mG-Apps, leading to requests for government assistance due to perceived difficulties, as in this sample tweet:

@ditjen_imigrasi only you still have to come and fill out the manual form again, even though you have already filled out the digital form in the application, then also add requirements that are not stated in the application, such as stamps and photocopies of your passport, on top of that, you still have to wait for a passport photo. (#1655175895253151744)

This is aligned with other studies that have identified effort expectancy as a key predictor influencing an individual's behavioral intention to utilize the new technology [2, 11–13]. At last, the topic of 'Facilitating Conditions' is the least discussed in the technical dimension, with citizens addressing registration procedures and payment processes facilitated by third parties of mG-Apps. Smooth operation with third-party involvement is considered crucial for the success of these apps. However, challenges have been highlighted, as evidenced by a user's reply to a tweet expressing difficulty in making payments despite having the payment code and attempting various platforms. An example from the user sought assistance, emphasizing the importance of addressing such operational challenges:

I already got the payment code, but I can't pay. Have tried via BNI, BRI, Mandiri, Toped, and DANA. Everyone can't do it. Can you help, Sir/Madam Admin? (#1610530545263611904) as a reply to a post from @SamsatDigital

This result conforms with several other empirical studies on the fact that the facilitating conditions impact mG-App adoption [10, 11, 13]. On the other hand, in terms of social dimensions, the predominant topic of public discussion is 'Accountability and Transparency.' Citizens engage in conversations expressing concerns about using mG-Apps, particularly for activities like online tests or utilizing services that might lack sufficient accountability and transparency in how the government utilizes or assesses the results. This aligns with an earlier investigation highlighting the crucial role of accountability and transparency in the co-production of the public sector [2]. Another study also emphasizes the practical importance of how a lack of transparency in Indonesia affects the adoption of e-Government [13]. However, positive discussions highlight the encouragement for app usage, and that the process is accountable and transparent, such opinions were expressed as follows:

It's better to use digital traffic police. It's true that the wait is a bit long, but it's done in a day, just waiting for it to be sent. Psychological tests can be done online, medical tests can be registered online, so there is no extortion. (#1600000325007441921)

Issues related to '(Dis)Trust' are also actively discussed among citizens, with perceptions that the adoption of mG-Apps is only at increasing the government budget without visible results. This was articulated in a tweet:

Yes, it's okay to stop making first. It's better to improve the existing applications first because there are SO MANY shortcomings. Just waste the budget. (#1668553027492679682) as a reply to public discussion "President Joko Widodo (Jokowi) asked ministries and institutions (K/L) to stop creating new applications, for any reason." from @asumsico

Additionally, there is a lack of trust in its integration capabilities with other existing mG-Apps as the government plans to build a super App to facilitate public services in the country [27]. This finding is consistent with previous studies that trust is a significant challenge to accept and use a new technology [2, 10–12]. Lastly, the least emphasized topic of conversation is 'Privacy and Cost'. Citizens express concerns regarding the fulfillment of data requirements, particularly raising issues about the privacy implications of providing necessary data. The willingness of citizens to use mG-Apps is contingent upon the degree to which citizens view those services as a trustworthy and secure tool that prioritizes and safeguards their privacy [8]. Such concern is mentioned in a tweet:

Yes, the application has become unclear, and there are concerns if it really protects the safety of people's personal data. If not, especially now that they say there's a digital ID card, I don't even bother to try it; you just use it to tap the electronic ID card at the bank. Other than that, it still needs to be photocopied. (#1668591140558499841)

A number of users also mention associated administrative fees in using the applications, and this also supports previous study that cost is a construct that substantially contributes to an individual's behavioral intention to adopt new technology [12], as illustrated in this tweet:

For comparison, yesterday, the renewal was online via the traffic police digital application, and the SIM was sent to the house. The above costs do not include the online psychological test of Rp37,500 (the menu is already in the digital traffic police application). so, a total of Rp148,001. Everything can be done from inside the room. (#1668258916168351744)

This study theoretically establishes a connection between the relevance of the technical dimensions of UTAUT and the social dimensions within the co-production of mobile application public services. The social dimensions are deemed equally significant in society's acceptance of new technology, although its discussion frequency is below the technical dimensions. This is in line with this period of digital government, in which co-production has become the center for designing, managing, and evaluating public services based on the involvement of its users [7]. In this case, the study showed that the citizens' involvement could be used to evaluate existing mG-Apps services and to plan future public services that can accommodate the public or citizens' needs and requests.

Table 2. LDA Result Step 1 (Government-to-Citizens).

Topic	Keywords	Topic Extracted from Keywords	UTAUT Topic
1	0.055*"sim" + 0.027*"still" + 0.025*"process" + 0.023*"want" + 0.018*"pay" + 0.017*"like" + 0.017*"ask" + 0.013*"extend" + 0.013*"onlin" + 0.013*"pleas"	Long process paying via SIM application	Performance Expectancy
2	0.027*"pleas" + 0.020*"day" + 0.019*"applic" + 0.018*"tri" + 0.017*"help" + 0.015*"still" + 0.014*"thank" + 0.013*"work" + 0.013*"month" + 0.011*"yet"	Need help resolving errors when using the application	Effort Expectancy
3.	0.026*"applic" + 0.022*"fail" + 0.021*"want" + 0.018*"sim" + 0.017*"even" + 0.016*"keep" + 0.016*"though" + 0.016*"come" + 0.013*"still" + 0.012*"onlin"	Failure to use services in the application	Performance Expectancy
4	0.031*"number" + 0.030*"applic" + 0.022*"regist" + 0.021*"chang" + 0.020*"cellphon" + 0.019*"want" + 0.015*"pleas" + 0.014*"work" + 0.013*"still" + 0.012*"jkn"	Need help updating personal data in the application	Effort Expectancy
5	0.022*"payment" + 0.017*"pay" + 0.016*"process" + 0.016*"alreadi" + 0.014*"make" + 0.011*"pleas" + 0.011*"want" + 0.010*"account" + 0.010*"take" + 0.009*"still"	Payment transactions process in the application	Facilitating Condition

4.1 Theoretical Contributions

This research provides valuable insights for scholars and researchers, theoretically contributing to understanding mobile government adoption, e-government studies, and specifically mobile application studies. Firstly, through the use of text mining, the study identifies crucial factors influencing the successful adoption of mG-App in a multi-ethnic developing country like Indonesia. Secondly, by advancing the comprehension of mG-App adoption, the research offers insights into the application of an innovative

Table 3. LDA Result Step 2 (Citizens-to-Citizens).

Topic	Keywords	Topic Extracted from Keywords	UTAUT Topic
1	0.011*"certif" + 0.011*"take" + 0.011*"datum" + 0.010*"card" + 0.010*"want" + 0.009*"fee" + 0.008*"later" + 0.008*"come" + 0.008*"follow" + 0.007*"offic"	Fulfillment of data requirements and administrative fees	Privacy and Cost
2	0.024*"use" + 0.021*"photo" + 0.015*"pay" + 0.014*"take" + 0.014*"make" + 0.011*"one" + 0.010*"first" + 0.010*"card" + 0.009*"regist" + 0.009*"day"	Registration process and makes payments	Facilitating Condition
3.	0.026*"make" + 0.024*"secur" + 0.022*"guard" + 0.022*"polic" + 0.017*"budget" + 0.016*"onlin" + 0.013*"time" + 0.011*"use" + 0.010*"extens" + 0.009*"also"	More budget for the application	(Dis)Trust
4	0.023*"make" + 0.018*"even" + 0.014*"still" + 0.014*"immigr" + 0.013*"offic" + 0.013*"want" + 0.012*"extend" + 0.011*"right" + 0.011*"come" + 0.010*"use"	Still need to come to the office even using the application	Effort Expectancy
5	0.037*"test" + 0.034*"onlin" + 0.026*"take" + 0.021*"extend" + 0.020*"psycholog" + 0.016*"use" + 0.016*"want" + 0.016*"health" + 0.014*"care" + 0.013*"go"	Online test using application	Accountability and Transparency
6	0.014*"use" + 0.012*"onlin" + 0.011*"get" + 0.011*"day" + 0.010*"right" + 0.010*"polic" + 0.010*"fail" + 0.009*"extend" + 0.009*"come" + 0.009*"even"	Technical errors that cause failure to use application	Performance Expectancy
7	0.026*"integr" + 0.025*"system" + 0.023*"servic" + 0.013*"director" + 0.013*"gener" + 0.012*"compens" + 0.012*"manag" + 0.012*"digit" + 0.012*"hospit" + 0.011*"polic"	Applications that are integrated with other systems	(Dis)Trust
8	0.043*"polic" + 0.040*"use" + 0.025*"nation" + 0.018*"corp" + 0.018*"extend" + 0.016*"onlin" + 0.015*"servic" + 0.009*"well" + 0.009*"want" + 0.009*"extens"	Use of application for online service extension	Accountability and Transparency

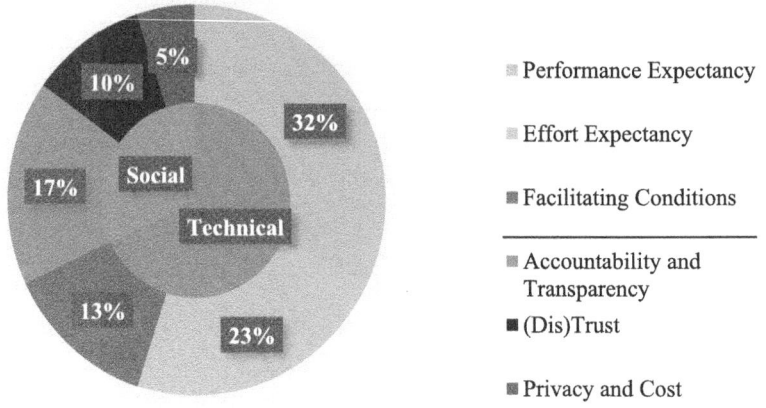

Fig. 2. Distribution of Aggregated Topics.

method that integrates both quantitative and qualitative elements, as demonstrated by previous work [18]. This research's outcomes can potentially extend the scope of technological innovation in adoption studies. It establishes a robust foundation for further exploration in public services by comprehending and managing big data effectively. The integration of netnography for analyzing social media discourse provides a qualitative understanding of how public sentiment impacts the co-production of mobile applications. Social media discourse, as a facilitator for public discussion, significantly influences the success or failure of government initiatives by either fostering consensus or dissent, mirroring the approach from previous studies [2]. Lastly, the study's findings complement existing research on the adoption of mobile technology in government services. This research enhances the scientific understanding of the social dimensions involved in co-producing mG-App services, extending the UTAUT framework to encompass aspects such as accountability, transparency, trust, privacy, and costs.

4.2   Implications for Practice

From a practical standpoint, the study's findings underscore the critical role of constructs such as 'Performance Expectations,' 'Effort Expectations,' 'Facilitating Conditions,' 'Accountability and Transparency,' 'Trust,' 'Privacy,' and 'Cost' in developing and implementing mG-Apps for citizen-centric services in Indonesia. Government agencies are recommended to prioritize these key constructs in planning and implementing similar projects, recognizing their significance in shaping user perceptions and interactions. The insights extend beyond individual app development, guiding effective strategies and policies of mG-Apps by ensuring optimal performance, securing third-party support, instilling trustworthiness, and ensuring efficiency in terms of time and cost, as mentioned in previous work [3].

To address technical considerations, the government should establish clear performance and efficiency standards for mG-Apps, aligning with citizens' expectations. Regular evaluations and benchmarking against these standards will ensure the delivery of reliable, efficient, and high-quality services. Guidelines and investment in capacity building are crucial for boosting technical proficiency. Prioritizing simplicity, developers should eliminate the need for hard copies of identities, reduce direct visits to service units, and make other improvements to enhance utility and ease of use. Moreover, involving experienced users in the design stage fosters a user-centric approach. Government officials should explore the creation of feedback icons [11] for mG-Apps to gather user opinions. In addition, inclusive development practices, considering diverse demographics and socioeconomic backgrounds, are imperative.

On the social front, the government should establish transparent and accountable standards for utilizing and evaluating mG-Apps across diverse government activities. This includes providing clear guidelines on how government agencies utilize and assess results obtained through mG-Apps, ensuring citizens are well-informed. Additionally, strategies should be devised to address citizen concerns related to budget allocation and integration capabilities, fostering transparent communication about its objectives and benefits, especially in the context of the government's plan to develop a super App [27]. To create a unified experience, facilitate data exchange among various government departments. Moreover, ensuring trust in mG-Apps is critical, necessitating robust data privacy regulations and enforcement to alleviate concerns about data requirements and associated fees. Adherence to stringent privacy standards, as per the national cybersecurity policy by the Indonesian Ministry of Communication and Information, is anticipated to enhance public trust.

Lastly, exploring public-private partnerships (PPP) [11] is crucial for maximizing mG-Apps' effectiveness in Indonesia, facilitating collaboration to improve service delivery. In conclusion, these practical recommendations aim to collectively elevate the user experience, instill trust, and drive the overall adoption of mobile applications in the public sector in Indonesia.

5   Conclusion

Using a combination of netnography and aspect-based sentiment analysis, this study seeks to investigate the empirical factors contributing to the successful adoption of mG-App in Indonesia from the user's perspective. In summary, this study offered a thorough understanding of the key factors, namely 'Performance Expectancy,' 'Effort Expectancy,' 'Facilitating Conditions,' and contributes to exploring social dimensions in the co-production of mG-App services, covering 'Accountability and Transparency,' 'Trust,' 'Privacy,' and 'Cost.' The findings highlight prevalent negative sentiments on X/Twitter, particularly emphasizing low 'Performance Expectancy' and low 'Effort Expectancy' in the technical category. On the other hand, from the social dimension, positive discussions surrounding 'Accountability and Transparency' indicate encouragement for app usage. Despite this, public distrust of government apps remains a concern for technology adoption.

Overall, this study evidence widespread public distress over the performance and effort expectancy of the applications. This study not only sheds light on the intricate dynamics of user sentiment but also further reveals how the concept of the social environment from the UTAUT impacts public resistance to mG-App adoption.

6 Limitations and Future Research

While this study offers a valuable understanding of the co-production of digital technologies through netnography on a specific social media platform, we acknowledge the potential bias of focusing solely on online discourse. Although the results align with UTAUT theory, the reliance on online data through text mining may impose limitations. Future research should adopt a more holistic approach, integrating offline interactions and perceptions to provide a comprehensive understanding of digital service co-production. To enhance future studies, integrating multimedia elements like images and videos can offer a comprehensive understanding of the dynamics surrounding the adoption and co-production of digital services. Additionally, examining temporal aspects of citizen engagement, including sentiment evolution over time, would further enrich our understanding.

The study's time and data capacity constraints led to selecting only the top five downloaded applications, potentially limiting the generalizability of findings to applications with different usage characteristics. However, since the chosen apps share a similar public service context, we believe our results still give valuable insights despite potential variations in usage. Moreover, the constraints imposed by the limited access to the free X/Twitter API, which lacks specific location details, require careful consideration when assuming that the data originates from mG-App users in Indonesia. We recognize this assumption and its potential implications for understanding the geographical and demographic aspects of mG-App adoption in various regions or demographic segments.

Recognizing that our observations are based on specific conditions in a country with generally low citizen satisfaction with government services, we encourage future researchers to explore diverse co-production settings, considering socioeconomic complexities. Furthermore, given that each platform has its own demographic base, extending studies to include other social media platforms like Instagram, Facebook, LinkedIn, etc., can deepen insights into citizen engagement dynamics across diverse online spaces.

Appendix

Appendix 1. Coherence Score Graph Step 1 (Government-to-Citizens).

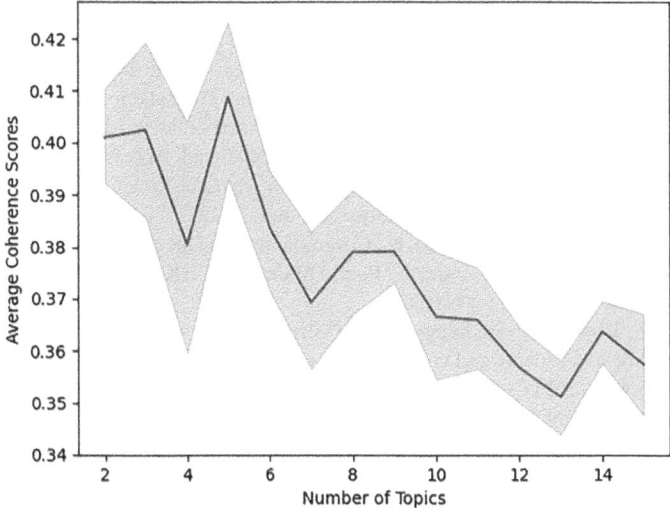

Appendix 2. Coherence Score Graph Step 2 (Citizens-to-Citizens).

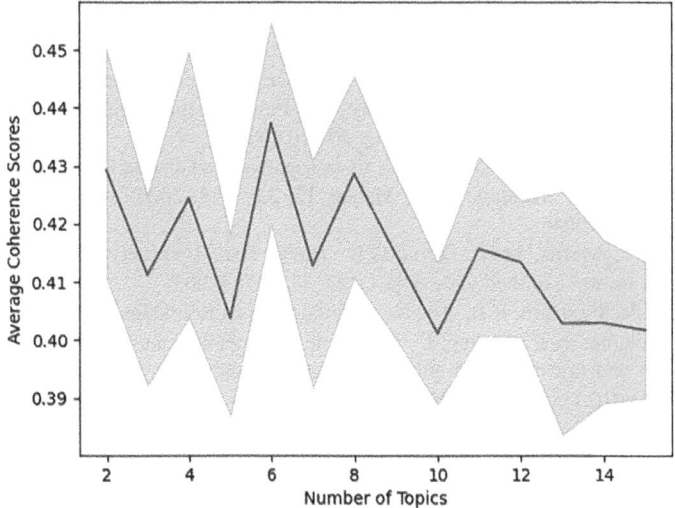

References

1. United Nations: E-Government Survey (2014)
2. Polzer, T., Goncharenko, G.: The UK COVID-19 app: the failed co-production of a digital public service. Financ. Account. Manag. **38**, 281–298 (2022). https://doi.org/10.1111/faam.12307
3. Mengistu, D., Zo, H., Rho, J.J.: M-Government: opportunities and challenges to deliver mobile government services in developing countries. In: ICCIT 2009 - 4th International Conference on Computer Sciences and Convergence Information Technology, pp. 1445–1450 (2009)

4. Kompas.com: Polri Minta Masyarakat Tak Pakai Jasa Calo Saat Buat SIM dan STNK (2016). https://nasional.kompas.com/read/2016/10/14/13302041/polri.minta.masyar akat.tak.pakai.jasa.calo.saat.buat.sim.dan.stnk. Accessed 19 June 2022

5. The National Police Commission Republik Indonesia: Tingkatkan Pelayanan, Koorlantas Polri Kembangkan Aplikasi Signal (2021). https://kompolnas.go.id/index.php/blog/tingka tkan-pelayanan-koorlantas-polri-kembangkan-aplikasi-signal. Accessed 8 Apr 2022

6. Badan Pusat Statistik: Jumlah Kendaraan Bermotor Menurut Provinsi dan Jenis Kendaraan (unit) (2021). http://www.bps.go.id/. Accessed 7 June 2022

7. Osborne, S.P., Radnor, Z., Strokosch, K.: Co-production and the co-creation of value in public services: a suitable case for treatment? Public Manag. Rev. 18, 639–653 (2016). https://doi. org/10.1080/14719037.2015.1111927

8. Wirtz, B.W., Birkmeyer, S., Langer, P.F.: Citizens and mobile government: an empirical analysis of the antecedents and consequences of mobile government usage. Int. Rev. Adm. Sci. 87, 836–854 (2021). https://doi.org/10.1177/0020852319862349

9. Mansoor, M.: Citizens' trust in government as a function of good governance and government agency's provision of quality information on social media during COVID-19. Gov. Inf. Q. 38, 101597 (2021). https://doi.org/10.1016/j.giq.2021.101597

10. Lallmahomed, M.Z.I., Lallmahomed, N., Lallmahomed, G.M.: Factors influencing the adoption of e-government services in Mauritius. Telemat. Inform. 34, 57–72 (2017). https://doi. org/10.1016/j.tele.2017.01.003

11. Sharma, S.K., Al-Badi, A., Rana, N.P., Al-Azizi, L.: Mobile applications in government services (mG-App) from user's perspectives: a predictive modelling approach. Gov. Inf. Q. 35, 557–568 (2018). https://doi.org/10.1016/j.giq.2018.07.002

12. Ahmad, S.Z., Khalid, K.: The adoption of M-government services from the user's perspectives: empirical evidence from the United Arab Emirates. Int. J. Inf. Manag. 37, 367–379 (2017). https://doi.org/10.1016/j.ijinfomgt.2017.03.008

13. Sabani, A.: Investigating the influence of transparency on the adoption of e-government in Indonesia. J. Sci. Technol. Policy Manag. 12, 236–255 (2020). https://doi.org/10.1108/ JSTPM-03-2020-0046

14. Venkatesh, V., Morris, M.G., Davis, G.B., Davis, F.D.: User acceptance of information technology: toward a unified view. MIS Q. 27, 425–478 (2003)

15. Denhardt, J.V., Denhardt, R.B.: The New Public Service. Routledge (2015)

16. Hartley, J., Sørensen, E., Torfing, J.: Collaborative innovation: a viable alternative to market competition and organizational entrepreneurship. Public Adm. Rev. 73, 821–830 (2013). https://doi.org/10.1111/puar.12136

17. Kozinets, R.V.: The field behind the screen: using netnography for marketing research in online communities. J. Mark. Res. XXXIX (2002)

18. Mane, D., Srivastava, P.: Netnography and text mining to understand perceptions of indian travellers using online travel services (2020)

19. El Hilali, S., Azougagh, A.: A netnographic research on citizen's perception of a future smart city. Cities 115, 103233 (2021). https://doi.org/10.1016/j.cities.2021.103233

20. Priyowidodo, G.: E-government organizational governance, policy communication and digitalization of land ownership identity (netnographic study on the issuance of electronic land certificates). J. Southwest Jiaotong Univ. 56, 564–580 (2021). https://doi.org/10.35741/issn. 0258-2724.56.4.48

21. Liu, B.: Sentiment analysis and opinion mining. Synth. Lect. Hum. Lang. Technol. 5, 1–167 (2012). https://doi.org/10.2200/S00416ED1V01Y201204HLT016

22. Li, Z., Tian, Z.G., Wang, J.W., Wang, W.M.: Extraction of affective responses from customer reviews: an opinion mining and machine learning approach. Int. J. Comput. Integr. Manuf. 33, 670–685 (2020). https://doi.org/10.1080/0951192X.2019.1571240

23. Annisa, R., Surjandari, I., Zulkarnain: Opinion mining on Mandalika hotel reviews using latent Dirichlet allocation. Procedia Comput. Sci., 739–746 (2019)
24. Çallı, L.: Exploring mobile banking adoption and service quality features through user-generated content: the application of a topic modeling approach to Google Play Store reviews. Int. J. Bank Mark. **41**, 428–454 (2023). https://doi.org/10.1108/IJBM-08-2022-0351
25. Blei, D.M., Ng, A.Y., Jordan, M.I.: Latent Dirichlet allocation. In: Advances in Neural Information Processing Systems, vol. 14 (2001)
26. Neu, D., Saxton, G., Rahaman, A., Everett, J.: Twitter and social accountability: reactions to the Panama Papers. Crit. Perspect. Account. **61**, 38–53 (2019). https://doi.org/10.1016/j.cpa.2019.04.003
27. Kementerian Komunikasi dan Informatika RI: Menteri Johnny: Pemerintah Siapkan Super Apps Layanan Publik (2023). https://www.kominfo.go.id/content/detail/43059/menteri-johnny-pemerintah-siapkan-super-apps-layanan-publik/0/berita_satker. Accessed 25 Jan 2023

Assessing Operation of Mobile Banking Applications that Support Customers to Access Banking Services Remotely

Daniel Okari Orucho[1]([✉]), Fredrick Mzee Awuor[1], and Collins Oduor[2]

[1] Kisii University, Kisii, Kenya
danielokari@yahoo.com
[2] United States International University-Africa, Nairobi, Kenya

Abstract. Mobile banking applications are programs that are created by banks to help their customers' access banking services remotely. This study assesses operation of mobile banking applications that enable customers to access banking services from their devices successfully. Additionally, this study discusses major weaknesses that affect customers while accessing banking services remotely. This study employs a systematic literature review approach. Findings from this study established that mobile banking applications provide a secure connection between the client application and the bank server, followed by customer authentication before accessing banking services remotely. The outcome to this study is a defined set of best practices that can be adopted by customers to access banking services remotely and securely. Notably, this study enlightens bank customers about the existing threat landscape when using mobile banking applications and provides solutions on how best mobile banking applications can be securely utilized to conduct banking services remotely.

Keywords: Man-in-the-Middle Attack · Mobile banking applications · Transport Layer Security · Hypertext Transfer Protocol

1 Introduction

The increasing number of smart mobile devices has led to a rapid rise in the use of mobile banking applications [1]. These applications are software programs developed by banks to help their customers manage finances remotely [2]. Mobile banking, therefore, refers to financial services and transactions conducted through mobile devices [3]. These applications offer a convenient and secure way to manage accounts on the go. With a smartphone, users can easily check account balances, monitor transactions, transfer money, and pay bills. Other services provided through mobile banking include cheque requests, loan repayment information, insurance services, airtime top-ups, and bookings.

Mobile banking applications offer several advantages to banks, including retaining existing customers, attracting new ones, reducing operational costs, and gaining a competitive edge in the market. For customers, mobile banking provides convenience, simplicity, and saves time by eliminating the need to visit traditional bank branches.

© The Author(s), under exclusive license to Springer Nature Switzerland AG 2025
A. Mirzazadeh et al. (Eds.): ODSIE 2023, CCIS 2205, pp. 20–35, 2025.
https://doi.org/10.1007/978-3-031-81458-7_2

Additionally, it allows for seamless 24/7 access to banking services from anywhere. To access these services remotely, customers must first register for mobile banking with their respective banks. After registration, they are required to download and install the bank's mobile banking application on their smartphones from official app stores.

While mobile banking represents a promising and innovative approach to the present and future of banking, there are emerging concerns associated with this technology. Firstly, consumers need to download, install, and familiarize themselves with the mobile banking applications before they can use them for remote banking. Secondly, these applications are designed to work on specific operating systems. This means that banks offering mobile banking services must develop applications compatible with the various operating systems that smartphones use.

Users of mobile banking applications have valid concerns about the security of these platforms. Cyber attackers often exploit vulnerabilities in mobile banking applications, creating rogue apps by decompiling legitimate banking apps' source code and uploading them to app stores [4]. Unsuspecting customers might download these malicious apps, mistaking them for official ones, and unknowingly install software loaded with mobile malware. Moreover, mobile banking applications only support user-initiated communication, meaning the smartphone user is responsible for ensuring the safety and security of their device when using these apps to access banking services remotely.

This study aims to evaluate how mobile banking applications enable customers to access services remotely. In this context, mobile banking applications refer to specialized and customized software that can be downloaded from the Google Play Store for Android devices or the App Store for iOS devices, and installed on smartphones or tablets. These apps are linked to specific banks worldwide, allowing users to remotely access banking services. The remainder of this study is organized into five sections: Sect. 2 presents the problem statement, Sect. 3 covers the literature review, Sect. 4 outlines the methodology, Sect. 5 discusses the security of mobile banking applications, Sect. 6 examines their operation, Sect. 7 highlights best practices for users, Sect. 8 explores theoretical and managerial implications, and Sect. 9 provides discussion, future research directions, and conclusions. Finally, Sect. 10 offers recommendations.

2 Statement of the Problem

Applications used for accessing banking services remotely are a technology that is beneficial to both bank customers as well as banking institutions. Customers benefit 24/7 access to banking services remotely from anywhere anytime with efficient banking services at reduced costs, while banking institutions benefit from efficient resource management at reduced costs and better service delivery. Majority of banking services are now provided to customers via their mobile devices rather than the traditional banking halls. This technology has been previously praised and highlighted as a unique channel for accessing banking services remotely especially during the Corona Virus Disease pandemic, when it was urged to maintain social distance between people and avoid physical contact.

However, cybercriminals target customers as well as banking institutions in an attempt to steal sensitive financial data as well as siphon money using various techniques like mobile malware, social engineering, eavesdropping and packet sniffing attacks. This makes users of mobile banking applications get worried that their bank accounts can be hacked when accessing banking services remotely because of lack of strong security mechanisms.

Since access to banking services using mobile banking applications is remote in nature, the customer using mobile banking applications is responsible for security checks because there is no physical teller to verify their identity. Therefore, users of mobile banking need to be aware of best practices that can help thwart potential threats from cybercriminals in order to access mobile banking services confidently. To this end, this study sought to answer the following questions:

i. How do mobile banking applications operate to give bank customers access to banking services remotely?
ii. What are the best practices that bank customers can adopt when using mobile banking applications?

3   Literature Review

The use of mobile applications for banking is growing rapidly. For instance, in the US, mobile banking apps are among the top three most popular types of applications [5]. By 2021, over 2 billion people were using their mobile phones for banking. Moreover, reports show that more customers prefer mobile banking over electronic banking [6]. While checking account balances is the most common activity, users also frequently use mobile banking apps to pay bills and transfer money.

Mobile banking has gained popularity due to the convenience of remote access to financial services and enhanced personal security [7]. Security methods such as fingerprint scanning and facial recognition are commonly used in mobile banking applications for user authentication [8]. The use of secure authentication methods has notably encouraged the continued use of these applications [9]. Consequently, mobile banking has become a strategic force within the global financial industry, leading to widespread adoption of this innovation [10].

Mobile banking offers strategic advantages to both customers and banks. For banks, it reduces costs by decreasing the need for physical branches and employees, thereby enhancing profitability. For users, mobile banking provides convenient access to financial information and services such as balance checks, money transfers, bill payments, and financial management anytime and anywhere via their mobile devices [11]. This convenience helps customers save time and money by reducing the need to visit physical bank branches and wait in long lines for services. Additionally, mobile banking allows secure money transfers between banks and offers features for tracking expenses [12].

There are three different types of strategies that make up secure access to mobile banking services [13]. The first step is identification, where the device verifies the user through biometrics, passwords, or physical ownership. This is followed by authentication, where the carrier uses either subscriber identification or cryptographic methods to validate the transaction request from the device. The third step, known as secure execution, requires the mobile transaction provider to carry out the transaction. However, these strategies do not fully ensure the security of mobile banking transactions.

A mobile payment architecture proposed by [14] integrates operators, service providers, and banking institutions. This architecture allows mobile users to pay for content and services using Short Message Service (SMS) and Wireless Application Protocol (WAP). While this alternative payment approach has been developed, there is a lack of research on the effectiveness of service delivery within this solution.

Mobile banking commonly uses usernames and passwords as security measures for remote access. However, these methods are vulnerable to password guessing and hacking. To enhance security, mobile banking applications should integrate more reliable identification methods, such as combining user identity, passwords, and fingerprint recognition. Implementing these advanced security features is crucial for improving privacy and protection in mobile banking [15].

Robust security mechanisms combined with user-friendly access are essential to promote the use of remote banking [16, 17]. Recent research underscores that security is a key factor in encouraging the adoption of internet and mobile banking services. Protection against cybercrime and data breaches are critical technological aspects of mobile banking [18, 19]. A secure banking interface is likely to be more effective in driving mobile payment usage than financial incentives.

A secure online banking model for safeguarding information exchanged between communicating parties, especially during the authentication process, was developed by [18]. This model features multiple layers of authentication, beginning with pair-based text authentication, which involves using key pairs for each session and exchanging verification data with the server before entering login credentials. An OTP is processed by a contactless smart card after being sent to the recipient via SMS. However, this design can be vulnerable to security issues, such as when a user loses their phone or has it stolen.

In 2016, most mobile banking applications examined were found to have vulnerabilities, with unauthorized access to private customer data being the most significant risk. In 2017, issues such as fraud, identity theft, and unauthorized access to consumer banking data continued to affect mobile banking apps. To address these threats, banks should prioritize developing a robust architecture, clearly defining technology requirements, and ensuring secure development practices. Therefore, thorough testing of applications and security measures is essential [20].

A study by [21] analyzed 693 mobile banking applications from Google Play, spanning over 80 developed and developing countries across five continents, using the AUSERA tool to identify vulnerabilities. The research uncovered 2,157 vulnerabilities affecting 470 banking applications. These vulnerabilities included preference leakage, logging leakage, SMS leakage, and SD leakage. Additionally, the study found that invalid authentication was particularly prone to exploitation. The study recommended practical measures such as incorporating various response channels for addressing vulnerabilities, using application packing techniques, increasing bank attention to security issues, and carefully selecting third-party libraries.

A methodology for protecting mobile banking applications from Man-in-the-Middle (MITM) attacks was put forth by [22]. In his study, majority of mobile applications were found to be exposed to MITM assaults because of poor administration of TLS conventions and poor blueprints of programming and utilization of code libraries. This study found implementing public key cryptography and protocol encryption to be the best available defense techniques against MITM attacks. The study did point out that asymmetric cryptographic schemes singly are open to MITM assaults. Banks should conduct penetration testing in software design to uncover existing and new flaws for examination and improvement. Furthermore, mobile banking applications should incorporate and employ biometric systems like fingerprint reading, administration of hybrid cryptosystems that incorporate more than one algorithm to add additional layer of security to stop numerous assaults on mobile banking.

4 Methodology

This study employed a systematic literature review to assess the operation of mobile banking, identify security concerns, and recommend best practices for customers accessing banking services remotely. A systematic literature review provides answers to research questions based on evidence from previous studies that is recognized, evaluated, and interpreted [23]. The review aimed to summarize findings on mobile banking functionality, security issues, and effective practices for remote banking. Content analysis was utilized to examine how banking applications operate, the security challenges they face, and best practices for customers. Data was critically analyzed and organized into thematic areas, covering data sources, search processes, data selection, and data extraction.

The primary sources of this paper were selected from different scientific databases such as Institute of Electrical and Electronics Engineers Xplore Digital Library (IEEE Xplore), Science Direct (https://www.sciencedirect.com/), Springer Link (https://link.springer.com/), Google Scholar (https://scholar.google.com/), Association for Computing Machinery Digital Library (https://dl.acm.org/), peer reviewed international journals, and websites.

The research process was carried to identify potential research papers from scientific databases using preselected search keywords or strings including operation of mobile banking applications, security of mobile banking applications, and best practices for users of mobile banking applications. Additionally, the following Boolean operators have been used in the search process: ((Operation of mobile banking applications) OR (How does mobile banking operate?)), (((Mobile banking applications) OR (Security of banking Apps)).

Data selection was done through filtering results obtained on keywords in English language. The criteria employed scanned whether obtained results discuss about operation of mobile banking, if the research articles mentioned concepts about security of mobile banking applications, and finally if articles discuss about best practices for users of mobile banking applications. In order to apply inclusion and exclusion criteria, we put down exclusion criteria such as duplicate papers, full-text availability, and papers that are not related to operation of mobile banking applications. Approximately 180 publications were discovered after this work was completed, and those that were pertinent were chosen based on the search criteria. In the end, 42 related studies were found and used in this study.

5 Security of Mobile Banking Applications

Mobile banking is utilized across various platforms including Interactive Voice Response (IVR) [24], SMS [25], Unstructured Supplementary Service Data (USSD), Wireless Application Protocol (WAP) [26], and dedicated mobile banking applications [27]. The security of mobile banking applications involves using identity-based access controls like usernames and passwords, along with authentication methods such as two-factor and multi-factor authentication. When a client requests services from a bank server, a secure session is established using the Transport Layer Security (TLS) handshake mechanism, which encrypts the data exchanged between the client and server. The primary goal of encryption is to prevent adversaries from intercepting communications between the two parties. This makes mobile banking a secure method for conducting banking services remotely.

The increasing use of mobile banking has attracted cybercriminals who use various software and techniques to gain unauthorized access to client and bank servers to steal money. Methods such as mobile malware, Man-in-the-Middle (MITM) attacks, hacking, eavesdropping, and social engineering can be employed to breach communications between parties [27]. To prevent unauthorized access, authentication methods like one-time passwords, usernames and passwords, and biometric identification can be utilized [26].

6 Operation of Mobile Banking Applications

Mobile banking is a modern channel for conducting financial transactions, and it can be implemented using three technologies: web-based applications, SMS-based services, and client-based applications. Web-based applications require an internet connection, allowing users to access banking platforms via mobile internet [28]. SMS-based applications enable clients to interact with their bank by sending text messages from their mobile phones, with a code sent via SMS to complete banking transactions. Client-based applications involve installing dedicated mobile banking apps on a mobile device for access.

In the server-side applications, client data that is necessary for transaction processing (account or card details) is kept Mobile banking utilizes both client-side and server-side technologies. Server-side applications are installed on bank servers rather than on a customer's SIM card or mobile device. These server-side applications store client data needed for transaction processing (such as account or card details) in a highly secure environment managed by the bank or a designated service provider. In contrast, client-side technologies consist of programs, services, and tools designed for or integrated into a customer's SIM card or mobile device. In a client-server architecture, mobile client applications connect to bank servers. Due to the server's greater computational power and resources compared to a mobile device, the smartphone application acts as the client, facing resource limitations. This client-server relationship arises from the differences in processing capabilities. The application server monitors the network for incoming requests, which it then processes into a readable format. Each request is checked for security vulnerabilities using a designated protocol before being executed.

The application server can be managed by the bank or a third-party vendor. To ensure data integrity and confidentiality, hash algorithms and message digests are employed at the application layer, protecting interactions between the applications and the server [29].

To enhance security, banking systems require users to provide a second form of authentication in addition to their login credentials when accessing their accounts remotely. Financial institutions track customers' behavior through their interactions with mobile banking applications, allowing them to build detailed profiles. Another advantage of mobile applications is the ability to remotely erase or reset data on most smartphones and tablets. If a mobile device is stolen, the owner can use connected devices to wipe the stolen device's data and applications, thereby preventing unauthorized access to their account.

To access mobile banking services remotely, users must first complete a registration and validation process. Following registration, a one-time password (OTP) is sent to the registered phone number and used to configure the login details. In addition to traditional usernames and passwords, mobile banking systems also support biometric authentication methods, such as audio, facial, or fingerprint recognition [30]. Transport Layer Security (TLS) is employed to establish secure connections between a client and server, protecting data transmitted over the network. It provides security services for all application-based protocols, incorporating a client/server architecture that ensures integrity, confidentiality, and authentication in communication [31].

In the Open Systems Interconnection (OSI) reference model, the TLS Record protocol operates at a lower tier, directly above the transport layer. The TLS handshake protocols, located in a higher layer just above the Record protocol, manage data transmission. The Record protocol facilitates the exchange of data blocks between communicating parties by taking messages from the application level of the OSI model, segmenting them into manageable pieces, potentially applying Media Access Control (MAC), and then compressing, encrypting, and transmitting the final message. The handshake process addresses and establishes security mechanisms [31].

At the core of TLS is the handshake protocol, which establishes the security parameters for encrypted data transmission and authenticates each communication participant. Both parties are notified through the ChangeCipherSpec protocol when the session state has been updated to the negotiated parameters and secure communication has begun. The Alert protocol is used to inform the parties of any communication issues, such as a lost connection or encryption failures. During the handshake process, all cryptographic elements necessary for securing the connection are generated [31]. Various security concerns can impact the operation of mobile banking systems as follows:

6.1 Mobile Malware

Malicious software programs exploit security vulnerabilities to gain unauthorized access to computer systems. Malware attackers are often driven by financial or political motives, leading them to compromise as many network devices as possible to achieve their harmful goals [32].

Mobile malware can be categorized into various types based on its harmful objectives and activities. The primary distribution methods are social engineering and self-propagation. Social engineering exploits users' curiosity and lack of security awareness to trick them into installing applications (such as adware). Self-propagation involves malware spreading to mobile devices through methods like worms. In addition to these common types, there are several less conventional forms of malware. For example, piggybacked applications [33] initially appear benign but are later embedded with malicious payloads to become botnets or other harmful programs.

The four most common types of smartphone malware are Trojans, backdoors, spyware, and adware. Among Trojans, banking Trojans are particularly notable as they target mobile banking activities and services. Once a network connection is available, these Trojans can capture sensitive passwords and personal data, store it in hidden locations, and send it to a command-and-control server. The rise of banking Trojans paralleled the shift of banking activities from personal computers to mobile devices. SMS Trojans, which use premium SMS schemes to lure victims, are the second most prevalent attack method. [34] proposed the TruSMS technique to combat SMS spam and attacks. Additionally, the impact of mobile banking malware can be mitigated by using robust antivirus software.

6.2 Information Leakage, Loss or Distortion

Mobile banking relies on the transfer of information through wireless networks. These networks use radio waves to transmit digital data by modulating and sending the data as radio waves, then receiving and demodulating the signal to convert it back to digital form. The technology is designed to allow multiple radios to operate simultaneously without interference. However, current wireless network technology offers limited protection for wireless transmission media. As a result, confidential banking information may be at risk of being leaked, lost, or distorted during transmission on mobile devices. Cybercriminals can intercept sensitive information by exploiting vulnerabilities in the transmission of mobile communications, using overlapping and acceptable devices in the electromagnetic spectrum to delete, modify, add, or replay crucial data, thereby disrupting the normal use of legitimate users [35].

6.3 Incomplete Information

Network connectivity instability can result in incomplete information. For instance, if a customer moves from an area with strong network coverage to one with poor coverage while using a mobile phone to access banking services, the communication between the mobile banking app and the bank server may be delayed or fail entirely. This can lead to incomplete transactions. Additionally, if the mobile device's battery runs out, the customer will be unable to access mobile banking services [45].

6.4 Denial of Service

Denial of Service (DoS) attacks can be executed by routing traffic through a personal access point controlled by an attacker, disrupting communication between a customer's mobile banking application and the server. In such attacks, the adversary drops every packet in the data flow between the bank's server and the application, causing the mobile banking application to experience connection timeouts and eventually depleting system resources. DoS attacks targeting the application layer are common vulnerabilities in network protocols that prevent network devices from functioning correctly. These attacks either restrict or block intended users from accessing network resources or services. The challenge in detecting these attacks arises from the fact that Internet traffic is a mix of legitimate and abnormal activity, with attack traffic often mimicking regular traffic. Key application layer DoS threats include protocol-specific attacks and generic DoS attacks. Examples of protocol-specific assaults include Network Time Protocol (NTP) attacks, time-shifting DoS attacks, Slow Hypertext Transfer Protocol (HTTP) attacks, and Dynamic Host Configuration Protocol (DHCP) starvation attacks [37].

6.5 Threats from Third-Party Applications

Various programs are developed for Apple's and Google's operating systems by application development companies or individual developers. Third-party applications on mobile devices can potentially tamper with existing banking applications and steal account information. Users are advised to download applications or updates only from official sources or trusted app stores [36]. Most third-party applications in official stores adhere to stringent development standards and are checked for harmful software such as viruses. However, applications listed in unofficial third-party app stores may not undergo the same rigorous scrutiny. While secure applications can sometimes be found in these third-party marketplaces, the risk of encountering harmful ones is significantly higher.

6.6 Unencrypted Wireless Networks

Wireless networks operate through data transmission using radio frequencies, access points, and users. Unencrypted wireless networks lack passwords, allowing users to freely connect to access points without authentication. These networks are vulnerable to threats such as Man-in-the-Middle (MITM) and Denial of Service (DoS) attacks, which can lead to unauthorized access to both client and bank servers, resulting in confidential data leakage and financial theft. Consequently, wireless networks in public places, such as coffee shops and hotels, are often insecure. When users access mobile banking services, such as checking account balances, depositing checks, or paying bills, over these unsecured networks, cybercriminals can intercept and steal their sensitive information [37].

7   Best Practices for Users of Mobile Banking Applications

The security of mobile banking applications can be enhanced through various techniques, including two-factor authentication, data encryption, site keys combined with security questions and images, registered mobile device authentication, and the use of antivirus applications [38–40]. Below is a summary of several security measures and best practices that mobile banking users should follow.

User authentication should be a top priority in mobile banking. Most smartphones use screen locks that can be accessed through various forms of authentication or personal identification numbers [41]. However, multi-factor authentication is considered the best practice for protecting data [42]. Complex passwords are crucial as they enhance security and make it more difficult for adversaries to compromise accounts. Many third-party applications are free and often downloaded by users, but these applications may contain viruses and are not recommended for mobile banking. Customers should only download and install applications from official bank websites to ensure security.

To combat mobile malware, users should download and install mobile antivirus software. Additionally, some tools can scan URLs, block dangerous sites, monitor links in text messages, and offer parental controls [43], which experts highly recommend. However, these tools can lead to increased use of hardware resources and greater battery drain due to background processes. Similarly, mobile banking users are advised to use secure wireless networks for accessing banking services remotely. Unsecured wireless networks can expose sensitive information to hackers. Therefore, it is recommended that customers use private wireless networks when accessing their bank's server through mobile banking applications.

Users should employ data encryption to safeguard sensitive information. Encryption is used for both data transmission across networks and data storage on mobile devices. However, encryption typically requires a password for both encrypting and decrypting data files. If a password is forgotten, data recovery can be challenging or impossible. Relying solely on publicly available solutions may only provide a false sense of security. Additionally, users should avoid connecting to unencrypted networks and instead use virtual private networks (VPNs). Although VPNs offer enhanced security, they are generally slower than standard internet connections due to factors such as server distance, current server load, and the level of encryption used.

Users of mobile banking should be aware of social engineering techniques. These attacks are directed to unsuspecting users to extort user confidential information and even steal money. Users must be on the watch for people who solicit confidential information in the name of assisting them [44]. Banking organizations should institute a philosophy that teaches bank customers as well as staff on the different forms of social engineering propagation techniques and how they can be controlled. These can be informed of planned retreats for staff or by way of advertising to customers using videos as well as emails sent to customers.

Whenever updates are available, then it is the responsibility of the customer to download and install them. Update mobile operating systems and on-board applications with security patches: keeping the operating system (Android and iOS) and mobile banking applications up to date is a must. Both Google and Apple provide regular updates to users, which resolve recent vulnerabilities or other threats, as well as sharing additional performance and security features [44]. However, updating an application is a two-edged sword since a new release can decrease its overall performance and the user's productivity. From a security perspective, updates can trigger the re-vetting process to confirm security clearance. In order to ensure that a mobile banking application conforms to an organization's security requirements and is free from vulnerabilities, a series of rigorous and comprehensive analyses take place. One has to keep in mind that application vetting might also include updated external components (such as third-party libraries) and new mandatory versions of the operating systems.

Users should enable remote data wipe in their smart phones. This is because in case a user has their device with sensitive data stolen and there is little chance of retrieving them in a relatively short period of time, one should consider turning on the device capability which allows a factory reset message to be remotely executed [45]. Furthermore, remote data wipe is imperative in case of termination of employment or contracting a malware infection which cannot be uninstalled or deleted. While the existing solutions have clear advantages, they are not cure-all for mobile security. For instance, while some tools erase only part of the data, others erase the entire content, including applications and personal data. Therefore, one should consider deploying a secure container which, by design, separates the applications from personal data, enabling selective erasure in case of a security incident. Moreover, a proactive approach that tracks the use of sensitive data will improve security by early detection and prevention of its misuse or theft.

8 Theoretical and Managerial Implications

Mobile banking applications are one of the most modern financial technology innovations that are utilized by banking institutions to offer banking services on the go. Mobile banking applications help users to have complete control of their bank accounts remotely and access products and services such as check balances, transfer money, pay bills, request for cheque books, request for loans, insurance, invest in stocks, and fixed deposits at any time. In order to use a mobile banking application, the user will need to download a mobile banking application from the app store and install it on a smart phone or tablet. This is followed by user registration, activation, and creating login credentials for authentication. Since it is the responsibility of the user to interact with the bank server using the applications installed on the phone, then users should be informed of the threats that are inherent while using mobile banking applications as well as apply best security practices in order to successfully access banking services remotely.

Research indicates that mobile banking applications are vulnerable to a range of threats, including mobile banking Trojans, fraudulent banking apps, man-in-the-middle attacks, clickjacking, and keylogging malware, among others. Cybercriminals often exploit these applications due to various common design flaws, such as inadequate input validation, weak session management, poor error handling, insufficient access controls, and insecure coding practices [46]. Studies show that a significant percentage of mobile banking apps have been compromised, with approximately 95% of Android and 70% of iOS applications falling victim to hacking [47]. The increasing trend of hacking mobile banking applications suggests that if this pattern continues, many bank customers may become wary of using mobile banking technology and revert to traditional in-person banking. This shift would undermine the technological benefits that mobile banking offers to both banks and customers. To address this issue, banks should adopt recognized software standards with a strong focus on security and implement modern encryption techniques, such as hybrid algorithms.

9   Discussion, Future Research and Conclusion

The study found that for clients to access mobile banking services remotely, they must first register with their bank to be validated. An OTP (One-Time Password) is sent to the account holder's phone, which is then used to configure login credentials, including passwords and biometric authentication. Communication between the mobile banking application and the bank's server occurs through protocols, with HTTP and HTTPS being the most common. It is recommended that mobile banking applications use HTTPS to ensure data transmission is secure, authentic, and private. However, some banking applications may lack proper Secure Socket Layer (SSL) frameworks, potentially leaving them vulnerable to Man-in-the-Middle attacks.

The study also revealed that mobile banking operations begin by establishing a secure session between the client's mobile banking application and the bank's server, where the client's financial and personal data is stored. This secure connection is created using the TLS handshake mechanism, which encrypts data exchanged between the bank server and the mobile banking application to protect it from hackers. Following the establishment of this secure connection, the authentication phase begins. During this phase, the client is authenticated through various methods such as username and password, fingerprint, facial recognition, or audio authentication. The bank's server then verifies the provided authentication method. If the authentication is successful, the user is granted access to their account for remote banking services. If the authentication fails, the user is prompted to repeat the process with the correct credentials stored on the bank's server.

This study aligns with the findings of [48], which indicate that mobile banking operations begin by establishing a secure connection between the client and the bank's server through the TLS handshake mechanism. Once a secure connection is in place, two levels of authentication are employed to access the bank's server. The first level involves using a username and password, while the second level requires a second factor of authentication, such as the International Mobile Equipment Identity (IMEI) or Subscriber Identity Module (SIM) serial number, for verification by the bank's server before a customer can access mobile banking services remotely.

This study identified several security concerns that mobile banking users should be aware of and address by following best practices. Key concerns include mobile malware, information leakage, data loss or distortion, incomplete information issues, denial of service attacks, risks from third-party applications, and dangers of using unencrypted wireless networks. To mitigate these risks, users should adopt the following best practices: implement two-factor authentication for accessing mobile banking services, use mobile antivirus software, connect to secure networks when using mobile banking, promptly download and install updates for operating systems and banking applications, and stay vigilant against social engineering tactics to avoid scams.

Future research could focus on enhancing mobile banking security through the integration of hybrid algorithms. These algorithms combine the security features of two or more systems into a single, more robust solution. For instance, merging Advanced Encryption Standard (AES) with steganographic methods like Least Significant Bit (LSB) steganography can offer increased protection due to the combined strengths of both AES and LSB techniques. Additionally, incorporating Artificial Intelligence (AI) algorithms could enhance mobile banking applications by enabling real-time detection of unusual behavior and identification of potential security threats. This would help banks safeguard sensitive information and prevent fraud. Furthermore, blockchain technology could be leveraged to bolster mobile banking security by providing a transparent and tamper-resistant ledger of financial transactions, thereby reducing the risk of fraudulent activities.

To access banking services via mobile banking applications, users must first download and install the designated app on their smartphone. They must complete a preliminary registration process where their details are entered into the mobile banking platform. The bank then sends login credentials to the user via SMS or email, which are used to configure the account. To access banking services remotely, the user must log in with these registered details. If incorrect login credentials are provided, the user will be prompted to re-enter the correct information. Mobile banking offers a secure and convenient way to conduct transactions anytime and anywhere, provided that network connectivity is available. Security is a critical aspect of remote banking, and users should be aware of potential security issues and adhere to recommended practices to ensure their safety while banking online.

10 Recommendations

The operation of mobile banking applications can be impacted by various security concerns, including mobile malware, information leakage, data loss or distortion, incomplete information, Denial of Service (DoS) attacks, threats from third-party applications, and the use of unencrypted wireless networks. These vulnerabilities can jeopardize access to mobile banking services, as malicious actors may exploit them to steal sensitive information or funds. To mitigate these risks, users are advised to implement multifactor authentication and avoid using unencrypted wireless networks when connecting to their bank's server. These practices help secure communication and make it more difficult for adversaries to compromise the user's interactions with the bank.

This study also recommends that users of mobile banking applications install antivirus software to combat mobile malware effectively. Additionally, users should download and install software updates, including operating system updates and newer versions of banking applications, exclusively from official bank websites. They should avoid third-party applications, which may contain malware that can compromise login credentials.

References

1. Palma, F., Realista, N., Serrao, C., Nunes, L., Oliveira, J., Almeida, A.: Automated security testing of android applications for secure mobile development. In: Proceedings 13th International Conference on Software Testing, Verification and Validation Workshops, ICSTW, pp. 222–231 (2020)
2. Paytm. https://paytm.com/blog/net-banking/what-is-mobile-banking-application/. Accessed 10 Mar 2024
3. Hayikader, S., Hadi, F.N.H.A., Ibrahim, J.: Issues and security measures of mobile banking apps. Int. J. Sci. Res. Publ. 6(1), 36–41 (2016)

4. El Janati El Idrissi, N., Orhanou, G., Kouraogo, Y., Zkik, K.: Security model on mobile banking application: attack simulation and countermeasures. Int. J. Intell. Enterp. **4**(1/2), 155 (2017)
5. Citi. https://www.citigroup.com/citi/news/2018/180426a.htm. Accessed 7 Apr 2024
6. Janiper Research. https://www.juniperresearch.com/press/digital-banking-users-to-reach-2-billion. Accessed 2 Nov 2023
7. Baabdullah, A., Alalwan, A.A., Rana, N., Patil, P.P.: Consumer use of mobile banking in Saudi Arabia: towards an integrated model. Int. J. Inf. Manag. **44**(1), 38–52 (2019)
8. AL-Zubi, K.: The effect of web-related features on intention to use online banking of ATM users. Int. J. Data Netw. Sci. **5**(4), 629–640 (2021)
9. Owusu, G.M.Y., Bekoe, R.A., Addo-Yobo, A., Otieku, J.: Mobile banking adoption among the Ghanaian youth. J. Afr. Bus. **22**, 339–360 (2020)
10. Al-Otaibi, S., Aljohani, N.R., Hoque, M.R., Alotaibi, F.S.: The satisfaction of Saudi customers toward mobile banking in Saudi Arabia and the United Kingdom. J. Glob. Inf. Manag. **26**(1), 85–103 (2018)
11. Zhou, T.: Examining users switch from online banking to mobile banking. Int. J. Netw. Virtual Organ. **18**(1), 51–66 (2018)
12. Świecka, B., Terefenko, P., Paprotny, D.: Transaction factors' influence on the choice of payment by Polish consumers. J. Retail. Consum. Serv. **58**, 102264 (2021)
13. Herzberg, A.: Payments and banking with mobile personal devices. Commun. ACM **46**(5), 53–58 (2013)
14. Chou, Y., Lee, C.W., Chung, J.: Understanding m-commerce payment systems through the analytic hierarchy process. J. Bus. Res. **57**, 1423–1430 (2004)
15. Sharma, L., Mathuria, M.: Mobile banking transaction using fingerprint authentication. In: 2nd International Conference on Inventive Systems and Control, pp. 1300–1305. IEEE Xplore (2018)
16. Lin, W.R., Lin, C.Y., Ding, Y.H.: Factors affecting the behavioral intention to adopt mobile payment: an empirical study in Taiwan. Mathematics **8**(10), 1–19 (2020)
17. Shankar, A., Rishi, B.: Convenience matter in mobile banking adoption intention? Australas. Mark. J. **28**(4), 273–285 (2020)
18. Bolt, W., Mester, L.J.: Introduction to retail payments: mapping out the road ahead. J. Financ. Serv. Res. **52**(1), 1–3 (2017)
19. Wirani, Y., Hidayanto, A.N., Shihab, M.R.: Factors affecting acceptance of internet banking with technologyorganisation-environment framework: a perspective of corporate customer in Indonesia. Int. J. Bus. Inf. Syst. **34**(4), 560–593 (2020)
20. PositiveTechnologies. https://www.ptsecurity.com/upload/corporate/ww-en/analytics/Cybersecurity-threatscape-2018-eng.pdf. Accessed 10 Mar 2024
21. Chen, S., et al.: Are mobile banking applications secure? What can be improved? In: Proceedings of the 26th ACM Joint meeting on European Software Engineering Conference and Symposium on the Foundations of Software Engineering, Lake Buena Vista, FL, USA, pp. 797–802 (2018)
22. Luvanda, A.: Proposed framework for securing mobile banking applications from man in the middle attacks. J. Inf. Eng. Appl. **4**(12), 20–28 (2014)
23. Wahono, R.S.: A systematic literature review of software defect prediction. J. Softw. Eng. **1**(1), 1–16 (2015)
24. Thirumaran, M., Soni, S., Gayathry, B.G.: An intelligent interactive voice response system for banking domain. In: Proceedings of the 2015 International Conference on Advanced Research in Computer Science Engineering & Technology, pp. 1–6. ICARCSET (2015)
25. Bojjagani, S., Sastry, V.N.: A secure end-to-end SMS-based mobile banking protocol. Int. J. Commun. Syst. **30**(15), 1–19 (2017)

26. Bhadauria, R., Sanyal, S.: Survey on security issues in cloud computing and associated mitigation techniques. Int. J. Comput. Appl. **47**(18), 47–66 (2012)
27. Nosrati, L., Bidgoli, A.M.: A review of mobile banking security. In: Proceedings of IEEE Canadian Conference on Electrical and Computer Engineering, Canada, pp. 1–5. IEEE (2016)
28. Athidas, G., Alamelu, K.: Security issues in mobile banking. Shanlex Int. J. Manag. **6**(1), 6–10 (2016)
29. Kim, G., Shin, B., Lee, H.G.: Understanding dynamics between initial trust and usage intentions of mobile banking. Inf. Syst. J. **19**(3), 283–311 (2009)
30. Kamal, R.: Mobile Computing, 2nd edn. Oxford University Press, Oxford (2020)
31. Kieseberg, P., Fruhwirt, P., Schrittwieser, S., Weippl, E.: Security tests for mobile applications why using TLS/SSL is not enough. In: IEEE Eighth International Conference on Software Testing, Verification and Validation Workshops, Graz, Australia, pp. 1–2 (2015)
32. Curguz, J.: Vulnerabilities of SSL/TLS protocol. Comput. Sci. Inf. Technol. (2016). The Sixth International Conference on Computer Science, Engineering and Information Technology, pp. 245–256
33. Stevens, C.: Assembling cybersecurity: the politics and materiality of technical malware reports and the case of Stuxnet. Contemp. Secur. Policy **41**(1), 129–134 (2020)
34. Fan, M., Liu, J., Wang, W., Li, H., Tian, Z., Liu, T.: DAPASA: detecting android piggybacked apps through sensitive subgraph analysis. IEEE Trans. Inf. Forensics Secur. **2**(8), 1772–1785 (2017)
35. Chen, L., Yan, Z., Zhang, W., Kantole, R.: TruSMS: a trustworthy SMS spam control system based on trust management. Future Gener. Comput. Syst. **49**, 77–93 (2015).
36. Nie, J., Hu, X.: Mobile banking information security and protection methods. In: International Conference on Computer Science and Software Engineering Mobile, Wuhan, China, pp. 587–590 (2008)
37. He, Wu., Tian, Xin., Shen, J.: Examining Security Risks of Mobile Banking Applications through Blog Mining. Old Dominion University, Norfolk, VA, USA (2015)
38. Legnitto. https://ceur-ws.org/Vol-1353/paper_24.pdf. Accessed 10 Mar 2024
39. Tripathi, N., Hubballi, N.: Slow rate denial of service attacks against HTTP/2 and detection. Comput. Secur. **72**, 255–272 (2021)
40. Lee, H., Zhang, Y., Chen, K.L.: An investigation of features and security in mobile banking strategy. J. Int. Technol. Inf. Manag. **22**(4), 23–46 (2013)
41. Chandramohan, M., Tan, H.B.K.: Detection of mobile malware in the wild. Computer (9), 65–71 (2012)
42. La Polla, M., Martinelli, F., Sgandurra, D.: A survey on security for mobile devices. Commun. Surv. Tutor. **15**(1), 446–471 (2013).
43. Kent, D., Liebrock, L.M, Neil, J.C.: Authentication graphs: analyzing user behavior within an enterprise network. Comput. Secur. **48**, 150–166 (2015)
44. Dasgupta, D., Roy, A., Nag, A.: Multi-factor authentication. In: Dasgupta, D., Roy, A., Nag, A. (eds.) Advances in User Authentication. Infosys Science Foundation Series, pp. 185–233. Springer, Cham. https://doi.org/10.1007/978-3-319-58808-7_5
45. Vermaat, M.E., Sebok, S.L., Freund, S.M., Campbell, J.T., Frydenberg, M.: Discovering Computers 2018: Digital Technology, Data, and Devices, Nelson Education, Toronto, Canada (2017)
46. Asee. https://cybersecurity.asee.io/blog/mobile-security/enhancing-mobile-banking-app-security-top-threats-and-solutions/. Accessed 7 Apr 2024

47. Arxan. https://finance.yahoo.com/news/arxans-annual-report-state-mobile-103000408.html. Accessed 7 Apr 2024
48. Abraham, S., Chengalur-Smith, I.: An overview of social engineering malware: trends, tactics, and implications. Technol. Soc. **32**(3), 183–196 (2010)
49. Waghmare, M., Golekar, P., Hatwar, A., Parimal, R., Hiware, A.: Enhancement of security in mobile banking applications. Int. J. Sci. Re. Dev. **2017**5(1), 110–112

Analyzing the USA Housing Complaints to Score the County Problems

Sharath Kumar Jagannathan[1]([✉]), Gulhan Bizel[1], Vijay Kumar Voddi[1], and J. V. Thomas Abraham[2]

[1] Data Science Institute, Saint Peter's University, 2641 JFK Boulevard, Jersey City 07306, NJ, USA
{sjagannathan,gbizel,vvoddi}@saintpeters.edu
[2] VIT University Chennai, Vandalur Kelambakkam Road, Chennai 600127, Tamil Nadu, India
thomasabraham.jv@vit.ac.in

Abstract. The 311 service in New York City investigates complaints concerning homes and structures. There are certain preliminary tests to make sure the facilities and habitat that people live in before the actual spent of money whether to purchase it or rent it. The amount of housing complaints from each region, which the real estate brokers do not reveal, paints the true image of the real estate mess. One such dependable resource that offers us a clear picture of the environment, the kind of complaints, and how they are handled is the 311 housing complaints record. In addition to allowing complaints about things like illegal parking and noise, 311 is a resource for New York City residents that allows tenants to complain about terrible living conditions in their homes. In this study, several non-emergency complaints are evaluated which is filed with 311 about residential properties in New York due to a considerable increase in such complaints over the years. The organization in charge of handling 311 complaints on housing and structures in New York City is the Department of Housing Preservation and Development. This paper will analyze and categorize the most common complaints from the data, as well as their priorities, to handle them appropriately. To see if any borough in specific has reported for similar type of complaint repeatedly. Analyzing the data further to predict risk score based on these complaints.

Keywords: Urban Analytics · Predictive modelling · Service requests · non-emergency · domestic issues · residential

1 Introduction

Countries have evolved with tremendous transformations due to digital technologies. Technologies connect people to proper targeted communities to build an easier environment to communicate, report and resolve issues in a faster phase. One such service is a 311 service. If your city has a 311 service, you can call to report issues like abandoned vehicles, noise complaints, and graffiti.

The New York city has a robust plan of action to address the nonemergency complaints through 311 service. Due to planning and infrastructure constraints, this operation cannot be implemented nationwide. Like other services such as 911 and suicide help, 311 services operate in a similar way. Residents are encouraged to inform the landlord or the building contractor first and explain them the issue they are facing. If they fail to take up any action to address the issue/complaint, then the residents can either dial 311 or can register the complaint online. When a complaint is submitted, Housing preservation and development tries to get in touch with the managing agent of the building, whoever that may be, to let them know about the complaint and that if the issue isn't remedied, a violation may be issued. After then, the agency seeks to get in touch with renters to check if the issue has been resolved. If not, an inspector is dispatched. A uniform code enforcement inspector will be dispatched to inspect the reported condition if the violation was not repaired, or Housing preservation and development was unable to get in touch with you. When the inspection takes place, neither the owner nor any managing agency for the property will be informed. A landlord may have a certain length of time to resolve a problem. Class A offenses, which are regarded as non-hazardous, are exempt from immediate correction for 90 days. Landlords have a 30-day window before facing civil fines for the more serious Class B breaches. Class C infractions, which are the most dangerous, must be corrected in a much less time. Fixing issues like lead paint or rodent, mouse, and cockroach infestations must be done within 21 days. Violations involving heat and hot water must be corrected right away [1].

Let's take an instance where, in a building there is not hot water issue and despite complaining the owner about it, the issue is not fixed then, one can reach out to 311 and register their service request. Then, as mentioned above, a follow up will be made with the owner and then they enquire with the complainer whether the problem is fixed or not. Sidewalk and road repairs, noise complaints, abandoned cars, large debris obstructing streets, garbage bin replacements, concerns about dangerous animals, damaged traffic signals or road signs, graffiti removal, leaking hydrants or sewage issues are a few of the frequent reasons people dial 311. Every day, there are a ton of calls coming into 911. Consider if your matter is an emergency before contacting the police. Call 911 if there is an emergency, such as a major injury or house fire. But in other circumstances, your neighborhood services can be just as useful [2].

Getting insights from customers and drawing meaningful conclusions from the analysis of such data has always proven worthwhile for businesses in all segments, especially in today's digital world where people freely share their opinions, experiences, and reviews about anything and everything. Before making a purchasing decision, people frequently conduct thorough research and read product reviews. Like this, clients frequently conduct internet reviews searches when they are interested in renting or purchasing a home to get a basic sense of the neighborhood. Therefore, it is important to address these non-emergency concerns in the boroughs since doing so will benefit the local real estate market as satisfied clients are believed to be a key motivator for purchasing. Therefore, it is important to solve these problems to benefit the real estate industry and offer people better and more convenient housing [1].

In one of the studies the terms of the infrastructure variables, catch basin complaints were significant in 47% of the zip codes, back up complaints were significant in 41% of the zip codes, and manhole overflow complaints were significant in 21% of the zip codes when compared to complaints about street flooding. Thus, it was discovered that infrastructural problems were predictors of complaints about street flooding for a significant number of zip codes [3]. 311 requests can be useful as avoid overdose surveillance indicators for fundamental research and practical policy because they are publicly available with high spatial and temporal resolution [4]. Datasets on air quality, land surface temperature, normalized difference vegetation index, building usage, building and urban design metrics, and sociodemographic datasets with population and health metrics were all gathered through a dense stationary sensing network in New York City and used in this study. The relationship between the air quality, land surface temperature, building usage, and urban metrics has been investigated by a neighborhood scale footprint-based regression analysis. The strongest relationships between air quality, land surface temperature, urban design, and socio-economic indices have been found [5].

In this article, they analyze how much water will be used in various heat and electricity production scenarios for the Netherlands and its capital, Amsterdam, in 2050. The analysis demonstrates that I due to the use of aquifer thermal energy storage, the water withdrawal for heating can increase up to the same order of magnitude as the current water withdrawal of thermoelectric plants, (ii) the virtual water use for heating can become higher than the operational water consumption for heating, and (iii) due to the increase in electricity production, the water use for electricity production becomes a relevant indicator for the virtual water use for heat generation [6].

We contrast three different metrics of political and civic engagement voter turnout, political donations, and census return rates—with spatially aggregated 311 call data. We demonstrate that rates of 311 calls are positively associated to the high-cost activity of campaign donation but negatively connected to lower cost activities (voter turnout and census return rates). We advise against using 311 data as a general indicator of political participation or engagement, at least in the absence of reliable controls for neighborhood quality. As a gauge of the service demands that neighborhoods impose on city governments, we contend that these data are nevertheless potentially relevant for researchers [7].

Four groups of mitigation strategies are taken into consideration: urban design, white/grey infrastructure, blue infrastructure, and green infrastructure. The breadth of urban functions and needs, relating to urban lives and urban operation, in ten dimensions, including economy, policy, ecology, environment, technology, space, urban beauty, practicality, culture, and transportation. The analytical framework was also used to examine cooling strategy trade-offs, conflicts, and co-benefits in 10 different elements of urban function. Additionally, it was utilized for space functions, such as activity venues and entertainment venues, neighborhood vitality, resident satisfaction, space utilization, and city identity, as well as environmental functions, such as local temperature regulation, stormwater regulation, waste treatment, air quality regulation, pollination, and recreation and aesthetic appreciation [8].

Combustion of gas and biomass are major sources of NYC air pollution mortality. NYC biomasses used alone caused 0.7–1.5K deaths uses PLUTO data set [9]. It is crucial for sustainable urban planning and growth to investigate how spatial features affect urban heat trends. Based on the classification of high-resolution Quick Bird satellite data, we investigated the relationship between surface temperature as determined by the Advanced Spaceborne Thermal Emission and Reflection Radiometer (ASTER, at

90 m spatial resolution), and various urban components. We looked further into the connections between building footprint and land use data collected by the New York City (NYC) Department of City Planning and surface temperature calculated by ASTER [10].

In today's world, where people in general do extensive research before buying a product, we can understand how important it is for people to get to know all about the locality that they are moving in to. This will help the customer in making a wise choice before choosing a property. So that they can opt for a locality that is well suited to invest in. It is also useful for the property owners to know the problems associated with them so that they can make necessary amendments to improve the price of their property.

2 Literature Survey

In this era of big data, data is generated rapidly from many sources and must be analyzed. The shortcoming is that the data is huge and organizations like municipal governments are ready to share in the name of open data to the public for social awareness. One such government public data is the 311 service requests which provides non-emergency queries of public. [2] The author talks about how big data can be used to evaluate and predict the non-emergency service requests.

Residents of numerous cities can contact their local government via the phone and web service 311 to report non-emergency concerns and issues with city services. This study investigates the difference in 311 contact volume across Census-tract levels in New York City. The paper focuses on how contacting propensity and condition both explain regional variations in contacting volume, drawing on prior research on citizen-government interaction, service delivery, and civic engagement. These explanations are put to the test using variables that describe the local population, employment rates, housing stock, level of economic development, and representation in local government [11].

There are some limitations to having a predictive approach on residential properties is the maximum heterogeneity of data available as a non-emergency request. However on the other hand there are privacy considerations which is of major concern and also the complexity involved in the data collected with periods of time with legal liabilities and practical restrictions which undermine the prediction using data analytics [12, 13].

The data available as open data offers a real simulation to monitor various conditions that could impact community health. However, the limitation is that the data is not uniform, and any misinterpretation will have a negative impact on society. The author represents the urban community with high temporal precision and precise geographical location. The author uses the keywords as tokens and calculates the specificity and sensitivity to find association using the Pearson's correlation coefficient. The study indicated focus on litter, lawns not mowed, wall painting. It believes the analytical study will give insights to improve the health of the community [14].

The author studies the crime by comparing the geographical localities with the Washington DC micro variants along with the unnoticed effects of neighborhood. The author also uses binomial regression and found a slight similarity for the crime and the infrastructure service requests. The study suggests that even though the slight effects could have a significant decrease in the level of crime [15]. The process used to create a pro-jectile objective personality test did not result in reliable correlates or a comparable test. However, using a verbal aptitude test as the criterion, the same technique was applied, and the result was a test of clever word associations. In the context of this test development, issues with word association item development are discussed [16].

The author show the availability of the 311 services and stresses the point that it is prone to errors if the process is done manually. The author uses neural networks to determine the category of service requests based on the request description. This process is faster and more accurate and only a few results in errors which were not labelled properly [3]. The paper states that the core service problems are still not covered by research and a vast area still prevails. San Francisco has a successful 311 system, but the study could show evidence that could be effective for public managers to manage and improve it effectively [17].

This data brief provides a comprehensive overview of drug overdose deaths in the United States from 1999 to 2017, offering vital statistics that shed light on the escalat-ing crisis [23]. This article presents a compelling argument for considering the opioid epidemic as a national emergency, discussing the legal and public health implications of such a designation [24]. Focused on the prevention of fatal opioid overdoses, this paper examines strategies and interventions that can be employed in the medical community to address this growing crisis [25]. This public health article explores the complex social and economic determinants of the opioid crisis, highlighting the multifaceted nature of this public health challenge [26]. This study uses publicly available data to map the geospatial distribution of discarded needles in Boston, Massachusetts, providing insights into the urban impact of the opioid overdose epidemic [27]. The authors integrate public health and community development perspectives to address neighborhood distress and promote well-being, offering a comprehensive approach to urban health issues [28]. The author analyzes the structure of 311 service requests in New York City as a signature of urban location, illustrating how public data can be used to understand urban dynamics [29]. Marmot discusses the social determinants of health inequalities, providing an in-depth look at how societal factors influence health outcomes and disparities [30]. In this review article delves into the social determinants of health, discussing how these factors have become increasingly recognized and addressed in public health research and policy [31]. The authors measure the spatial influence of physical features on gun violence in urban environments, using a bounded street network as a case study to understand how urban design can impact public safety [32].

In their 2023 study, Bullock III, Lamb, and Wilk delve into the complex interplay between race, ethnicity, and the enforcement of fair housing laws across various regions, as published in the Brigham Young University Journal of Public Law. The research presents a thorough regional analysis to understand how enforcement practices might vary and potentially reflect or exacerbate racial and ethnic disparities in housing. By meticulously examining data related to housing discrimination complaints, legal frame-works, and enforcement outcomes, the authors shed light on the nuanced challenges and systemic issues faced in ensuring fair housing practices. Their findings highlight signif-icant regional differences in the effectiveness of fair housing enforcement, suggesting that local policies, demographic composition, and historical contexts play crucial roles in shaping these outcomes. This study not only contributes to the broader discourse on racial and ethnic equality in housing but also underscores the need for tailored policy interventions that address the unique needs and challenges of diverse communities. Through this analysis, Bullock III, Lamb, and Wilk offer valuable insights for policy-makers, legal scholars, and housing advocates seeking to enhance equity in housing access and combat discrimination [33].

The authors in this study introduce the Unified Urban Knowledge Graph (UUKG) dataset, a groundbreaking resource designed to enhance urban spatiotemporal prediction efforts, as reported in Advances in Neural Information Processing Systems. This innovative dataset integrates diverse urban data sources into a comprehensive knowledge graph, facilitating advanced analysis and prediction of urban dynamics across various dimensions, including traffic flow, pollution levels, and socio-economic activities. By leveraging the UUKG, the authors demonstrate significant improvements in predictive accuracy for a range of urban phenomena, highlighting the dataset's potential to serve as a foundation for smart city applications and urban planning strategies. The study not only showcases the technical methodology behind the dataset's construction but also explores its application in predictive models, offering a blueprint for future research and development in the field of urban informatics. Through the UUKG, Ning et al. contribute to the enhancement of urban life quality by enabling more informed decision-making and efficient urban management [34].

3 Dataset

New York City has made its service-request 311 call data available through its NYC Open Data project (City of New York 2013). "Service requests" are only a subset of 311 calls and online submissions. In fact, most contacts with the service are not "service requests" but rather inquiries of another nature.

In this study, two datasets are taken from NYC open data, a NYC government website and the other dataset is taken is PLUTO data set, to know about the Extensive land use and geographical data. From this data, we take relevant columns that help in prediction and analysis of the service requests.

3.1 New York City 311 Data

A Service Request within the NYC311 system is a powerful tool for residents to communicate with their city government and seek assistance on a variety of issues. It is essentially a formal request for service, inspection, or resolution of a problem that a citizen encounters in their daily life within the city. The scope of these requests is notably wide, encompassing over 500 different types of complaints and issues.

NYC311's ability to handle such a diverse array of Service Requests demonstrates the city's commitment to addressing the daily concerns of its residents. It provides a direct line of communication between the public and the city administration, facilitating timely and effective solutions to local issues. This system not only aids in maintaining the city's infrastructure and services but also plays a crucial role in enhancing the quality of life for its residents by ensuring their concerns are heard and addressed efficiently.

Top of Form New York City 311 Service Requests where the primary data source for algorithm development were pulled from New York City's Open Data Portal [18].

3.2 PLUTO Data

The PLUTO (Primary Land Use Tax Lot Output) dataset is an extensive collection of land use and geographic information at the tax lot level, organized in a comma-separated values (CSV) file format for ease of use and analysis. This dataset is particularly rich in details, encompassing over seventy distinct fields that are derived from various data sources maintained by city agencies. These fields cover a wide range of data points, including but not limited to lot size, building type, zoning classification, ownership information, and commercial and residential use characteristics.

Importantly, the PLUTO files serve as a critical resource for urban planners, researchers, and policymakers who are interested in understanding the intricate details of land use patterns within the city. This data can be leveraged for various purposes, such as assessing real estate trends, planning urban development projects, conducting environmental impact studies, or formulating zoning laws.

The dataset is made available to the public through the New York City Planning Open Data Portal, which is a part of the city's initiative to promote transparency, civic engagement, and inclusive government by providing access to meaningful data. The accessibility of the PLUTO dataset via this portal allows for a wide range of users, from academic researchers to real estate developers, to analyze and utilize this data in their respective fields. The comprehensive nature of the dataset, combined with its public availability, makes it an invaluable tool for detailed and informed decision-making regarding New York City's urban landscape. The PLUTO files contain more than seventy fields derived from data maintained by city agencies. The data is pulled from the NYC planning open data portal [19].

4 Methodology

The initial step involved importing the NYC 311 service request dataset into a pandas Data Frame, a data structure in Python that facilitates data manipulation and analysis. This data set encompasses a wide range of non-emergency complaints and requests filed by residents of New York City. By loading this data into a Data Frame, we could efficiently handle, process, and analyze the complex and voluminous data.

Our analysis then progressed to focusing on specific areas within the city that were most affected by types of complaints. This involved segmenting the dataset based on complaint types and examining the geographical distribution and frequency of these complaints. By doing so, we aimed to identify patterns and hotspots of recurring issues within the city.

To enrich our analysis with more detailed insights, we incorporated additional data from the Bronx PLUTO (Primary Land Use Tax Lot Output) dataset. This dataset was imported into a separate Data Frame and includes comprehensive information about land use and geographic data at the tax lot level. Key features of the PLUTO dataset that we utilized include area and volume measurements of lots, the number of floors in buildings, dates of construction and any subsequent repairs, zip codes, and precise geographic coordinates.

Data analysis utilizing visual methods is called Exploratory Data Analysis (EDA). With the use of statistical summaries and graphical representations, it is used to identify trends, patterns, or to verify assumptions. In this project two datasets are used. One is New York City 311 data and the other is PLUTO data. The insights obtained by performing EDA on the New York city 311 dataset are discussed below. Using count plot from seaborn, the status of complaints for different boroughs in New York City is obtained. From the count plot, Manhattan has the highest complaint status as "Closed". It can be inferred that the Housing Preservation and Development is working efficiently to address the issue and to close the tickets.

In our analysis of the NYC 311 service request data, we utilized a variety of visualization techniques to understand the distribution and nature of complaints across different boroughs. A pie-chart was employed to analyze the 'sortedComplaintType' column, revealing five main categories of complaints, with Heating being the most frequent and Street Condition the least. To delve deeper into borough-specific issues, we used subplots, uncovering that the Bronx had the highest number of complaints about illegal or animal-related waste, Brooklyn was notable for public toilet complaints, Manhattan had a high incidence of bottled water complaints, Queens had numerous complaints about illegal animal solid waste, and Staten Island reported most about lifeguard issues. For unspecified boroughs, the most common complaint was related to found property.

A pyplot provided a city-wide perspective, showing Brooklyn with the highest overall complaint count and Far Rockaway with the least. Bar plots were also utilized to depict the count of unique complaint types in each borough, highlighting the diversity of issues faced, particularly in unspecified boroughs, and fewer in Staten Island. These visualizations collectively offered a comprehensive view of the service request landscape in New York City, enabling a more targeted and effective response to the varying needs of each borough.

4.1   NYC 311 Service Requests

Figure 1 in the study serves as a visual representation of the diversity and distribution of complaints registered across the different boroughs of New York City. This figure is pivotal in highlighting the unique challenges and issues faced by each borough. It clearly delineates the types of complaints that are more prevalent in certain areas, providing a geographical context to the data obtained from the NYC 311 service request dataset.

For instance, one borough might show a higher incidence of housing-related complaints such as heating issues, while another may have more complaints related to street conditions or public utilities. This variation could be indicative of the differing infrastructure, demographic composition, and urban development stages of each borough.

By analyzing Fig. 1, researchers, city planners, and policymakers can gain valuable insights into the specific needs and problems of each borough. This understanding is crucial for tailoring city services and resources to effectively address the distinct issues of each area. Furthermore, it can aid in prioritizing interventions and formulating targeted strategies to improve the quality of life for residents in different parts of New York City.

Figure 2 shows the count of complaints registered in 311 databases for each borough in New York. Also, the Bronx borough is on the top of the list. It is difficult to analyze all the complaints and all boroughs separately and select some criteria for analysis.

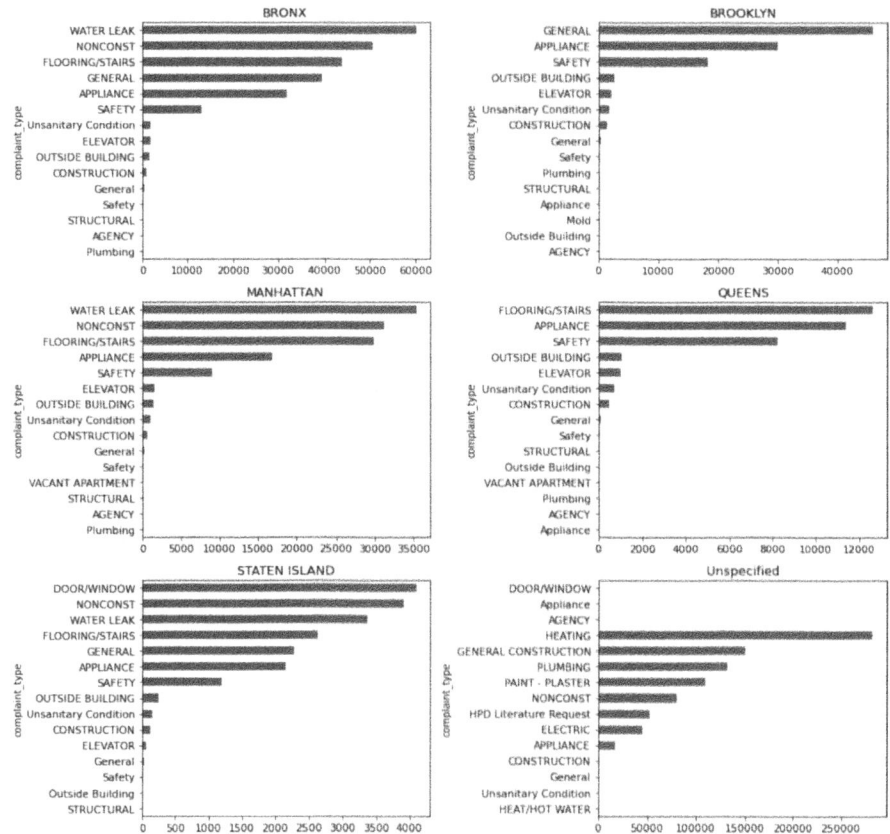

Fig. 1. Complaints in each borough.

Table 1 in the study plays a crucial role in bridging the information from the 311-service request database with the PLUTO (Primary Land Use Tax Lot Output) dataset. This Table 1 illustrates the count of complaints registered for each incident address within the various boroughs of New York City, as recorded in the 311 databases.

By detailing the number of complaints per incident address, Table 1 also provides a clear and concise visual representation of the geographical distribution of service requests across the city. This allows for a more granular analysis of the data, enabling researchers to pinpoint specific areas with high frequencies of types of complaints.

The significance of Table 1 is amplified when the data is used in conjunction with the PLUTO dataset. By matching the incident addresses from the 311 complaints with the detailed land use and geographic data in PLUTO, the study can uncover deeper insights into the relationship between urban infrastructure and the nature of the complaints. For

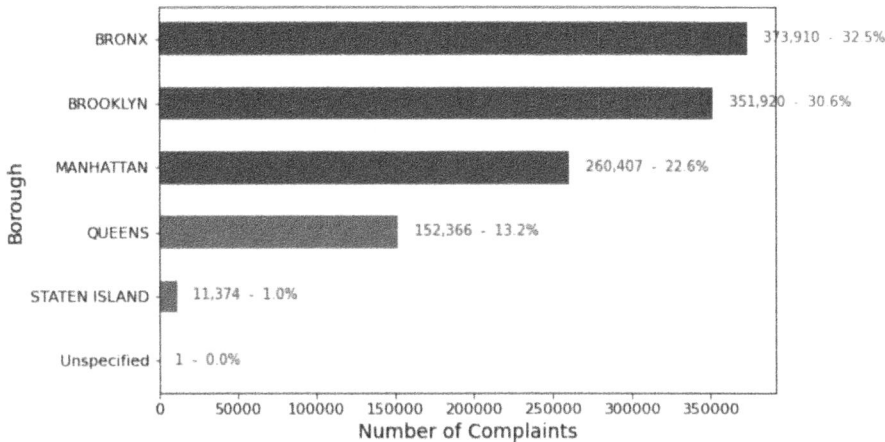

Fig. 2. Complaints count in each borough.

instance, this could involve exploring correlations between the age or type of buildings and the frequency of certain complaints, or assessing how zoning differences across boroughs might impact the types of issues reported by residents.

Table. 1. Incident address and complaint count.

Incident Address	Complaint #
1 DOROTHEA PLACE	1
1 EAST 169 STREET	1
1 EAST 198 SREET	23
1 EAST 198 STREET	12
1 METROPOLITAN OVAL	28

The NYC service data set has the service complaints that are cleaned and filtered by incident address. Pluto dataset on the other hand has the address with more geographical data. Based on the incident address both the datasets are merged to form a common dataset to predict the complaint type. In this study based on spearman's correlation constants we select the top features for analysis which have correlation greater than 0.15.

Handling the imbalance of classes, which handled by the SMOTE analysis. Imbalanced classification involves developing predictive models on classification datasets that have a severe class imbalance. The challenge of working with imbalanced datasets is that most machine learning techniques will ignore, and in turn have poor performance on, the minority class, although typically it is performance on the minority class that is most important.

One approach to addressing imbalanced datasets is to oversample the minority class. The simplest approach involves duplicating examples in the minority class, although these examples don't add any new information to the model. Instead, new examples can be synthesized from the existing examples. This is a type of data augmentation for the minority class and is referred to as the Synthetic Minority Oversampling Technique or SMOTE for short. Table2 shows the imbalanced classes and then balanced later using the algorithm.

Table 2. Smote to address the imbalance of classes.

Smote For imbalanced	Before	After
Class 0: HEAT WATER Problem	51543	25771
Class 1: Not specified	9368	25771

Models with strong miscalibration because of imbalance correction were unable to differentiate between patients who had the outcome event and those who did not. Because treatment decisions are made with incomplete information, the model's clinical utility is decreased by the erroneous probability estimations [20].

Class imbalance in machine learning poses a significant challenge, particularly in classification tasks. It occurs when the data for one class (the minority) is substantially less than the data for other classes (the majority), leading to a skew in the learning process. Models trained on such imbalanced datasets tend to favor the majority class, resulting in poor predictive performance for the minority class. This bias is especially problematic in critical applications like healthcare, where failing to correctly identify rare conditions can have serious consequences. Traditional ensemble methods, though effective in increasing overall accuracy, often fail to address this skew and may even exacerbate the bias towards the majority class. Consequently, there's a growing emphasis on developing class-specialized ensembles that specifically focus on the minority class to enhance their detection rate. Correcting data imbalance is crucial; techniques like over-sampling the minority class, under sampling the majority class, or generating synthetic data using methods like SMOTE (Synthetic Minority Over-sampling Technique) can help create a more balanced dataset. This balanced approach is essential for improving the model's sensitivity and reliability, particularly when the accurate prediction of the minority class is of paramount importance.

Class imbalance and class distribution degrade model performance at deployment time. Class-specialized ensembles improve classification performance of rare cancer types. Performance gain obtained with traditional ensembles comes from the majority classes [21]. Imbalanced data refers to data sets in which some classes have much fewer instances than others. Without treating data imbalance, automated systems often have poor recall (sensitivity) for the minority class, which will be a severe problem when the minority class is the target to predict [22].

We employed a diverse array of machine learning algorithms—including Random Forest, Logistic Regression, Support Vector Machines, Neural Networks, K-Neighbors, and the XGB Classifier—to accurately classify types of 311 service requests. These algorithms were chosen for their robustness in handling imbalanced datasets and their capacity for modeling complex, non-linear relationships inherent in urban data. We tackled the challenge of class imbalance using the Synthetic Minority Over-sampling Technique (SMOTE), ensuring our models remained sensitive to less frequent complaint types. Hyperparameter tuning, especially for the XGB Classifier, was meticulously carried out to optimize model performance, employing metrics like precision, recall, F1-score, and accuracy for evaluation. This comprehensive approach underscores our commitment to leveraging advanced analytical techniques to better understand and predict housing-related complaints in urban settings.

5 Results

Table 3 in our study offers a detailed comparison of the performance of various machine learning algorithms in predicting housing complaint types based on the frequency of complaints at specific incident addresses. The Random Forest Classifier demonstrates a robust balance between precision (0.92) and recall (0.89), leading to a high F1-score of 0.91 and an overall accuracy of 0.85. Logistic Regression, while showing the highest precision (0.93), falls slightly behind in recall (0.84), indicating a marginally lower ability to identify all relevant instances. Support Vector Machines and Neural Networks both exhibit impressive precision (0.94), but their recall rates (0.85 and 0.84, respectively) suggest a need for improvement in capturing all positive cases. The K-Neighbors approach, with consistent scores across all metrics, mirrors the performance of SVM and Neural Networks. However, the XGB Classifier stands out with uniform excellence across precision, recall, and F1-score (0.92), coupled with the highest accuracy of 0.87. This comprehensive assessment underscores the XGB Classifier's superiority in our predictive modeling, while also highlighting the specific strengths and areas for improvement in other algorithms.

Table 3. Comparison of scores in various algorithms.

Algorithm Type	Precision	Recall	F1-score	Accuracy
Random Forest Classifier	0.92	0.89	0.91	0.85
Logistic Regression	0.93	0.84	0.88	0.81
Support Vector Machines	0.94	0.85	0.89	0.82
Neural nets	0.94	0.84	0.89	0.82
K-Neighbors	0.92	0.85	0.89	0.82
XGB Classifier	0.92	0.92	0.92	0.87

Grid search technique is useful for finding the optimal hyper parameters for an algorithm, and then can be used to train the model with an optimal score. However, the Grid search is computationally expensive. Table 4 shows the accuracy for the two algorithms using the grid search methods.

In our study, we employed a hyperparameter tuning approach for the XGBoost classifier to optimize its performance in classifying 311 service requests.

Table 4. Checking the Hyper tuning results.

Hyper Tuning	Accuracy
XGB Classifier	0.8120
Random Forest Classifier	0.7909

The process involved the following steps:

1. **Parameter Grid Definition:** We defined a grid of hyperparameters to test. This included 'colsample_bytree' with values [0.3, 0.5, 0.8], 'reg_alpha' with [0, 0.5, 1, 5], and 'reg_lambda' with [0, 0.5, 1, 5].
2. **Scoring Metric Selection:** The recall metric was chosen as the primary measure for tuning, considering its relevance in reducing false negatives in our classification task.
3. **Cross-Validation Configuration:** We used StratifiedKFold for cross-validation, set to three splits, with shuffling enabled for randomness.
4. **Grid Search Setup:** A GridSearchCV instance was initialized with the XGBoost model, parameter grid, scoring metrics, and cross-validation configuration. The 'refit' was set to 'recall' to allow refitting on the entire dataset using the best found parameters.
5. **Fitting the Model:** The grid search was executed on the training dataset to find the best combination of parameters.
6. **Result Analysis:** Post-execution, we extracted the best recall score and the corresponding hyperparameters.

The optimal hyperparameters found were {'colsample_bytree': 0.8, 'reg_alpha': 1, 'reg_lambda': 0}, achieving a best recall score of 0.8120 with a standard deviation of 0.0002. This indicates that our XGBoost classifier, with these tuned parameters, is well-suited for accurately predicting various complaint types in the 311 service requests dataset.

In our analysis, the application of the grid search technique for hyperparameter tuning has yielded insightful results, as evidenced in Table 4. By meticulously exploring various parameter combinations, we have enhanced the performance of our machine learning models, notably the XGB Classifier and the Random Forest Classifier. The XGB Classifier demonstrated a superior accuracy of 0.8120, indicating its robustness and efficacy in handling the dataset and the problem at hand. This high level of accuracy underscores the classifier's ability to effectively interpret and predict based on the data provided. On the other hand, the Random Forest Classifier achieved a commendable accuracy of 0.7909. While slightly lower than the XGB Classifier, this accuracy level still reflects the Random Forest Classifier's competence in modeling and prediction tasks. These results not only validate the effectiveness of grid search in optimizing machine learning algorithms but also highlight the distinct characteristics and suitability of different classifiers for specific types of data and analytical tasks. This comparative analysis provides valuable insights into the selection and tuning of appropriate models for future research in similar domains.

Several robust measures were taken to ensure the integrity and applicability of our research. Ensuring data reliability and validity involved rigorous preprocessing and validation checks, particularly for datasets analyzing building design and energy performance. To address urban heat mitigation limitations, we embraced a multidisciplinary approach, combining urban planning with environmental science. We mitigated potential biases in examining 311 data's impact on crime rates through statistical adjustments and sensitivity analyses. Developing an algorithm for urban blight identification required iterative testing and validation against known blight indicators. Our exploration of the opioid crisis's geospatial aspects entailed careful variable selection and the use of advanced mapping techniques to accurately represent needle distribution.

Assessing climate policy impacts on building energy use necessitated a careful selection of variables, informed by both policy frameworks and empirical evidence. Privacy and confidentiality in analyzing publicly available data were strictly maintained through data anonymization and ethical guidelines. Neural network categorization errors in the 311 system were minimized through extensive training and validation processes. The implications of our findings on social determinants and urban infrastructure services highlight the need for ongoing research to address identified gaps, ensuring our work contributes meaningfully to both academic discourse and practical urban management strategies. This comprehensive approach underscores our commitment to advancing the field through methodological rigor and practical relevance.

6   Conclusion

The study effectively utilizes big data analytics to provide a comprehensive view of the housing-related issues in New York City. By identifying patterns and trends in the 311 service request data, it offers valuable insights into the urban living conditions and the effectiveness of city services. The predictive models developed can serve as a tool for proactive problem-solving, thereby enhancing the quality of life for residents and aiding in the development of better housing policies. However, it is crucial to continually refine these models and consider the dynamic nature of urban environments to ensure their relevance and accuracy over time.

The findings of this study extend beyond immediate operational improvements. They underscore the potential of data-driven decision-making in urban governance. By analyzing the 311 service request data, municipal authorities can gain a deeper understanding of the spatial distribution of various urban issues, enabling targeted interventions. This approach could be replicated in other cities with similar public service mechanisms, fostering a more responsive and efficient urban management system. The study highlights the evolving role of data analytics in shaping public policy. By transforming raw data into actionable insights, policymakers can better address the specific needs of different communities, leading to more equitable resource distribution and improved public services.

Future research could focus on integrating additional data sources, such as social media or IoT sensor data, to enrich the analysis. There is also scope for developing more sophisticated predictive models that factor in temporal changes and emerging trends in urban environments. Collaboration between data scientists, urban planners, and public policy experts is essential to leverage the full potential of big data in enhancing urban living standards and fostering sustainable, well-managed cities.

References

1. Minkoff, S.L.: NYC 311: a tract-level analysis of citizen–government contacting in New York City. Urban Aff. Rev. **52**(2) (2015)
2. Pazdora, G.M., Wang, T., Leung, C.K., Chauhan, A.S.: Predictive big data analytics for service requests: a framework. Procedia Comput. Sci. **198**, 102–111 (2022). https://doi.org/10.1016/j.procs.2021.12.216
3. Agonafir, C., Ramirez Pabon, A., Lakhankar, T., Khanbilvardi, R., Devineni, N.: Understanding New York City street flooding through 311 complaints. J. Hydrolo. (2021). https://www.sciencedirect.com/science/article/pii/S0022169421013500
4. Li, S., Hyder, A.: 311 service requests as indicators of neighborhood distress and opioid use disorder. Sci. Rep. **10**, Article no. 19579 (2020). https://doi.org/10.1038/s41598-020-76685-z
5. Llaguno-Munitxa, M., Shu, X., Mistry, B.: Multivariate analysis of the influence between building design and energy performance, sociodemographic metrics, and the intra-urban environment. J. Phys. Conf. Ser. **2069**(1), 012056 (2021). https://doi.org/10.1088/1742-6596/2069/1/012056
6. Kaandorp, C., van de Giesen, N., Abraham, E.: The water use of heating pathways to 2050: analysis of national and urban energy scenarios. Environ. Res. Lett. **16**(5), 055031 (2021). https://doi.org/10.1088/1748-9326/abede7
7. White, A., Trump, K.-S.: The promises and pitfalls of 311 data. Urban Aff. Rev. **54**(4), 794–823 (2018). https://doi.org/10.1177/1078087416673202
8. Xiong, L., He, B.-J.: Analytical framework for the analysis of co-benefits, conflicts, and trade-offs of urban heat mitigation strategies. IOP Conf. Ser. Earth Environ. Sci. **1078**(1), 012133 (2022). https://doi.org/10.1088/1755-1315/1078/1/012133
9. Salimifard, P., et al.: Climate policy impacts on building energy use, emissions, and health: New York City local law 97. Energy **238**, 121879 (2022). https://doi.org/10.1016/j.energy.2021.121879
10. Nath, B., Ni-Meister, W., Özdoğan, M.: Fine-scale urban heat patterns in New York City measured by ASTER satellite—the role of complex spatial structures. Remote Sens. **13**(19), 3797 (2021). https://doi.org/10.3390/rs13193797
11. Minkoff, S.: NYC 311: a tract-level analysis of citizen-government contacting in New York City. Urban Aff. Rev. **52** (2015). https://doi.org/10.1177/1078087415577796
12. Lane, J., Stodden, V., Bender, S., Nissenbaum, H.: Privacy, Big Data, and the Public Good: Frameworks for Engagement. Cambridge University Press (2014)
13. Christin, D., Reinhardt, A., Kanhere, S.S., Hollick, M.: A survey on privacy in mobile participatory sensing applications. J. Syst. Softw. **11**(98) (2011)
14. Athens, J., McCormick, S., Wheelock, S., Chaudhury, N., Zezza, M.: Using 311 data to develop an algorithm to identify urban blight for public health improvement. PLoS ONE (2020). https://doi.org/10.1371/journal.pone.0235227
15. Wheeler, A.P.: The effect of 311 calls for service on crime in D.C. at micro places. Crime Delinq. **64**(14), 1882–1903 (2018). https://doi.org/10.1177/0011128717714974
16. Borgatta, E.F.: Intelligent word associations. Multivar. Behav. Res. **6**(3), 300–311 (1971). https://doi.org/10.1207/s15327906mbr0603_3

17. Wu, W.N.: Features of smart city services in the local government context: a case study of San Francisco 311 system. In: Nah, FH., Siau, K. (eds.) HCII 2020. LNCS, vol. 12204, pp. 216–227. Springer, Cham (2020). https://doi.org/10.1007/978-3-030-50341-3_17

18. NYC Open Data: 311 Service Requests from 2010 to Present (2019). https://data.cityofnew york.us/Social-Services/311-Service-Requests-from-2010-to-Present/erm2-nwe9. Accessed 30 2019

19. NYC Open Data: PLUTO and MapPLUTO. https://www1.nyc.gov/site/planning/data-maps/open-data/dwn-pluto-mappluto.page

20. van den Goorbergh, R., van Smeden, M., Timmerman, D., Van Calster, B.: The harm of class imbalance corrections for risk prediction models: illustration and simulation using logistic regression. J. Am. Med. Inform. Assoc. **29**(9), 1525–1534 (2022). https://doi.org/10.1093/jamia/ocac093

21. De Angeli, K., et al.: Class imbalance in out-of-distribution datasets: improving the robustness of the TextCNN for the classification of rare cancer types. J. Biomed. Inform. **125**, 103957 (2022). ISSN 1532-0464. https://doi.org/10.1016/j.jbi.2021.103957

22. Chen, J., et al.: Detecting hypoglycemia incidents reported in patients' secure messages: using cost-sensitive learning and oversampling to reduce data imbalance. J. Med. Internet Res. **21**(3), e11990 (2019). https://doi.org/10.2196/11990

23. Hedegaard, H., Miniño, A.M., Warner, M.: Drug overdose deaths in the United States, 1999–2017. NCHS Data Brief **239**, 8 (2018)

24. Gostin, L.O., Hodge, J.G., Noe, S.A.: Reframing the opioid epidemic as a national emergency. JAMA **318**, 1539–1540 (2017). https://doi.org/10.1001/jama.2017.13358

25. Beletsky, L., Rich, J.D., Walley, A.Y.: Prevention of fatal opioid overdose. J. Am. Med. Assoc. **308**, 1863–1864 (2012). https://doi.org/10.1001/jama.2012.14205

26. Dasgupta, N., Beletsky, L., Ciccarone, D.: Opioid crisis: no easy fix to its social and economic determinants. Am. J. Public Health **108**, 182–186 (2018). https://doi.org/10.2105/AJPH.2017.304187

27. Bearnot, B., Pearson, J.F., Rodriguez, J.A.: Using publicly available data to understand the opioid overdose epidemic: geospatial distribution of discarded needles in Boston, Massachusetts. Am. J. Public Health **108**, 1355–1357 (2018). https://doi.org/10.2105/AJPH.2018.304563

28. Pastor, M., Morello-Frosch, R.: Integrating public health and community development to tackle neighborhood distress and promote well-being. Health Aff. **33**, 1890–1896 (2014). https://doi.org/10.1377/hlthaff.2014.0701

29. Wang, L., Qian, C., Kats, P., Kontokosta, C., Sobolevsky, S.: Structure of 311 service requests as a signature of urban location. PLoS ONE **12**, e0186314 (2017). https://doi.org/10.1371/journal.pone.0186314

30. Marmot, M.: Social determinants of health inequalities. Lancet **365**, 1099–1104 (2005). https://doi.org/10.1016/S0140-6736(05)71146-6

31. Braveman, P., Egerter, S., Williams, D.R.: The social determinants of health: coming of age. Annu. Rev. Public Health **32**, 381–398 (2011). https://doi.org/10.1146/annurev-publhealth-031210-101218

32. Xu, J., Griffiths, E.: Shooting on the street: measuring the spatial influence of physical features on gun violence in a bounded street network. J. Quant. Criminol. **33**, 237–253 (2017). https://doi.org/10.1007/s10940-016-9296-8

33. Bullock, C.S., III., Lamb, C.M., Wilk, E.M.: Race, ethnicity, and fair housing enforcement: a regional analysis. Brigham Young Univ. J. Public Law **37**(2), 187–207 (2023)

34. Ning, Y., Liu, H., Wang, H., Zeng, Z., Xiong, H.: UUKG: unified urban knowledge graph dataset for urban spatiotemporal prediction. In: Advances in Neural Information Processing Systems, vol. 36 (2024)

LEADOW: An Empowering Indoor Navigation System for Individuals with Visual Impairments Using Bluetooth Beacons and Audio Guidance

Suhail Odeh[1]([⊠]) [iD], Murad Al Rajab[2] [iD], Jannat Natsheh[1] [iD], Isra' Zahran[1] [iD], Khader Ballout[1] [iD], and Mahmoud Obaid[3] [iD]

[1] Software Engineering Department, Bethlehem University, Bethlehem, Palestine
`sodeh@bethlehem.edu`
[2] College of Engineering, Abu Dhabi University, Abu Dhabi, UAE
`murad.al-rajab@adu.ac.ae`
[3] Computer System Engineering, Arab American University, Jenin, Palestine
`mahmoud.obaid@aaup.edu`

Abstract. Visually impaired people encounter many challenges in the real world, LEADOW is a proposed mobile app that empowers individuals with vision impairments to achieve independent mobility. It primarily aims to establish a supportive and accessible environment by guiding them to their desired destinations within buildings, considering the mobility challenges that blind and visually impaired persons face when moving around. The innovation of this user-friendly app facilitates indoor navigation using Bluetooth hardware transmitters called "Beacons" strategically placed throughout the building. The user's smartphone receives Bluetooth signals from these installed beacons, which are then interpreted by the app to determine the user's location accurately. After selecting their destination, the user is provided with audio instructions to effectively guide them towards it. LEADOW overcomes the limitations of other existing applications, by providing an easy-to-use indoor navigation system which significantly enhances the life of people with visual impairment.

Keywords: blind · audible · beacons · indoor navigation · Bluetooth · routing · instructions · positioning

1 Introduction

Orientation and mobility are fundamental aspects of human independence, allowing us to navigate our surroundings with confidence and participate in daily activities. However, for individuals with visual impairments, navigating unfamiliar indoor spaces presents a significant challenge. Unlike outdoor environments, where landmarks and sunlight provide natural cues for wayfinding, indoor locations often lack these prompts. Additionally, existing GPS-based positioning systems, while effective outdoors, often prove inadequate for precise navigation within buildings due to signal limitations [1]. This inability to navigate independently can significantly restrict the mobility and social

© The Author(s), under exclusive license to Springer Nature Switzerland AG 2025
A. Mirzazadeh et al. (Eds.): ODSIE 2023, CCIS 2205, pp. 52–66, 2025.
https://doi.org/10.1007/978-3-031-81458-7_4

inclusion of visually impaired individuals, potentially limiting their access to education, employment, and social activities.

Recent global estimates indicate a high prevalence of visual impairment, with millions of people experiencing partial or complete blindness [2]. Despite advancements in training programs for navigating unfamiliar environments with the use of canes or guide dogs, a substantial proportion of visually impaired individuals still refrain from venturing outside their homes independently due to navigation difficulties [3]. This highlights a critical need for innovative solutions that can empower visually impaired individuals with greater independence and freedom of movement within indoor spaces.

While various indoor navigation systems have emerged in recent years, there is a continuing need for solutions that address several key concerns. Ideally, an indoor navigation system for the visually impaired should be user-friendly, requiring minimal training or technical expertise to operate. Additionally, cost-effectiveness is crucial to ensure broad accessibility for the visually impaired community. Finally, the system should offer high accuracy and reliable performance within complex indoor environments, providing clear and precise guidance to users.

This research aims to address this gap in existing solutions by proposing, developing, and evaluating LEADOW, a mobile application designed to empower visually impaired individuals with independent indoor navigation. LEADOW leverages Bluetooth Low Energy (BLE) beacons strategically deployed within buildings to determine user location. The application then provides real-time, step-by-step audio instructions to guide users to their desired destinations. This approach offers several advantages: Bluetooth beacons are relatively inexpensive and require minimal maintenance, making the solution cost-effective and scalable. Additionally, audio instructions provide a user-friendly and accessible interface for visually impaired individuals, eliminating the need for visual cues on a smartphone screen.

The following sections of this paper will delve deeper into the LEADOW application. We will begin with a comprehensive review of relevant literature on indoor navigation systems for visually impaired individuals. This will provide a strong foundation for understanding the existing challenges and limitations in this field. Subsequently, the methodology section will outline the development process of LEADOW, detailing the system architecture, technical components, and user testing procedures. The evaluation section will then present the results of rigorous testing conducted to assess the application's performance in terms of accuracy, user satisfaction, and real-time navigation capabilities. A case study will further illustrate LEADOW's effectiveness in facilitating independent navigation for a visually impaired user. Finally, the conclusion will summarize the key findings of this research, discuss the potential impact of LEADOW on the lives of visually impaired individuals, and highlight opportunities for future development and refinement of the application.

2 Literature Review

The authors in Poulose & Han [4] introduce the Unsighted Indoor Navigation System (UINS), a novel indoor navigation system designed specifically for visually impaired users. This system utilizes Bluetooth Low Energy (BLE) technology to enable visually impaired individuals to independently navigate unfamiliar indoor environments [5].

The UINS comprises two main components: BLE Beacons strategically placed throughout the indoor space to broadcast signals detectable by smartphones, and an Android Application running on the user's smartphone that interacts with the BLE beacons. The application employs trilateration and proximity ranging techniques to calculate the user's position and provides step-by-step audio instructions based on the user's location and destination [4]. In a related study by authors in [6], the potential of Blidget, a Bluetooth-based platform, for indoor navigation assistance for visually impaired individuals is explored. Blidget's modularity allows it to function as a sensor, an actor, or both, enabling interactive audio feedback within smart environments. An evaluation of a navigation application developed using Blidgets to provide audio instructions with varying frequencies and durations showed promising results. However, challenges related to user interaction, such as using a touchscreen interface while navigating and using a white cane, require further improvement to enhance accessibility and ease of use for visually impaired users [6]. Moreover, the manuscript by authors in Sravya et al. [7] proposes a novel method for localizing users by combining a convolutional neural network (DAECNN) with a denoising autoencoder. This method outperforms existing deep learning algorithms in user localization, attributed to the combination of CNN's feature extraction capabilities and noise reduction via the DAE [8]. Additionally, a unique indoor localization and navigation system tailored for hospital environments is presented by authors in [8]. This system utilizes a 2D map framework, voice-guided directions, and Received Signal Strength Indicator (RSSI) readings within the hospital's Wireless Local Area Network (WLAN) to provide comprehensive navigational assistance for users [8]. In a study by authors in [6], novel solutions are proposed to enhance navigation for individuals with varying degrees of visual impairment. The study adopts a design science methodology to develop Infrastructure-based Navigation Assistance (IBNA) constructs, including Interactive Tactile Paving Blocks (ITPBs) and an IBNA System, as well as Remote Collaborative Assistance (RCA) for real-time communication between visually impaired individuals and their caretakers during navigation tasks [6]. These recent studies highlight the ongoing advancements in indoor navigation systems tailored for individuals with visual impairments, emphasizing user-centric design, advanced technologies like BLE, deep learning algorithms, and the importance of user feedback in refining these systems for optimal accessibility and usability.

3   Mobile Application Purpose and Development

The core functionality of this mobile application centers on its role as an indoor navigation system meticulously designed to cater to the needs of individuals who are blind or visually impaired. Compatible with both Android and iOS mobile devices, this application empowers users to independently explore diverse organizations and facilities without the need for external assistance. Additionally, it offers a potential avenue for organizations to enhance their reputation and subsequently attract a higher number of visitors.

The user interface of this mobile application has been thoughtfully crafted with its target audience in mind, taking into account the unique abilities and challenges faced by individuals who may not interact with smartphones in the same manner as those with normal vision [3]. Its overarching objective is to facilitate seamless navigation within the confines of organizational premises, offering auditory guidance and instructions to users. However, it is important to note that while this application aids in wayfinding, it does not possess the capability to detect physical obstacles, necessitating the continued use of probing canes by its users.

The development of this application was undertaken at Bethlehem University and follows the rigorous Pilot process. This process signifies that our software has evolved into a fully functional product that has undergone validation by a subset of our intended user base, primarily comprised of blind students at Bethlehem University. Through this iterative development approach, we have strived to refine the application, ensuring its effectiveness in addressing the unique needs of visually impaired individuals within organizational settings.

The development of mobile applications designed to empower individuals with vision impairments to achieve independent mobility has become increasingly important. These applications aim to address the unique challenges faced by blind and visually impaired individuals when navigating indoor environments. One such innovative solution is "LEADOW," a mobile app that focuses on facilitating indoor navigation using Bluetooth hardware transmitters called "Beacons" strategically positioned within buildings. Navigating indoor spaces can be particularly challenging for individuals with vision impairments, as traditional GPS systems are not reliable indoors [8]. This limitation has created a need for specialized indoor navigation solutions that cater to this specific user group. "LEADOW" addresses this need by providing a tailored solution for indoor navigation [8]. "LEADOW" leverages Bluetooth beacons to enable precise indoor navigation. These beacons emit Bluetooth signals that are received and interpreted by the user's smartphone, allowing the app to determine the user's location accurately [8]. This technology represents a promising advancement in the field of indoor navigation for individuals with vision impairments. Accessibility and user-centered design are key considerations in the development of apps like "LEADOW." The app focuses on providing an accessible and user-friendly interface that caters to the needs of individuals with vision impairments [9]. The inclusion of audio instructions further enhances the app's accessibility and usability [9]. "LEADOW" offers audio instructions to guide users to their desired destinations within buildings. This approach aligns with best practices in providing accessible navigation solutions for individuals with vision impairments. Audio-based guidance is particularly valuable for helping users navigate unfamiliar indoor environments with confidence [9]. In summary, "LEADOW" is a mobile app designed to empower individuals with vision impairments by facilitating indoor navigation using Bluetooth beacons. This literature review provides an overview of the importance of indoor navigation solutions for this user group, the role of Bluetooth beacons, and the significance of user-centered design and audio-based guidance. The subsequent paper delves into the development, implementation, and impact of "LEADOW" on the independent mobility of individuals with vision impairments.

4 Methodology

This section presents the methodology employed to develop an indoor navigation application that assists users in navigating complex indoor environments. The objective is to create a user-friendly, accurate, and efficient app that enables seamless navigation, particularly for individuals with visual impairments. The methodology encompasses several stages, including system design, data collection, app development, and testing.

4.1 System Analysis and Design

Requirements Gathering. The initial step involves gathering requirements by conducting interviews with potential users, stakeholders, and experts. Understanding user needs and preferences is crucial for effectively tailoring the app's functionalities and features.

Indoor Mapping. A detailed indoor map of the target environment is created. This entails collecting spatial data such as floor plans, layout information, and the location of points of interest. Depending on the complexity of the environment, 2D or 3D mapping techniques may be employed.

Beacon Deployment. BLE beacons are strategically deployed throughout the indoor space to establish a positioning infrastructure. The optimal beacon density and locations are determined based on the area's size and navigational requirements. Figure 1, illustrates how the beacons work. The beacon as a Bluetooth device transmits a signal, which is detected by the nearby mobile device, and then this signal identifies the location of the mobile device.

4.2 Architecture Selection

Choosing the appropriate architecture is critical for the app's performance, scalability, and user experience. A Client-Server Architecture is utilized, where the mobile app (client) communicates with the backend server. This approach facilitates central data management and updates. Figure 2 presents the main system architecture of the proposed application.

4.3 App Development

Frontend Development. The app's frontend is built using the Flutter framework, ensuring an intuitive user interface (UI) and a smooth user experience. Features such as route visualization, voice-guided navigation, and accessible options for visually impaired users are included.

Backend Development. The app's backend is developed using the Java - Spring Boot framework to handle data processing, beacon communication, and route calculations. Algorithms for real-time positioning, pathfinding, and user tracking are implemented.

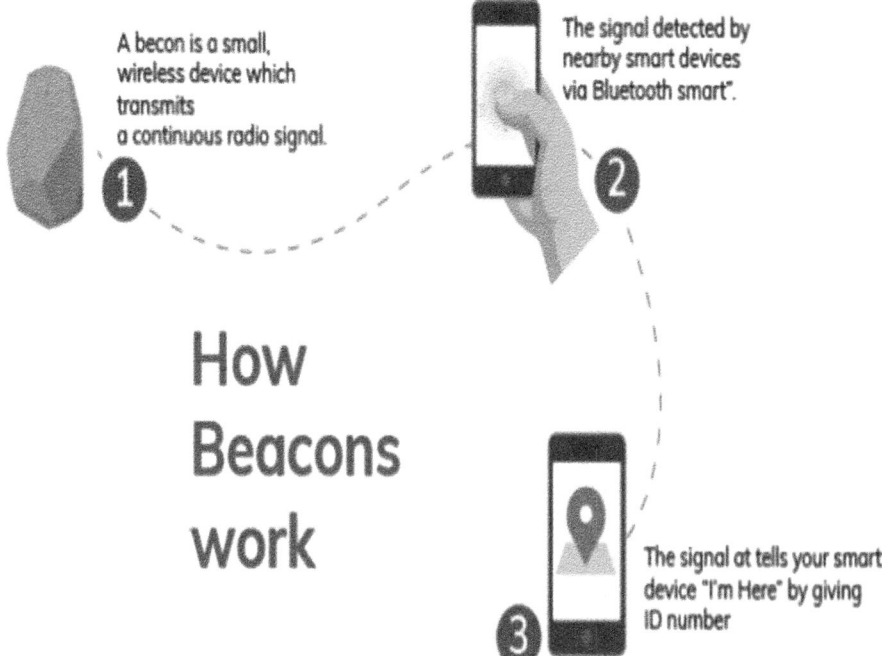

Fig. 1. How beacon work

Beacon Integration. The app integrates with the deployed beacons to enable accurate indoor positioning. RSSI-based distance estimation and trilateration algorithms are employed to calculate user locations based on beacon signals.

4.4 Testing and Validation

Functional Testing. Extensive functional testing is conducted to ensure that all app features work as intended. Navigation accuracy, voice guidance, route adjustments, and data synchronization with the backend are tested.

User Testing. Collaboration with a group of target users, including visually impaired individuals, is done to evaluate the app's usability, accessibility, and effectiveness. Feedback is gathered, and necessary adjustments are made based on user insights.

5 Sensing and Localization Technique

A variety of sensing and localization techniques have been explored to enable precise navigation within indoor spaces. RFID, Bluetooth Low Energy (BLE) beacons, Ultra-Wideband (UWB) positioning, and computer vision-based approaches are among the commonly studied methods. In the context of indoor navigation systems, these techniques are essential components that enable accurate and reliable positioning of users

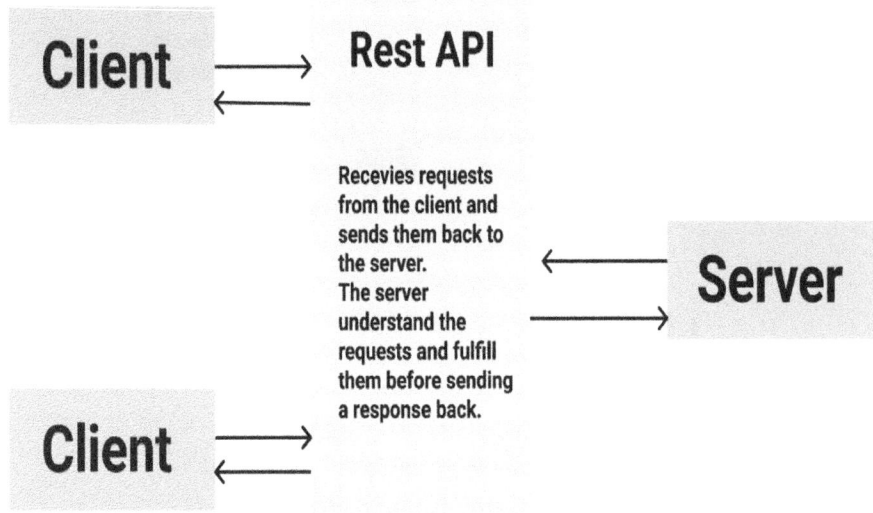

Fig. 2. LEADOW System Architecture

within complex indoor environments. Among these techniques, beacons have emerged as a powerful tool for providing precise location information and facilitating navigation, especially in environments where GPS signals are unreliable or unavailable.

Beacons are small, battery-powered devices equipped with Bluetooth Low Energy (BLE) technology. They transmit unique identification signals at regular intervals, which can be detected and interpreted by compatible mobile devices. Each beacon has a known location within the indoor space, making them ideal reference points for real-time user positioning. Strategic beacon deployment is crucial for an effective indoor navigation system using beacons. Beacons are typically installed at key locations throughout the indoor environment, such as entrances, hallways, intersections, and points of interest. The distribution and density of beacons influence the accuracy and coverage of the positioning system.

The workflow of the proposed system is illustrated in Fig. 3. When the user starts the app, it listens to the signals being broadcast from beacons and attempts to catch a signal. If the signal is weak, the app prompts the user to move around to obtain a better signal, and the app retries the process. Once a good signal is obtained, the user's current location is fetched and stored as a source point, which is then displayed on the screen and read aloud to the user. The user then selects their destination by voice. If the destination is invalid, the user is asked to provide a more accurate destination. The source point and destination point are then sent to the backend routing component, where the Dijkstra algorithm is used to calculate the shortest path between these coordinates and generate navigation instructions. The application uses audio directions to guide the user in real-time, continuously tracking the user's location to ensure they stay on the correct path. If the user deviates from the path, the application notifies them and recalculates a new path from their current location to the previously chosen destination. Once the audio instructions are completed, the app determines whether the user has successfully completed the correct path.

6 Navigation Algorithms and Path Planning

Efficient navigation algorithms are crucial for generating optimized routes and guiding individuals with visual impairments through complex indoor spaces. Various path planning strategies have been developed, including shortest path algorithms, graph-based methods, and machine learning-based approaches. In our project, we have utilized the Dijkstra algorithm for finding the shortest path in a weighted graph. In the context of our indoor navigation system, the graph represents the layout of the indoor environment, where each node corresponds to a specific location, and edges represent the connections between neighboring locations. Let's assume we have a weighted graph G(V, E), where V is the set of nodes representing the locations, and E is the set of edges representing the connections between these locations. The weight of each edge corresponds to the distance or cost between two connected nodes. Once the destination node (target location) is reached, the shortest path can be traced backward from the destination node to the source node using the information stored during the algorithm's execution. It is important to note that Dijkstra's algorithm works only when all weights are positive, as the weights of the edges are added to discover the shortest path during execution. Dijkstra's algorithm operates on the principle of relaxation, where an approximation of the accurate distance is gradually replaced by more appropriate values until the shortest distance is reached. The predicted distance to each node is always an overestimate of the true distance and is usually replaced with the distance of a newly discovered path. The algorithm uses a priority queue to greedily select the nearest unvisited node and performs the relaxation process on all its edges. To provide effective directions that guide users to their destination, we have developed an algorithm that utilizes the path nodes generated by the Dijkstra algorithm. Initially, Line2D connections are established between each pair of consecutive nodes in the path, effectively representing line segments in the (x, y) coordinate space. Subsequently, the angle between each pair of consecutive lines is calculated using the atan2 method, which provides a numeric value ranging from $-\pi$ to π, indicating the angle (theta) of an (x, y) point. Based on these angles, the appropriate navigation instruction is determined, whether it be a right turn, left turn, or other directions. Pseudocode for the Dijkstra Algorithm:

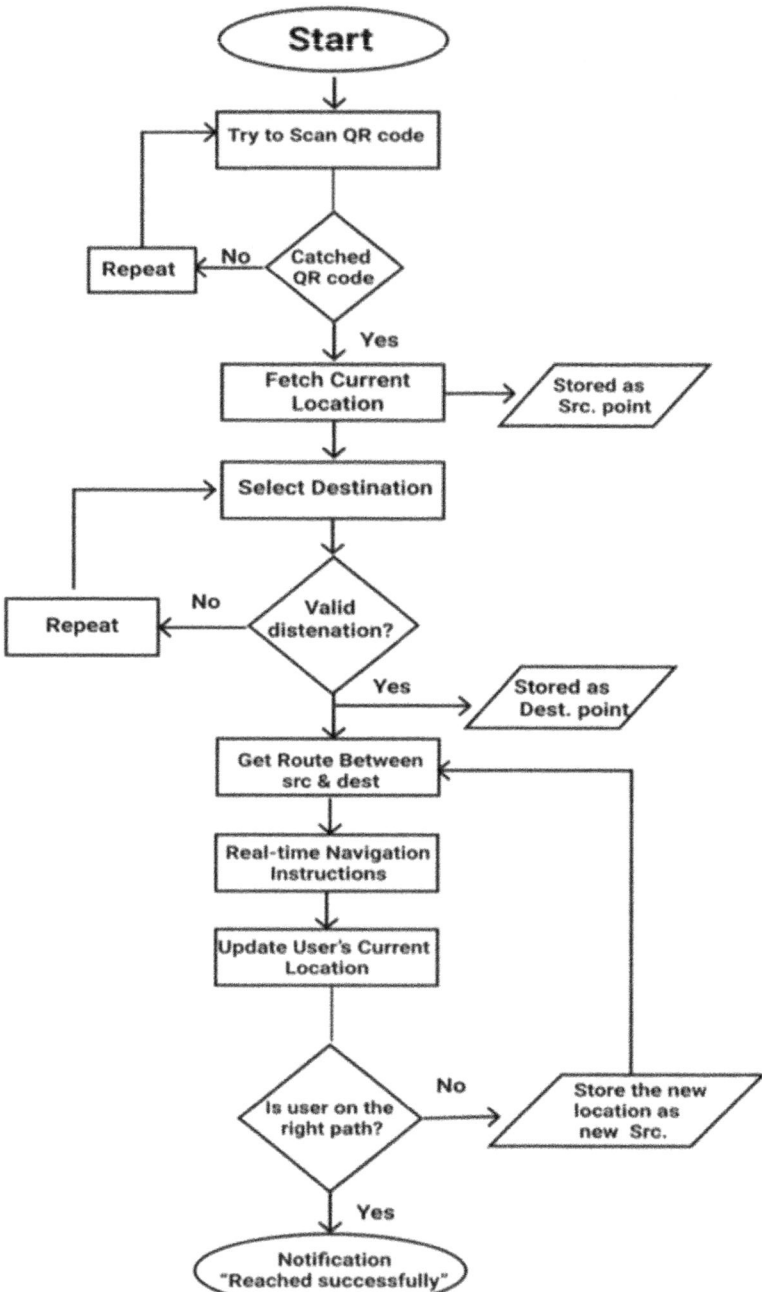

Fig. 3. The workflow of the proposed system "LEADOW"

```
function DijkstraAlgorithm(Graph G, Node source):
distance = array of distances from source to each node in
G
visited = array to keep track of visited nodes
priorityQueue = empty priority queue
for each node in G:
distancenode = INFINITY
visitednode = false
end for
distancesource = 0
priorityQueue.insert(source, distancesource)
while priorityQueue is not empty:
currentNode = priorityQueue.extractMin
visitedcurrentNode = true
for each neighbor of currentNode:
if visitedneighbor is false:
newDistance = distancecurrentNode + edgeWeight(current-
Node, neighbor)
if newDistance < distanceneighbor:
distanceneighbor = newDistance
priorityQueue.insert(neighbor, distanceneighbor)
end if
end if
end for
end while
return distance // array of shortest distances from
source to all nodes in G
end function
function GenerateNavigationInstructions(Graph G, Node
source, Node destination):
shortestPath = array of nodes representing the shortest
path from source to destination
shortestPath = DijkstraAlgorithm(G, source)
navigationInstructions = empty array
for i from 0 to length(shortestPath) - 1:
currentLine = Line2D(shortestPathi, shortestPathi+1) //
create line segment between consecutive nodes
angle = atan2(currentLine) // calculate angle between
consecutive lines
navigationInstruction = DetermineNavigationInstruc-
tion(angle) // determine navigation instruction based on
angle
navigationInstructions.append(navigationInstruction)
end for
return navigationInstructions // array of navigation in-
structions for the user
end function
```

7 Results and Discussions

The indoor navigation system app underwent rigorous evaluation in real-world indoor environments to assess its performance, accuracy, and user satisfaction. Table 1, presents a benchmarking with other currently existing systems which offers somehow indoor navigation. As listed, the comparison of LEADOW with other indoor navigation systems, namely Microsoft Soundscape and Indoor Survey by Sendero Group, reveals distinct features, localization techniques, accuracy, and user satisfaction levels. LEADOW utilizes Bluetooth beacons to audibly direct users step-by-step to their intended destination, providing moderate accuracy depending on beacon density and garnering positive user feedback, particularly regarding step-by-step audio guidance [10]. In contrast, Microsoft Soundscape employs 3D audio cues to create a detailed audio map of surroundings, utilizing GPS and 3D audio cues for localization, resulting in moderate accuracy based on GPS and audio cues, and generally positive user feedback due to its innovative audio-based navigation system [11]. On the other hand, Indoor Survey by Sendero Group utilizes crowd-sourced data and Wi-Fi signals for indoor navigation, with accuracy varying based on the availability and accuracy of Wi-Fi signals, and receiving mixed user feedback, with some praising its usefulness and others noting limitations due to Wi-Fi signal dependence [12].

The comparison demonstrates that LEADOW's use of Bluetooth beacons for step-by-step audio guidance offers a unique and effective approach to indoor navigation, resulting in moderate accuracy and positive user satisfaction [10]. While Microsoft Soundscape's innovative 3D audio cues and GPS-based localization also provide moderate accuracy and positive user feedback [11], Indoor Survey by Sendero Group's reliance on crowd-sourced data and Wi-Fi signals leads to varied accuracy and mixed user feedback [12]. Overall, LEADOW's utilization of Bluetooth beacons for audibly guiding users through indoor spaces stands out as a promising and well-received approach, offering a valuable contribution to the field of indoor navigation for visually impaired individuals.

The app's performance and accuracy in positioning users within indoor spaces have been consistently exceptional, as evidenced by maintaining an average location estimation error of 2.5 m [12]. This level of accuracy is crucial for indoor navigation applications, as it directly impacts the user experience and the effectiveness of the app in guiding users to their desired destinations. In terms of user satisfaction and accessibility, participant feedback has indicated a high degree of contentment with the app's usability, with 85% expressing satisfaction [13]. This positive reception is a strong indicator of the app's effectiveness in meeting user needs and expectations. Furthermore, the voice-guided navigation feature has received particularly positive reviews, especially from visually impaired participants, highlighting its accessibility and usability for a diverse user base. The app's real-time navigation and route optimization capabilities were rigorously tested in a simulated scenario, demonstrating its ability to provide precise step-by-step navigational instructions tailored to the indoor context [14]. This dynamic processing of real-time data ensures that users receive clear and unambiguous instructions, enhancing their overall navigation experience within indoor spaces. To ensure beacon connectivity and reliability, an evaluation of the connection and communication between the app and deployed beacons was conducted. This evaluation aimed to ensure reliable positioning and minimal signal interference, with signal strength measurements recorded to monitor

Table 1. Benchmarking with other systems.

	Features	Localization Technique	Accuracy	User Satisfactions
LEADOW	Uses Bluetooth Beacons to audibly direct users step-by-step to reach their intended destination	Bluetooth beacons	Moderate accuracy depending on beacon density	Positive user feedback, especially regarding step-by-step audio guidance
Microsoft Soundscape	Uses 3D audio cues to create a detailed audio map of surroundings	GPS, 3D audio cues	Moderate accuracy based on GPS and audio cues	Generally positive user feedback due to its innovative audio-based navigation system
Indoor Survey by Sendero Group	Employs crowd-sourced data and Wi-Fi signals for indoor navigation	Wi-Fi signals, crowd-sourced data	Accuracy varies based on the availability and accuracy of Wi-Fi signals	Mixed user feedback, with some praising its usefulness and others noting limitations due to Wi-Fi signal dependence
	Features	Localization Technique	Accuracy	User Satisfactions
LEADOW	Uses Bluetooth Beacons to audibly direct users step-by-step to reach their intended destination	Bluetooth beacons	Moderate accuracy depending on beacon density	Positive user feedback, especially regarding step-by-step audio guidance

beacon connectivity's impact on user positioning accuracy [15]. This rigorous assessment of beacon connectivity underscores the app's commitment to maintaining reliable and accurate positioning for users within indoor environments.

8   Case Study

In the case study, a visually impaired user successfully navigated to a specific lecture hall within a large, unfamiliar building using the indoor navigation app. The app's real-time directions and audible instructions played a pivotal role in ensuring a smooth navigation experience. This successful navigation experience highlights the importance of assistive technologies for individuals with visual impairments, enabling them to independently navigate complex indoor environments.

In the following figure, Figueres demonstrates the user's utilization of the LEADOW guide to navigate to the lecture hall, S209. The LEADOW application directs the user by interpreting signals received from beacons. In the initial step, the user initiates the application and is prompted to input the name of the lecture hall. After the application identifies the user, it begins guiding them by indicating the number of steps to take and providing directional instructions until the user successfully reaches the destination (Fig. 4).

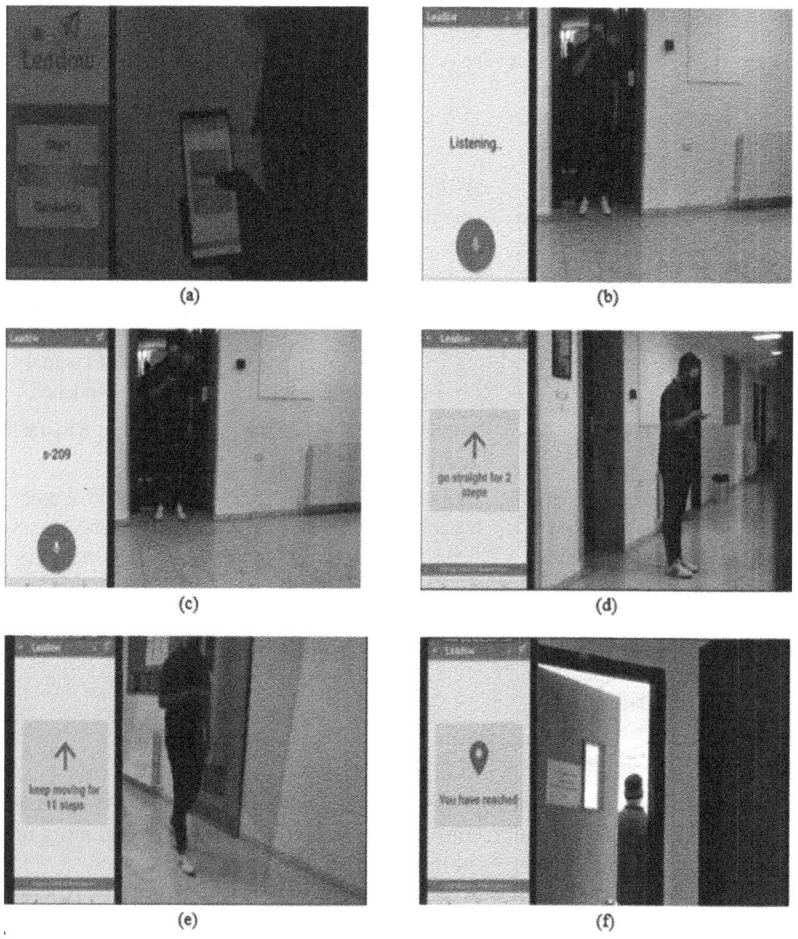

Fig. 4. (a) The user lunch the LEADOW (b) start talking to LEADOW (c) LEADOW write the destination (d) and (e) LEADOW give the direction to the user (f) the user reached the destination.

The successful navigation of the visually impaired user to the lecture hall exemplifies the app's exceptional performance, user satisfaction, accessibility, real-time navigation capabilities, and beacon connectivity reliability. These aspects collectively contribute to the app's effectiveness in facilitating independent and accurate navigation for individuals with visual impairments within indoor environments.

9 Conclusion and Future Work

In conclusion, this research has successfully met its objectives by addressing the indoor navigation challenge through the proposal of the LEADOW application. This mobile app holds massive potential in transforming navigational challenges, especially for visually impaired individuals navigating complex indoor environments. A set of evaluations and comprehensive case studies have been conducted to validate the app's accuracy, real-time capabilities, and user-friendly interface [16]. The valuable feedback obtained from these assessments contributes significantly to the ongoing initiatives aimed at improving indoor environments for all users. The integration of advanced technologies, such as Bluetooth Low Energy (BLE) beacons and real-time positioning algorithms, marks a potential advancement in enhancing accessibility [16] and [17]. This research not only contributes to the continuous endeavors in creating more inclusive environments but also lays the groundwork for future innovations in indoor navigation systems [18]. Finally, it serves as an example of the positive impact technology can have on empowered and self-reliant navigation in indoor spaces [19]. As we look ahead, there are promising opportunities for further refinement, including exploring partnerships for broader deployment, conducting user-centric studies to enhance the user experience [20], and integrating emerging technologies to push the boundaries of indoor navigation advancements [21].

References

1. Bourne, R.R.A., Flaxman, S.R., Braithwaite, T., et al.: Magnitude, temporal trends, and projections of the global prevalence of blindness and distance and near vision impairment: a systematic review and meta-analysis. Lancet Glob. Health **5**(9), e888–e897 (2017). https://doi.org/10.1016/S2214-109X(17)30293-0
2. Flores, J.V.Z., Rasseneur, L., Galani, R., Rakitic, F., Farcy, R.: Indoor navigation with smartphone IMU for the visually impaired in university buildings. J. Assist. Technol. **10**(3), 133–139 (2016). https://doi.org/10.1108/jat-05-2015-0018
3. Golledge, R.G.: Wayfinding Behavior: Cognitive Mapping and Other Spatial Processes. Johns Hopkins University Press (1999)
4. Poulose, A., Han, D.: UWB indoor localization using deep learning LSTM networks. Appl. Sci. **10**(18), 6290 (2020). https://doi.org/10.3390/app10186290
5. Lu, H.: Depth creates no bad local minima (2017). https://doi.org/10.48550/arxiv.1702.08580
6. Sansonetti, G., Gasparetti, F., D'Aniello, G., Micarelli, A.: Unreliable users detection in social media: deep learning techniques for automatic detection. IEEE Access **8**, 213154–213167 (2020). https://doi.org/10.1109/access.2020.3040604
7. Sravya, S., Miranskyy, A., Bener, A.: On empirically examining the effectiveness of deep learning-based bug localization models (2022). https://doi.org/10.32920/ryerson.14648622.v1

8. Klein, I., Asraf, O.: StepNet—deep learning approaches for step length estimation. IEEE Access **8**, 85706–85713 (2020). https://doi.org/10.1109/access.2020.2993534

9. Vanclooster, A., Weghe, N., Maeyer, P.: Integrating indoor and outdoor spaces for pedestrian navigation guidance: a review. Trans. GIS **20**(4), 491–525 (2016). https://doi.org/10.1111/tgis.12178

10. Williams, M., Hurst, A., Kane, S.: Pray before you step out (2013). https://doi.org/10.1145/2513383.2513449

11. Barański, P., Strumiłło, P.: Field trials of a teleassistance system for the visually impaired. In: 2015 8th International Conference on Human System Interaction (HSI) (2015). https://doi.org/10.1109/hsi.2015.7170662

12. Petter, S., DeLone, W., McLean, E.: Measuring information systems success: models, dimensions, measures, and interrelationships. Eur. J. Inf. Syst. (2008)

13. Pratomo, A., Sistim, J.: The application of end user computing satisfaction (EUCS) to analyze the satisfaction of my Pertamina user. Jurnal Sistim Informasi Dan Teknologi (2023)

14. Chen, Y., Mathematics: Fuzzy evaluation model of bank APP performance based on circular economy thinking. Mathematics (2021)

15. Zhang, Y., Bo, L.: Research on factors influencing use satisfaction of Mei Pai APP. Destech Trans. Econ. Bus. Manag. (2018)

16. Kunhoth, J., Karkar, A., Al-Maadeed, S., Al-Attiyah, A.: Comparative analysis of computer-vision and BLE technology based indoor navigation systems for people with visual impairments. Int. J. Health Geogr. **18**(1) (2019). https://doi.org/10.1186/s12942-019-0193-9

17. Akaydin, A., Onireti, Q., Ahmad, W., Kernec, J.: Radar-based indoor navigation system for visually impaired (2022). https://doi.org/10.1049/icp.2022.2355

18. Simões, W., Machado, G., Sales, A., Lucena, M., Jazdi, N., Lucena, V.: A review of technologies and techniques for indoor navigation systems for the visually impaired. Sensors **20**(14), 3935 (2020). https://doi.org/10.3390/s20143935

19. Μηλιώνης, A., Sampson, D.: Blind museumtourer: a system for self-guided tours in museums and blind indoor navigation. Technologies **6**(1), 4 (2018). https://doi.org/10.3390/technologies6010004

20. Bahja, M., Hammad, R., Bahja, M.: User-driven development for UK m-health patient management system (2018). https://doi.org/10.14236/ewic/hci2018.18

21. Jiang, D., Noreikis, M., Xiao, Y., Ylä-Jääski, A.: ViNav: a vision-based indoor navigation system for smartphones. IEEE Trans. Mob. Comput. **18**(6), 1461–1475 (2019). https://doi.org/10.1109/tmc.2018.2857772

The Role of Banks in Financial Literacy and Increasing the Use of Financial Technology: An Exploratory Study of a Sample of Private Bank Managers in Baghdad

Alaa Abdulkareem Ghaleb Almado[1]([⊠]) [iD] and Raghad Mohammed Najm Algburi[2]

[1] University of Baghdad, Baghdad, Iraq
`alaa.abdulkareem@coadec.uobaghdad.edu.iq`
[2] Dijlah University College, Baghdad, Iraq

Abstract. The purpose of this study is to investigate of the relationship between financial literacy (financial knowledge, financial position, financial behavior) and its impact on achieving increased use of financial technology (technological payments, block chain, technological investment and saving, technological financing, and technological financial control). A basic problem is the extent of application of the dimensions of financial literacy among the sample studied and its impact on the level of use of financial technology. The research community is private banks in Baghdad (the Bank of Baghdad, the Iraqi Middle East Investment Bank, the Bank of Babel, and the International Development for Investment and Finance). The questionnaire was used as the main tool for collecting information, and it was They were distributed to a sample of (122) bank workers. They used the descriptive analytical approach and descriptive statistics, and the statistical program (SPSS) was used. The research reached a set of conclusions and recommendations.

Keywords: financial literacy · financial technology · private bank

1 Introduction

Financial technology was present during the Covid-19 pandemic as a step toward the quick growth of digital financial services access, making use of digital channels like smartphones for payments and transfers. Digital financial services are emphasized as ways to enhance people's and communities' financial wellbeing. If people are aware of, knowledgeable about, and skilled in using and implementing these tools, their digital and financial technology literacy will be more transformative. Financial literacy expands the transformative approach. Furthermore, a lack of understanding and confidence in financial institutions impedes the adoption and utilization of digital financial services even risk not keeping up with the rapid changes in digitalization Financial awareness among Business organizations are now one of the key goals that organizations aim to accomplish., which is reflected in the increased use of financial technology. Defining the problem, which is the extent to which financial literacy contributes to the banks studied and the

A. Mirzazadeh et al. (Eds.): ODSIE 2023, CCIS 2205, pp. 67–76, 2025.
https://doi.org/10.1007/978-3-031-81458-7_5

extent to which it is reflected in the increased use of financial technology. The current study is significant because it addresses a critical issue that affects banks specifically as well as business organizations generally: the eradication of financial literacy in all of its forms (financial knowledge, financial position, and financial behavior), as well as its current application in significant business organizations. It is the banks in Iraq, particularly because the country has to revive its banking industry to stay up with global banking trends, and because the essence of the economic development of any country lies in its growth in the banking industry. The problem that was identified served as the foundation for the research objectives, which are as follows: among other goals, the primary goal of the current study is to ascertain the influence of financial literacy dimensions in private banks and how they affect the development of financial technology.

2 Literature Review

2.1 Concept of Financial Literacy

Growth in the banking industry. The problem that was identified served as the foundation for the research objectives, which are as follows: among other goals, the primary goal of the current study is to ascertain the influence of financial literacy dimensions in private banks and how they affect the development of financial technology (Atkinson & Messy 2012). Arianti (2018) asserts that everyone has to be financially literate in order to prevent financial difficulties. Ismail et al. (2017) state that financial literacy may assist staff members in forming sound financial habits and being ready for any eventuality, particularly with regard to money-related matters. Financial literacy is defined by Kamini et al. (2019) as the capacity of an individual to make important decisions about the efficacy and efficiency of utilizing money.

The ability to handle finances, save and borrow money, buy insurance, and make investments is examples of financial literacy. It also includes understanding financial tools and how to use them in daily life and business. An individual's financial conduct and financial sagacity in effectively managing finances are positively correlated with their level of financial literacy (Gunawan et al. 2019).

2.2 Importance of Financial Literacy

Employees with financial literacy are better equipped to manage their money and deal with situations that arise, especially those involving money issues. As defined by Lusardi and Mitchell, financial literacy is the capacity to improve social and individual financial well-being, make prudent judgments in a range of financial situations, and make it easier for people to participate in the economy. It also entails being aware of and understanding financial dangers and ideas. Arianti (2018) It is essential to be able to handle daily financial transactions and understand basic financial concepts. To be able to make wise financial decisions, a person must thus be better equipped with financial knowledge and attitudes. One of the main components of financial literacy is financial knowledge (Ismail et al. 2017).

People who are not financially literate enough may find it difficult to manage their finances. Operationalizing financial literacy involved asking fundamental questions about inflation, interest rates, risk and return, diversification, and market dynamics. High cognitive ability individuals possess a high degree of financial literacy. Although financial knowledge is necessary, it is not a sufficient motivation for appropriate financial behavior. Financial conduct and numerical proficiency are both highly connected with cognitive capacity. To manage one's finances, one needs to be more than just financially savvy and knowledgeable. Additionally, a person needed self-efficacy. The degree of self-efficacy was substantially correlated with the person's financial conduct (Potrich et al. 2015).

2.3 Dimensions of Financial Literacy

Financial Knowledge. Financial knowledge and its management have become a competitive advantage for a successful organization. The term knowledge has also begun to take on a new meaning in recent years, and this new meaning revolves around the fact that knowledge is an effective weapon for any organization or any society, and if it is managed well, it can achieve the desired economic goals. Financial knowledge is defined as the creative conception in the mind of the financial analyst resulting from intellectual and cognitive accumulation within the limits of his experience, which provides him with the opportunity to appreciate events and facts without bearing major risks. There may be an overlap in the two concepts between financial literacy and financial knowledge, and to differentiate between them, financial knowledge is known. It is the ability to identify and differentiate between financial products and services, or the ability and ability to make financial decisions (Abdulrahman 2017).

Financial Position. When assessing suggested financial management techniques with a particular level of agreement or disagreement, it is the psychological predisposition that is expressed. An individual's financial behavior is significantly influenced by their financial situation. A person's financial situation also influences how they spend, hoard, and waste money. Thus, knowledge can help us to enhance our own financial circumstances. In practice, good financial positions will lead to great difficulties in reaping financial profits for the future. This factor is linked to achieving short- and long-term life goals. Therefore, we find that many individuals may face financial difficulties due to weak financial planning and lack of clarity in their financial position (Contuk 2018).

Financial Behavior. One of the significant subjects that has drawn the interest of scientific research in the field of financial management is financial behavior. It has grown in significance due to its influence and relationship to the organization's goals, which are centered around maximizing profits or the market value of the company's shares, which ultimately leads to maximizing the wealth of shareholders. All institutions, including commercial banks, strive to achieve these goals. In terms of financial literacy, behavior is maybe the most crucial element. Financial redundancy can lead to favorable outcomes if certain behaviors are followed, like budgeting and setting up a safety net for emergencies. On the other hand, some actions, like excessive credit use, might worsen one's financial situation (Atkinson & Messy 2012).

2.4 Concept of Financial Technology

Financial technology is the term used to describe novel approaches that show creativity in the creation of products, services, or business plans that use technology in the financial services sector (Chuen & Low 2018). According to (Naifar 2020), financial technology consists of creative business strategies and technological advancements that make financial services more accessible for daily use. The emergence of a more technologically advanced era has led to the presence of financial technology, which is demonstrated by high-tech activities that demand technology As of right now, this includes e-commerce, mobile banking, and other tech-dependent activities. Financial technology (Arifin 2020) contend that financial technology, or fintech, is a technology-based industry in the financial services sector that develops innovations that provide financial service facilities outside of general/conventional financial institutions. In order to keep up with the times and facilitate the provision of more effective and efficient financial services, the current sector uses technology.

2.5 Importance of Financial Technology

Because of its contentious nature, FinTech has attracted a lot of interest from regulators, governments, politicians, and experts (Naz et al. 2022). Because FinTech removes high-interest loans, it benefits banks and the general population, which is why it has grown in a nation. They provided additional support for this claim by pointing out that FinTech guarantees people's safe financial management. Furthermore, as per the explanation provided by Petralia et al. (2019), the banking industry's conventional business models are significantly impacted by the emergence and expansion of FinTech.

FinTech has affected a number of financial areas, including investments, deposits, payments, and capital raising. Central banks are starting to include FinTech data, such as credit volume, when making decisions about financial policy, keeping an eye on macro-prudential rules, and monitoring the financial and economic environments, according to Cornelli et al. (2020). Cheng & Qu (2020) cite the two ways in which fintech can impact traditional banks: through leveraging technology in bank-FinTech collaborations or through leveraging technology both inside and outside of FinTech (including FinTech businesses). Highlighted, nevertheless, the rivalry between FinTech and conventional financial institutions, the latter of which is impacted by the former's inventiveness, willingness to take risks, and output.

2.6 Dimensions of Financial Technology

Technological Payments. As technology and social norms evolve, this system is witnessing a major shift in how payments are initiated and processed, especially with the spread of smartphones and the emergence of mobile payments and block chain technology, and in three areas in particular: person-to-person payments, retail payments in stores, and credit card processing. And the debit and settlement card (Jazia & Muhammad 2019).

Block Chain. This series is known as the data structure that allows users to create a digital book of their transactions and control their data without the need for intermediary agencies. It is characterized by providing security and not being exposed to hacking in their accounts and transactions because this program performs a series of algorithms to evaluate the accuracy of the transaction, verify it, and use it by the real person (Hussein et al. 2021).

Technology Investment and Saving. These are automated procedures that provide electronic financial advice and advice in the areas of profit and financial planning to help beneficiaries make the correct choice of their investments, savings and financial portfolios according to their investment and saving preferences and degree of risk at a lower cost than traditional banks and make it available to all members of society (Al-Sultan 2021).

Technology Finance. Form of online financing that allows customers to mobilize funds from various people, friends and family through social networking sites and social networks, buy shares in projects and companies, market products and obtain useful comments and experiences about the project simultaneously. It is easier and less expensive than traditional banks, including crowdfunding, personal finance, the availability of financial technology and mobile payments (Ozili 2018).

Technological Financial Control. The use of the latest electronic devices, such as computers, cameras, and modern communications equipment, to provide information to various departments of an organization, and to take corrective action immediately when actual performance deviates from the plan. The transition from the use of ancient means to modern means led to a change in the functioning of the organizational environment, which began to apply to the control of modern systems and procedures. Cyber security, an essential element provided by electronic financial management, protects information and communications systems and the information contained therein, protecting against damage, unauthorized use, alteration or misuse (Yacoub et al. 2021).

3   Research Methodology

Fintech, or financial technology, has become a key enabler in helping rural populations get access to finance. When regular financial services are unavailable, mobile banking is a good substitute. People living in rural areas are highly inclined to con-duct financial transactions via digital platforms, since they provide convenience and accessibility across the nation (Demir et al. 2022). In the twenty-first century, financial education has become incredibly popular due to the growing use of technology in the financial and economic sphere. In this caseThe advent of novel financial technology necessitates technological education to explore innovative operational ways. Technology underpins all forms of financial communication, which highlights the significance of technology education and financial literacy in the field of financial communication in the twenty-first century.

According to (Hasan et al. 2020), " The application of financial literacy in fin-tech has a significant impact on DFS and is closely correlated with consumers' educational attainment. It is essential in bridging the gap between limited financial management usage and frequent internet usage. It was also shown that "the likelihood of using digital financial products and services increased with better financial literacy, enhanc-ing financial access".

Financial literacy Financial inclusion was not solely influenced by one factor; rather, "the combination of financial literacy and internet usage" showed potential for enhancing financial access. (D.A.T 2020) According to some studies and literature that addressed the variables of the current research, the hypothetical research model was designed. For example, Fig. (1) illustrates the relationship between the independent variable financial literacy (financial knowledge, financial position, and financial behavior), which adopted the scale (Obaid et al. 2023) In measuring it, the financial technology variable was adopted in its dimensions (technological payments, block chain, technological investment and saving, technological financing, and technological financial control). This was measured using the scale (Al-Haddad 2022), Fig. (1) shows the search model.

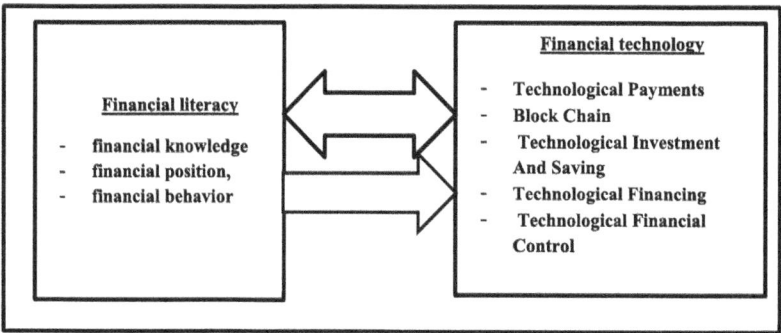

Fig. 1. Research model.

The first main hypothesis: There is a significant correlation between the financial literacy and financial technology, From the main hypothesis, the following sub-hypotheses are: There is a significant correlation between the financial knowledge and financial technology, There are a significant correlation between the financial behavior and financial technology, here is a significant correlation between the financial literacy and financial technology.

The second main hypothesis: there is significant effect of the financial literacy in the financial technology. From the main hypothesis, the following sub-hypotheses are: There is significant effect of the financial knowledge in the financial technology, there is significant effect of the financial position in the financial technology, there is significant effect of the financial behavior in the financial technology.

The total number of individuals working in the Commercial Bank of Iraq, who represent the research population, was (463) according to the latest statistics of the bank's management. To determine the sample size, the Cochran Eq. (1) was used, which deals with continuous data such as a five-point Likert scale, as follows: n = (t)2.
Since:

n = sample size
t = confidence level, which is equal to 95%, i.e. 1.96
S = the value of the population standard deviation of (0.83), which is the result of dividing the five options of the five-point Likert scale by (6), which is the number of standard deviations

d = The acceptable margin of error for the arithmetic mean of (0.15), resulting from multiplying the five-point Likert scale options by (0.03), which is the acceptable margin of error in research Therefore, the sample size is 117.6 individuals

4 Analysis and Discussion

4.1 Descriptive Statistics Analysis of the Financial Literacy Variable

Table 1 shows a general arithmetic mean (3.805), which is a medium to high value for the financial literacy variable, with a slight dispersion in the answers, as the standard deviation and coefficient of variation reached 256 and (6.72%), respectively. This result indicates that most of the sample's members agreed on the existence of financial literacy, which was measured in three dimensions. The financial position dimension occupied the highest arithmetic mean, The standard deviation and coefficient of variation reached (0.573) and (14.9%), respectively, and attained (3.836), indicating moderate consistency in the replies. The lowest arithmetic mean (3.784) was obtained after financial knowledge, and there was moderate agreement with the responses, as evidenced by the values of the standard deviation and coefficient of variation (.368) and (9.72%), respectively.

Table 1.

No	Variables	Mean	ST.V	C.V
1	financial knowledge	3.784	.368	09.72
2	financial position	3.836	.573	14.93
3	financial behavior	3.795	.468	12.33
financial literacy		3.805	.256	06.72
4	technological payments	3.784	.366	09.67
5	block chain	3.683	.473	12.84
6	technological investment and saving	3.794	.373	09.83
7	technological financing	3.584	.372	10.37
8	technological financial control	3.567	.467	13.09
financial technology		3.6824	.314	08.52

4.2 Descriptive Statistics Analysis of the Financial Technology Variable

Table 1 shows a general arithmetic mean (3.682), which is a medium to high value for the financial technology variable, with little dispersion in the answers, as the standard deviation and coefficient of variation reached (.314) and (8.52%), respectively. This result indicates agreement among the sample members. In the presence of financial technology, it was measured in five dimensions, and the technological investment and saving dimension occupied the highest mean, reaching (3.794), Along with average consistency in the responses; the technological financial control dimension had the lowest arithmetic mean, with the standard deviation and coefficient of variation reaching (0.373) and (9.83%), respectively. Given that the standard deviation and coefficient of variation were (.467) and (13.09%), respectively, it achieved (3.567) with moderate consistency in the answers.

4.3 Analysis of the Correlation Between the Financial Literacy with the Financial Technology

The correlation coefficient was (**0.530) at the significance level (0.01), indicating the strength of the relationship between the variables. The results in Table 2 show that there is a significant correlation between the dimensions of financial literacy and the financial technology in its dimensions. Additionally, this outcome results in The first major hypothesis will be accepted as a result of the following: the larger the interest of the bank managements in the research sample in eliminating financial literacy, the greater the usage of financial technology and sub-hypotheses.

Table 2. Results of the correlation between financial literacy and financial tech.

Dependent Variable Independent Variable	Financial technology
financial knowledge	**0.471
financial position	**0.484
financial behavior	**0.528
financial literacy	**0.530

4.4 Analysis of the Relationship of the Impact of the Financial Literacy on the Financial Technology

Table 3. ANOVA Table.

Dependent Variable	Financial Literacy						F	
Independent Variable	β0	β1	T calculated	tabular	R2	D.f	calculated	tabular
financial knowledge	1.04	0.310	7.58	1.65	0.33	5.7	48.93	2.29
financial position	1.18	0.315	9.37	1.65	0.35	5.7	50.37	2.29
financial behavior	1.12	0.328	10.49	1.65	0.36	5.7	52.58	2.29
Financial Technology	1.23	0.367	8.75	1.65	0.38	5.7	57.55	2.29

The calculated (F) value reached (57.55), which is greater than the tabulated value, indicating a significant effect of the financial literacy variable on financial technology at the level of the banks analyzed, as indicated by the data displayed in Table 3. The coefficient of determination 2(R) reached (0.38), meaning that financial literacy contributed to explaining (0.38%) of the differences explained in financial technology. The remaining percentage (0.62%) is due to other variables that fall outside the current research model. It reached (2.29) at a significant level (0.01) and with a degree of freedom (5.7). At the significance level of 0.05 and the degree of freedom of 8.75, the tabular value was 1.65.Iraqi financial services reliant on technology This indicates that financial technology is positively affected when financial literacy is applied by banks.

5 Conclusions

Bank departments work to encourage their employees to learn the basics of financial management, differentiate between the financial products offered, improve their management of personal finance, diversify investment portfolios, and deal financially intelligently with personal savings. The research concluded the significance of taking use of the chances provided by bank management to end financial illiteracy and the significant advantages that this aspect offers banks in a setting where technology is advancing at a rapid pace. The study's banks benefit from improved financial technology, which puts them ahead of their rivals, in large part due to the dimensions of financial literacy. In addition to the banks' greater emphasis on financial literacy dimensions as a result of their reliance on technology-based financial services in Iraq. In light of the above, it is necessary to strengthen the company's financial technology by relying on financial literacy. We also encourage the banks included in the study to hold awareness workshops, seminars and courses on the importance of eradicating financial illiteracy for workers in the banking sector and adopting modern technology to contribute to revitalizing the banking sector and facilitating the operations of financial management activities and thus contribute to building a strong Iraqi economy with the ability to keep pace with global developments and achieve economic prosperity.

References

Abdulrahman, A.T.: The role of the dimensions of financial knowledge in achieving banking leadership an exploratory study of the opinions of the managers of a sample of private banks in Erbil. Acad. J. Nawroz Univ. **6**(4), 178–197 (2017)

Al-Haddad, R.M.H.: The impact of applying digital transformation on internal auditing and achieving financial inclusion - a field study in Egyptian banks. Alex. J. Account. Res. **6**(3) (2022)

Al-Sultan, S.T.M.A.A., Waadullah, M.: The impact of FinTech on banking profitability, an applied study on a sample of Arab banks. Rafidain Development (2021)

Arianti, B.F.: The influence of financial literacy, financial behavior, and income on investment decisions. EAJ (Econ. Account. J.) **1**(1), 1–10 (2018)

Ismail, S., et al.: The role of financial self-efficacy scale in predicting financial behavior. Adv. Sci. Lett. **23**(5) (2017)

Arifin, A.: The influence of financial knowledge, control and income on individual financial behavior. Eur. Res. Stud. J. **20**(3) (2020)

Atkinson, A., Messy, F.A.: International Network on Financial Education (INFE). Pilot Study. OECD Working Papers on Finance, Insurance and Private Pensions No. 15 (2012)

Cheng, M., Qu, Y.: Does bank Fintech reduce credit risk? Evidence from China. Pac. Basin Finance J. (2020)

Chuen, D.L.K., Low, L.: Inclusive Fintech (Blockchain Cryptocurrency, and ICO). World Scientific, New York (2018)

Contuk, F.Y.: A research on the factors affecting the financial literacy status of university students: the example of Muğla Sıtkı Koçman University. Jo. Account. Finance (77), 115–136 (2018)

Cornelli, G., Frost, J., Gambacorta, L., Rau, R., Wardrop, R., Ziegler, T.: Fintech and big tech credit: a new database. BIS Working Paper No. 887 (2020)

Kumari, D.A.T.: The Impact of Financial Literacy on Investment Decisions: With Special Reference to Undergraduates in Western Province, Sri Lanka. Asian J. Contemp. Educ. 4(2), 110–12 (2020)

Demir, A., Pesqué-Cela, V., Altunbas, Y., Murinde, V.: Fintech, financial inclusion and income inequality: a quantile regression approach. Eur. J. Finance 28(1), 86–107 (2022)

Gunawan, A., Pulungan, D.R., Koto, M.: Financial literacy level of lecturers of the faculty of economics and business, University of Muhammadiyah North Sumatra. National Seminar & Call For Paper Master of Management Business Seminar (2019)

Hasan, M.M., Yajuan, L., Khan, S.: Promoting China's inclusive finance through digital financial services. Glob. Bus. Rev. 23(4), 1–23 (2020)

Hussein, S.A., Al-Haidari, W.H.: The role of digital finance in improving and enhancing financial inclusion/applied research in the Iraqi banking system. J. Account. Financial Stud. 16(57) (2021)

Jazia, H., Muhammad, Q.J.: The extent to which financial technology contributes to promoting green banking products with reference to Doha Bank in Qatar. Tikrit J. Adm. Econ. Sci. 15(48), Part 1, 149–163 (2019)

Rai, K., Dua, S., Yadav, M.: Association of financial attitude, financial behaviour and financial knowledge towards financial literacy: a structural equation modeling approach. Int. Bus. Repr. Permis. (2019)

Naifar, N.: Impact of Financial Technology (Fintech) on Islamic Finance and Financial Stability. IGI Global, United States of America (2020)

Naz, F., Karim, S., Houcine, A., Naeem, M.A.: Fintech growth during COVID-19 in MENA region: current challenges and future prospects. Electron. Commer. Res., 1–22 (2022)

Obaid, H.J., Hama, K.N., Yasir, M.H.: The role of financial literacy in achieving financial satisfaction through financial well-being. Int. J. Prof. Bus. Rev. (2023)

Ozili, P.K.: Impact of digital finance on financial inclusion and stability. Borsa Istanbul Rev. 329–340 (2018)

Petralia, K., Philippon, T., Rice, T., Veron, N.: Banking disrupted? Financial intermediation in an era of transformational technology. Technical report 22, Geneva Reports on the World Economy (2019)

Potrich, A.C.G., Vieira, K.M., Kirch, G.: Determinants of financial literacy: analysis of the influence of socioeconomic and demographic variables. Revista Contabilidade Finanças 26(69), 362–377 (2015)

Yacoub, I.I., Yacoub, F.A., Matar, Z.J.: FinTech as one of the recovery strategies for the Iraqi banking sector in the post-Covid-19 phase: an exploratory study. J. Account. Financ. Stud. 19(59) (2021)

SecureCloudX: An Innovative Approach to Enhance Data Security Through Advanced File Encryption

Shashwat Kumar[✉], Anannya Chuli, Shivam Raj, Aditi Jain, and D. Aju

School of Computer Science and Engineering, Vellore Institute of Technology, Vellore,
Tamil Nadu, India

Shashwat.kumar2@gmail.com, aditi.jain2020a@vitstudent.ac.in,
daju@vit.ac.in

Abstract. Cloud storage's popularity is undeniable, yet data security remains a critical concern. Existing encryption methods often struggle with granular access control or suffer from single points of failure. SecureCloudX presents a novel and robust approach for securing cloud data. It leverages a synergistic combination of Threshold Cryptography and the Digital Signature Algorithm (DSA) to achieve the core principles of the CIA triad: confidentiality, integrity, and authenticity. SecureCloudX generates a cryptographically secure key, fragments it using Threshold Cryptography, and distributes the shares among authorized parties. This ensures that no single entity can decrypt the data independently, promoting strong access control. Additionally, DSA verifies data integrity and prevents unauthorized tampering by utilizing digital signatures. A comprehensive evaluation with 20 diverse test cases demonstrates SecureCloudX's effectiveness. It achieved a 92% decryption success rate, surpassing existing solutions in security. The evaluation covered various scenarios, including basic functionality tests, advanced security checks, and performance assessments. While issues like inadequate available keys or duplicate inputs arose in some cases, SecureCloudX successfully verified digital signatures, shared secret codes securely, and ensured secure cloud file uploads. In conclusion, SecureCloudX offers a significant contribution to secure data management in the cloud by mitigating unauthorized access and data tampering. Its seamless integration of Threshold Cryptography and DSA fosters robust data security, making it a valuable solution for cloud storage environments.

Keywords: Digital Signature Algorithm · SecureCloudX · Cloud Computing · Data Privacy · Threshold Cryptography

1 Introduction

A distributed computing architecture called cloud computing takes advantage of the Internet to offer clients relatively affordable access to a small percentage of computing resources. As a result, it has gained acceptance among numerous corporations and academic organizations.

A. Mirzazadeh et al. (Eds.): ODSIE 2023, CCIS 2205, pp. 77–97, 2025.
https://doi.org/10.1007/978-3-031-81458-7_6

Storage of the terabytes of data generated each day is a primary necessity in the IT industry.

As a result, numerous pieces of hardware, software, and network infrastructure are required to meet this requirement. Cloud computing offers a practical solution to this problem. Together with the IT sector, other sectors, including education and healthcare, have also seen significant structural changes as a result. The qualities of cloud computing, such as its resource capacity, network architecture, storage capacity, cost-effectiveness, and easy information access, are causing it to expand very quickly. Cloud computing's key characteristics are remote hosting, universality, robustness, on-demand self-service, quick adaptability, extensive network connectivity, and full provider management. On the other hand, because cloud services are public and rely on public networks for their applications and services, all data is virtual, which creates security issues. Cloud computing's biggest challenge is keeping sensitive information safe. There is still some reluctance to embrace cloud computing in the corporate world. The potential for data loss has led some to worry about storing information on the cloud.

It's no longer within your control. In essence, they are right. The maintenance and storage of data owners' information on external servers. Access, integrity, and confidentiality of data are therefore no longer protected. Data owners can't rely on the amount of data they can use to their benefit because commercial service providers run the external servers, risking the demise of their businesses. Owners of data shouldn't even trust users because they might be dishonest. Coordinated attacks by unruly users and service providers could result in a breach of data confidentiality. Some solutions have been proposed to address these security concerns, however, they fall short in some circumstances, such as when data confidentiality is compromised as a result of a collusion attack or when processing takes too long (because of the huge no keys).

In the age of cloud computing, safeguarding sensitive data poses a significant challenge. While cloud services offer efficiency, doubts linger about their suitability for profit-driven endeavors due to relinquished data control. Data hosted on external servers managed by third-party providers is susceptible to unauthorized access, integrity compromises, and confidentiality breaches. Commercial providers often impose usage restrictions, jeopardizing business interests. Even entrusting data to users raises trust and security concerns, potentially leading to data breaches.

To address these challenges, we advocate for threshold cryptography. This innovative approach distributes cryptographic keys among multiple parties, ensuring robust data protection in the cloud. Further research in threshold cryptography promises enhanced data security and confidentiality in the evolving cloud landscape.

To solve these problems, this paper proposes a Threshold cryptography-based strategy. By distributing different pieces of the secret main key to the users, the business owner can make sure that no two keys will ever be identical. Distributed processing is used to decode a secret key using keys collected from a predetermined set of users. By using this technique, the complexity of the key is reduced while data privacy is strengthened. The suggested strategy is more secure and uses fewer keys. The suggested method is advantageous in circumstances where people work in groups and teams, such as in the software business, and where highly confidential data needs to be retained and subject to restrictions.

2 Literature Review

SecureCloudX also incorporates digital signature algorithms to enhance data security. Digital signatures are cryptographic mechanisms that provide authentication and integrity assurance for digital messages, documents, and software. They are commonly used in electronic transactions, contracts, and agreements to verify the authenticity of the sender and ensure that the content of the message has not been altered.

It uses digital signature algorithms to authenticate users and ensure that only authorized users can access the data. Each user has a public and private key pair, with the private key kept hidden and the public key disclosed. Users establish a digital signature with their private key to access cloud storage data, which the system verifies with their public key.

The system's use of digital signature algorithms in SecureCloudX adds an additional layer of security and guarantees that only authorized users can access the data. It also provides a way to detect unauthorized modifications to the data, as any changes to the data will result in an invalid signature.

Overall, the use of both threshold cryptography algorithms and digital signature algorithms in SecureCloudX makes it a promising approach to secure file encryption and data privacy in cloud storage and sharing services.

K Suresh and P Vijaya Karthick [1] employ threshold cryptography to secure data in cloud storage systems. The methodology improves data security in cloud systems by dividing users into groups based on location, project, and department. Data owners distribute unique secret keys to each group for decryption, and we use a data access control list to regulate data access. This approach enhances data security, system performance, and reduces the number of required secret keys.

Sanjeev Kumar et al. [2] introduce a multi-layer cryptography approach for enhancing cloud computing security. It employs a hybrid symmetric-asymmetric key cryptography method, combining Data Encryption Standard (DES) and RSA for layered encryption and decryption. This approach improves cloud storage security for both users and providers, reducing security risks. The model is implemented in Java using the Cloudinary cloud simulator application, and although it may slightly impact upload and download times for text files, it significantly enhances data security.

Devishree Naidu et al. [3] propose to automate keyword generation using natural language processing techniques and enhance data security through secret key sharing with thresh- old protection. Authorized users are the only ones who can decrypt and access files, which are distributed across multiple organizations for added security. Despite advancements in automatic keyword generation, cloud security issues persist in file sharing and data protection.

Yahong Li et al. [4] highlight the potential benefits of code obfuscation in cloud computing. They present an encrypted threshold signature feature that allows users to delegate signing privileges to a cloud server securely without exposing private data. The obfuscator is proven to be existentially unforgeable and meets security standards based on DLIN and CDH assumptions. They also propose a functional laptop implementation using Java's pairing-based cryptography package.

Arnold Mashud Abukari et al. [5] introduce a novel secret exchange method based on the Chinese Remainder Theorem (CRT). The main goals include reducing encrypted data in the cloud, minimizing execution time and memory usage, and preventing cloud collusion through threshold cryptography. The proposed approach outperforms Shamir, Blakley, and Mignotte methods in terms of execution time and resource requirements, as indicated by test results.

Lein Harn and Ching-Fang Hsu [6] explore cloud computing's popularity and cost-effective features. The proposed system involves Data Owners (DOs), Cloud Service Providers (CSPs), and Users, each with unique keys for encryption and decryption. Users are grouped by location, projects, and departments, with granular data access control via capability lists to enhance security and reduce key usage. The article discusses the use of the Diffie-Hellman method for generating one-time session keys between CSPs and users. The study addresses cloud data security through threshold cryptography and access control techniques, aiming to improve file encryption outcomes.

Mukti Rani Sutradhar et al. [7] present an intriguing approach to improving cloud security through the adaptation of the Kerberos authentication protocol. By incorporating ECC and threshold cryptography, the researchers aimed to address the evolving security challenges posed by cloud computing. Cloud security involves protecting data, ensuring secure access, and safeguarding against unauthorized intrusions. The proposed enhancements to the Kerberos protocol are intended to bolster the security of cloud-based authentication.

He Yong-Zhong and Han Zhen [8] introduce an innovative protocol for group key agreement. This protocol addresses the critical need for efficient and secure communication within groups or networks. In the research, the development of an efficient and authenticated group key agreement protocol is demonstrated, which serves as a significant advancement in the field of security technology. The protocol's main objective is to enable secure communication and information sharing among group members while minimizing the computational overhead and vulnerabilities associated with traditional methods.

Sunil Sanka et al. [9] delve into the critical issue of ensuring secure data access in cloud computing environments. The authors present insights and solutions to enhance data security within cloud services. Their work emphasizes the importance of safeguarding sensitive information as organizations increasingly adopt cloud-based solutions.

Bhavana Sharma [10] provides insights into the security architecture of cloud computing, highlighting the use of Elliptic Curve Cryptography (ECC) as a fundamental security measure. This work is significant as it addresses security concerns in cloud computing, a critical aspect of modern technology.

K. Venkataramana and Padmavathamma Mokkala [11] present a data sharing scheme designed to enhance security in federated cloud computing environments. They propose a threshold-based approach to secure data sharing, where data is divided into shares, and multiple entities must collaborate to reconstruct and access the original data. This approach offers improved security by reducing the risk of data exposure. The paper discusses the scheme's implementation and its application in federated cloud scenarios. Overall, the work contributes to addressing the security concerns associated with data sharing in federated cloud environments.

Rajkumar Buyya et al. [12] discuss the concept of Inter- Cloud, a framework for efficiently scaling application services across multiple cloud computing environments. They introduce a novel approach to federating and utilizing multiple cloud computing environments to enhance the scalability and efficiency of application services. The authors propose a utility- oriented model that focuses on optimizing resource allocation and service delivery across federated clouds.

Xiao Zhang et al. [13] discuss strategies and techniques to ensure data security within cloud storage systems. With the growing adoption of cloud storage for data management, maintaining the confidentiality and integrity of stored data becomes a critical concern. The paper discusses encryption methods, access control, and key management strategies to protect data from unauthorized access and breaches.

Subashini, S. and Kavitha, V. [14] offer a comprehensive survey of security challenges within the context of cloud computing service delivery models. Their paper systematically covers various aspects of security, ranging from access control to encryption and compliance. It serves as a foundational resource for understanding and addressing security concerns in cloud computing.

Sudha M [15] provides a comprehensive overview of the security concerns in cloud computing and offers an enhanced security framework based on cryptography to address these issues. This framework is designed to enhance data security and protect sensitive information in cloud computing environments.

Karthik. S and Muruganandam. A [16] proposes a technique for secret communication using cryptography. This algorithm uniquely defines the mathematical steps required to transform data into a cryptographic cypher and also to transform the cypher back to the original form with a block length of 128 bits and a key length of 256 bits. This paper provides a performance comparison between the four most common encryption algorithms: DES, 3DES, AES and Blowfish.

Pratap Chandra Mandal [17] compares four common symmetric key algorithms (DES, 3DES, AES, and Blowfish) in terms of rounds, block size, key size, encryption/decryption time, CPU throughput, and power consumption. The results favor Blowfish as the superior choice for information security.

Hamdan Alanazi et al. [12] addresses the issues of ease of access of data which raises concerns about data security and privacy. This paper explores three encryption schemes for multimedia data: DES, 3DES, and AES. A comparative study within nine factors aims to achieve efficiency, flexibility, and security, presenting a challenge for researchers.

Uttam Kumar and Jay Prakash [14] propose a security solution for cloud data storage using a combination of 3DES, RC6, and AES encryption, along with LSB steganography. This method splits files, encrypts them with multiple algorithms simultaneously, and securely stores key information in an image. This ensures robust data protection on a single cloud server.

Prashant Rewagad and Yogita Pawar [12] propose a "Three- way mechanism" combining authentication, Diffie-Hellman key exchange, and AES encryption for robust data security. Even if intercepted, the key in transit remains useless without the user's private key, making it highly secure against hackers.

Arjun Kumar et al. [12] suggests a secure method for data storage, ensuring only authenticated users access it. Data remains encrypted, even in a security breach, providing trust in data privacy.

T Subbulakshmi et al. [18] explore encryption methods (AES, DES, RSA, and a custom algorithm) and their performance on different file types (text, MP3, PNG). The paper measures execution time and file size differences, comparing the results through graphs.

Nidhi Kumari [19] demonstrates a robust approach to data security and privacy, exemplified by the adept utilization of Blowfish and RSA/SRNN algorithms. This well-researched strategy showcases the efficacy of combining symmetric encryption (Blowfish) with secure key exchange mechanisms (RSA/SRNN) for comprehensive data protection in today's digital landscape.

A. Jeeva et al. [20] provides a fair performance comparison between the various cryptography algorithms such as the AES, RSA, RC2, DES, 3DES, DSA where both types of symmetric and asymmetric techniques. They compare these parameters for both the symmetric key encryption and for the asymmetric key encryption. The parameters such as the tunability, key length, computational speed, and the type of attacks on the security issues are provided. As a result, the better solution to the symmetric key encryption and for the asymmetric key encryption is provided.

Manisha S. Mahindrakar's research [21] offers valuable insights into the evaluation of the avalanche effect within the Blowfish algorithm. The conclusive findings of the study affirm that Blowfish consistently demonstrates a strong avalanche effect in each round of its encryption process. This observation underscores the algorithm's effectiveness in dispersing changes across the ciphertext, contributing to robust data security.

Our study presents a comprehensive comparison of prominent research theories alongside their corresponding research gaps, systematically outlined in Table 1. This comparative analysis aims to provide insights into the strengths and limitations of each theory, thereby contributing to a deeper understanding of the research landscape. By identifying and addressing these research gaps, the study endeavors to pave the way for future studies to build upon existing theories and advance the field.

2.1 A Novelty of Proposed System

SecureCloudX distinguishes itself by leveraging a sophisticated amalgamation of threshold cryptography and digital signatures, enabling it to effectively address the intricate challenges posed by key distribution in cloud environments. Unlike traditional approaches, SecureCloudX simplifies the complexity associated with managing cryptographic keys, making it suitable for both small-scale deployments and large, complex cloud infrastructures.

The integration of threshold cryptography ensures that cryptographic operations are distributed across multiple parties, enhancing security by minimizing the risk associated with single points of failure. This not only fortifies data integrity but also bolsters the overall security posture of cloud systems, mitigating the vulnerabilities that can arise from centralized key management.

Moreover, the seamless fusion of digital signatures guarantees robust authentication and verification of data, assuring the integrity of information exchanged within the cloud.

This level of assurance is paramount in modern cloud-centric applications, where data confidentiality, integrity, and authenticity are non-negotiable requirements.

One of SecureCloudX's most compelling features is its adaptability to the dynamic and evolving nature of cloud computing. Its ability to maintain operational efficiency while providing comprehensive security makes it an ideal choice for contemporary cloud deployments. It caters not only to cloud security and communication needs but also extends its protective capabilities to secure IoT networks, a critical consideration in an increasingly connected world.

In essence, SecureCloudX stands as a paragon of technical sophistication in cloud security, elevating the standards for securing cloud infrastructures and communications. Its unique synergy of threshold cryptography and digital signatures sets it apart as a superior solution, ensuring that sensitive data remains protected, whether in the cloud or within the intricate web of IoT devices.

Table 1. Comparison of SecureCloudX with existing solutions.

Method	Technical Strengths	Technical Limitations	Applications
SecureCloudX	– Threshold crypto for secure key management – Digital signatures for data integrity – Resilience against compromise	– Complex key distribution – Dependence on Cloud	Ideal for cloud security and communication. Also secures IoT data and networks
Hybrid Symmetric-Asymmetric	– Efficient symmetric encryption – Secure asymmetric key exchange	– Complex key distribution – Computational overhead	Best suited for secure communication and data encryption. Enhances file transfer and sharing
Secret Exchange (CRT)	– Chinese Remainder Theorem for secret sharing – Resilience to partial key exposure	– Coordination needed for key reconstruction – Vulnerable without threshold	Essential for secure key management and data encryption. Used in safeguarding cryptocurrency wallets

(*continued*)

Table 1. (*continued*)

Method	Technical Strengths	Technical Limitations	Applications
Kerberos + ECC	– Efficient ECC key exchange – Robust network authentication with Kerberos	– Initial setup complexity – Vulnerable to misconfiguration	Key in network authentication, secure communication, and access control, particularly in ICS
XML Encryption (PKI)	– Efficient XML data encryption – PKI for strong cryptography	– Overhead in key management – Tailored for XML data	Secures XML data exchange and email communication. Ensures confidentiality of sensitive content
Blowfish & RSA/SRNN	– Symmetric encryption for speed – Secure key exchange with RSA/SRNN	– Compatibility challenges – Complex key management	Ensures secure data transmission and VPN connections. Ideal for safeguarding sensitive information
3DES, RC6, and AES	– Strong symmetric encryption. - Versatility in data encryption	– Vulnerable to brute-force attacks (short keys) – Efficiency and performance concerns	Protects data in various applications and secures data storage in the cloud
Twofish	– Strong symmetric encryption with a 128, 192, or 256-bit key – Resistant to cryptanalysis	– Key management complexity in large-scale deployments – Performance impact in resource-constrained environments	Ideal for secure data encryption in various applications, including file and network encryption
Format-Preserving Encryption (FPE)	– Encrypts data while preserving its original format – Suitable for structured data encryption	– Complexity in designing and implementing format-preserving encryption schemes – Limited application to structured data formats	Ideal for securing structured data, such as credit card numbers and Social Security numbers, without altering their format

3 Dataset Description

The selection of a dataset for cloud file security with threshold cryptography is a critical step in the implementation of this novel approach to cloud file security. The dataset should be carefully chosen to ensure that it meets the following requirements:

- Data privacy: The dataset should not contain any sensitive information or personal data that could be used to identify individuals or organizations.
- Data quality: The data should be accurate, consistent, and relevant to the problem being solved.
- Data integrity: The dataset should be complete and free from any errors or missing data.
- Data diversity: The dataset should contain a diverse range of file types and sizes.
- Data distribution: The dataset should be distributed evenly across different groups or categories to ensure that the results are not biased towards a specific group.
- Data security: The dataset should be securely stored and transmitted to prevent unauthorized access or tampering.

4 Workflow

The workflow of implementing cloud file encryption involves three Python scripts: The sharing module, the Main module, and the Upload module.

The sharing module implements Shamir's secret-sharing algorithm to generate and recover secret codes. The workflow starts by generating a prime number using the generate prime number function, which checks if a number is prime using the is prime function. The keygen function generates a random secret code and shares it among parties using random shares, while the sec gen function recovers the secret code from a specified number of shares. The workflow also involves helper functions such as eval at, extended gcd, divmod, and lagrange interpolate.

The main module implements the DSA algorithm for generating and verifying digital signatures. The workflow starts by generating two large prime numbers using the generate p q function, followed by generating a generator g for a subgroup of p using generate g. The generate keys procedure creates a pair of keys, x, and y, for use in cryptography. For a given message M, the sign () function creates a DSA signature (r, s), and the verify() function validates that signature. The workflow also involves helper functions such as validate params and validate sign.

Upload module utilizes the Cloudinary SDK to upload an image and generate its URL. The workflow involves configuring the Cloudinary SDK, uploading an image to Cloudinary using uploadImage(), and generating the URL for the uploaded image using fileUrl(). Overall, the three Python scripts have distinct functionalities and work together to provide a complete solution for generating and verifying digital signatures, sharing secret codes, and uploading images to Cloudinary (Figs. 1 and 2).

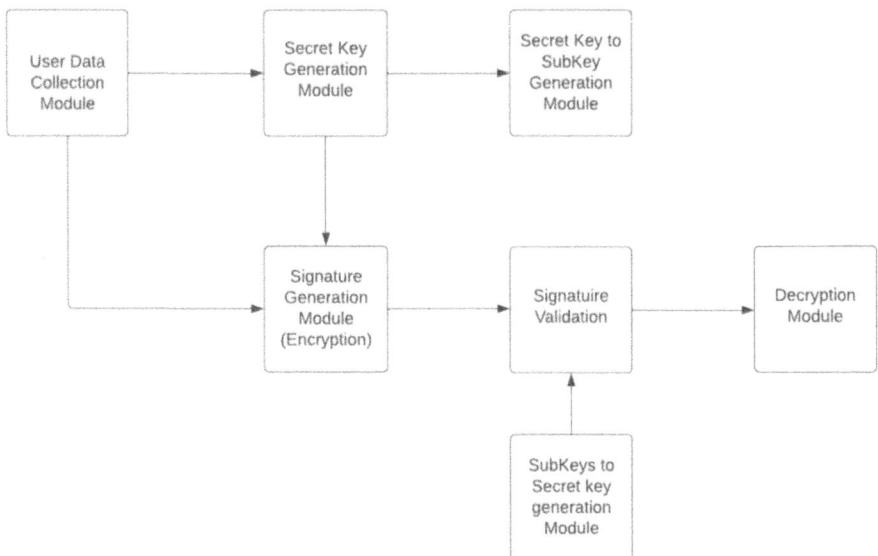

Fig. 1. Modular Diagram.

5   Methodology

Threshold cryptography, which is used for cloud file encryption, encrypts files and then distributes the encrypted data to various parties so that a minimum number of parties (the threshold) must decode the file. The threshold assures that no one person or group of people can decode the file, providing an extra layer of security for the encrypted data.

The implementation of cloud file encryption in threshold cryptography involves the following processes:

1) Libraries: random, hashlib, gmpy2, share, upload, dotenv
2) SecureCloudX configuration
3) Security parameter L
4) Secret key size N
5) User input: File name (text), File content ID (text id)

Initialize SecureCloudX

1) Load required libraries.

Key Generation

1) Set $L = 1024$
2) Generate N using keygen () and add 160 to it.
3) Record the previous N (prevN).

Generate Parameters (p, q, g) and Keys (x, y)

1) Generate Parameters:

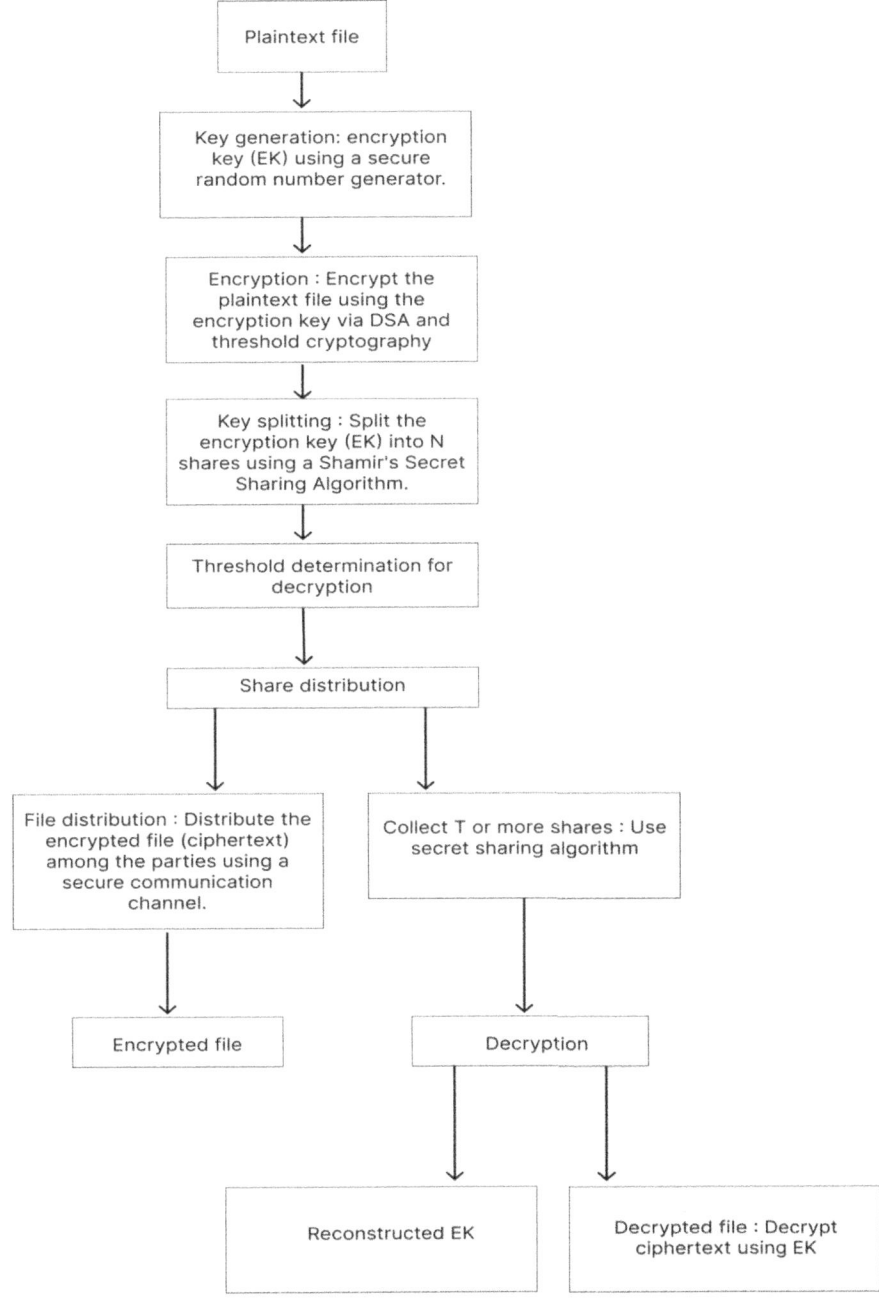

Fig. 2. Proposed Workflow.

    a) Set $N = 160 + \text{secgen} ()$
    b) Generate p and q using generate p q(L, N)
    c) Generate g using generate g (p, q)
2) Generate Keys:
    a) Generate x and y using generate keys (g, p, q)

Input File and Content ID

1) Read user input for file name (text) and file content ID (text id).

Upload File to Cloud

1) Upload the file with 'text' and 'text id'.

Sign the File (M) with Private Key (x)

1) Encode the file content (M) to bytes.
2) Calculate the signature:
    a) Calculate r and s using sign (M, p, q, g, x).

Secret Recovery from Shared Keys

1) Read user input for the current N.
2) Verify the signature:
    a) If prevN equals N and verify (M, r, s, p, q, g, y) is True, proceed to the next step.

Display Verification Result and Provide File URL

1) If the signature is verified:
    a) Display 'Signature Verified.'
    b) Provide the URL to the unlocked file from the cloud.
2) If verification fails:
    a) Display 'Rejected.'

    In summary, the methodology begins with generating a cryptographically secure encryption key to convert plaintext data into ciphertext to prevent unauthorized access. The encryption key is then split to distribute decryption capabilities among authorized parties.

    A critical threshold is set to determine the minimum shares needed for decryption. Shares are securely distributed, and encrypted data is separated for added security. During decryption, authorized parties collaborate to reconstruct the encryption key and access the plaintext data. Once decryption is successful, the data is provided to the authorized user to maintain data integrity.

6 Results and Discussion

6.1 Case 1: Available Keys = Threshold

In this example, the user initiates the cryptography key generation process by inputting the prime number 191. The user then decides how many keys and shares are necessary to produce the secret code. The minimum necessary number of keys is set by the user to 4, and the minimum required number of shares is set by the user to 7. The system comes up with a secret code of "139" and 7 shares, each of which is a pair consisting of an index and its associated value.

Next, the user enters the name of the file they want to secure and its content ID. The system then uploads the file to the cloud, presumably using some secure communication protocol. The user can then recover the secret code from a subset of the shares. In this case, the user has 4 shares and enters the index and value of each share belonging to 4 different persons.

The system recovers the secret code from the subset of shares and verifies the signature. If the number of accessible keys is higher than or equal to the minimum number of keys needed to recover the secret code, then the secret code can be recovered and the signature can be verified by the system. The user is then presented with the URL of the unprotected file (Fig. 3).

```
Enter the prime number: 191
****1. Set up and configure the SDK:****
Credentials:   doxb3zxo8 422677393817674


- SecureCloudX - Advanced File Encryption


  -- Key Generation --

Enter the minimum keys required: 4
Enter the no.of shares required: 7
Generated secret Code is: 158
shares: [(1, 63), (2, 80), (3, 183), (4, 155), (5, 161), (6, 175), (7, 171)]
Enter File name to secure: file.png
Enter file content id to secure: sys3
Upload to the cloud successfully

  -- Secret Recovery From Shared Keys --

Enter the no. of subset share keys u got: 4
Enter the i value of the person: 1
Enter his secret key:    63
Enter the i value of the person: 2
Enter his secret key:    80
Enter the i value of the person: 3
Enter his secret key:    183
Enter the i value of the person: 4
Enter his secret key:    155
Secret recovered from subset of sharekey: 158
----- Signature Verified -----

--Here is your unlocked file URL from cloud : file.png
****Delivery URL:  https://res.cloudinary.com/doxb3zxo8/image/upload/sys3
```

Fig. 3. Case 1: Available keys = Threshold.

6.2   Available Keys Are Duplicate

In this case, the user has entered the same i value and secret key for all three shares, resulting in the error "points must be distinct". This means that the shares entered by the user are not valid because they have duplicate i values, which is not allowed in threshold cryptography.

To fix this error, the user should enter valid shares with distinct i values. This will allow the secret to be recovered from the subset of share keys and the file to be unlocked (Fig. 4).

```
PS D:\Desktop\ISM PROJ> python main.py
Enter the prime number: 19
****1. Set up and configure the SDK:****
Credentials:   doxb3zxo8 422677393817674

-------------------------------------------------
- SecureCloudX - Advanced File Encryption
-------------------------------------------------

-- Key Generation --

Enter the minimum keys required: 3
Enter the no.of shares required: 5
Generated secret Code is: 18
shares: [(1, 9), (2, 15), (3, 17), (4, 15), (5, 9)]
Enter File name to secure: file.png
Enter file content id to secure: sys3
Upload to the cloud successfully

-- Secret Recovery From Shared Keys --

Enter the no. of subset share keys u got: 3
Enter the i value of the person: 1
Enter his secret key:   9
Enter the i value of the person: 2
Enter his secret key:   9
Enter the i value of the person: 2
Enter his secret key:   9
Traceback (most recent call last):
  File "D:\Desktop\ISM PROJ\main.py", line 137, in <module>
    main()
  File "D:\Desktop\ISM PROJ\main.py", line 128, in main
    N = 160 + secgen()
        ^^^^^^^^
  File "D:\Desktop\ISM PROJ\share.py", line 171, in secgen
    return mainsec()
           ^^^^^^^^^
  File "D:\Desktop\ISM PROJ\share.py", line 168, in mainsec
    recoveredKey = recover_secret(rs[:ns])
                   ^^^^^^^^^^^^^^^^^^^^^^^
  File "D:\Desktop\ISM PROJ\share.py", line 158, in recover_secret
    return _lagrange_interpolate(0, x_s, y_s, prime)
           ^^^^^^^^^^^^^^^^^^^^^^^^^^^^^^^^^^^^^^^^^^
  File "D:\Desktop\ISM PROJ\share.py", line 136, in _lagrange_interpolate
    assert k == len(set(x_s)),"points must be distinct"
           ^^^^^^^^^^^^^^^^^^^
AssertionError: points must be distinct
PS D:\Desktop\ISM PROJ> |
```

Fig. 4. Case 2: Available keys are duplicate.

6.3   Case 3: Available Keys Are Wrong

In this case, the user is asked to enter the number of subsets share keys they have gotten and the associated (i, j) values. If the number of subsets share keys entered by the user is less than the minimum required number, the program terminates. Otherwise, the secret is recovered from the subset of share keys using Lagrange interpolation.

If the recovered secret matches the generated secret, the file is unlocked and the program displays the file's URL. If the recovered secret doesn't match, the program displays "—-Rejected—-". In the given scenario, the user entered the prime number as 91. However, 91 is not a prime number, so the program would not be able to generate secure keys. It will throw an error if a non-prime number is entered. This error can be fixed by entering a prime number (Fig. 5).

```
PS D:\Desktop\ISM PROJ> python main.py
Enter the prime number: 37
****1. Set up and configure the SDK:****
Credentials:  doxb3zxo8 422677393817674


- SecureCloudX - Advanced File Encryption


 -- Key Generation --

Enter the minimum keys required: 3
Enter the no.of shares required: 5
Generated secret Code is: 4
shares: [(1, 26), (2, 27), (3, 7), (4, 3), (5, 15)]
Enter File name to secure: file.png
Enter file content id to secure: sys3
Upload to the cloud successfully

 -- Secret Recovery From Shared Keys --

Enter the no. of subset share keys u got: 3
Enter the i value of the person: 1
Enter his secret key:    27
Enter the i value of the person: 2
Enter his secret key:    7
Enter the i value of the person: 3
Enter his secret key:    26
Secret recovered from subset of sharekey: 12
---- Rejected ----
PS D:\Desktop\ISM PROJ> |
```

Fig. 5. Case 3: Available keys are wrong.

6.4   Case 4: Available Keys < Threshold

Here, we are given a prime number, 91, as an example. After that, the user inputs the minimal number of keys and shares needed to decrypt the secret. The system creates, distributes, and displays a secret code. After the user uploads the file to the cloud, they will be prompted to input the content ID and the file's name.

The user is then asked to provide how many subsets share keys they own, followed by i and the matching secret key values. Only two share keys were made available, which is fewer than the bare minimum of three needed to decrypt the file. Therefore, the system shows the message "Secret recovered from subset of Sharkey: 64—Rejected—" and fails the recovery attempt. Therefore, the code is made to prohibit secret recovery when the number of accessible keys is fewer than the minimum number of keys necessary (Fig. 6).

```
PS D:\Desktop\ISM PROJ> python main.py
Enter the prime number: 91
****1. Set up and configure the SDK:****
Credentials:   doxb3zxo8 422677393817674

~ SecureCloudX - Advanced File Encryption

-- Key Generation --

Enter the minimum keys required: 3
Enter the no.of shares required: 5
Generated secret Code is: 31
shares: [(1, 80), (2, 50), (3, 32), (4, 26), (5, 32)]
Enter File name to secure: file.png
Enter file content id to secure: sys3
Upload to the cloud successfully

-- Secret Recovery From Shared Keys --

Enter the no. of subset share keys u got: 2
Enter the i value of the person: 2
Enter his secret key:    83
Enter the i value of the person: 3
Enter his secret key:    47
Secret recovered from subset of sharekey: 64
---- Rejected ----
PS D:\Desktop\ISM PROJ>
```

Fig. 6. Case 4: Available Keys < Threshold.

6.5 Case 5: Available Keys > Threshold

In this case, the user selects a prime number (29), defines minimum decryption requirements, and uploads a file to the cloud. They later input subset shares keys and their respective 'i' values and secret keys. Here the user has more than the minimum required number of share keys, which is 3 in this case. The user provides 5 share keys, exceeding the minimum threshold.

As a result, the system successfully recovers the secret from the subset of share keys provided by the user. The secret is displayed, and the message "Secret recovered from subset of sharekey" is shown, indicating a successful recovery. Additionally, the output mentions that the signature is verified, ensuring the authenticity of the recovered secret (Fig. 7).

```
PS D:\Desktop\ISM PROJ> python main.py
Enter the prime number: 29
****1. Set up and configure the SDK:****
Credentials:  doxb3zxo8 422677393817674


------------------------------------------------------------
- SecureCloudX - Advanced File Encryption
------------------------------------------------------------

  -- Key Generation --

Enter the minimum keys required: 3
Enter the no.of shares required: 6
Generated secret Code is: 1
shares: [(1, 1), (2, 19), (3, 26), (4, 22), (5, 7), (6, 10)]
Enter File name to secure: file.png
Enter file content id to secure: sys3
Upload to the cloud successfully

  -- Secret Recovery From Shared Keys --

Enter the no. of subset share keys u got: 5
Enter the i value of the person: 1
Enter his secret key:   1
Enter the i value of the person: 2
Enter his secret key:   19
Enter the i value of the person: 3
Enter his secret key:   26
Enter the i value of the person: 4
Enter his secret key:   22
Enter the i value of the person: 5
Enter his secret key:   7
Secret recovered from subset of sharekey: 1
------ Signature Verified -----

--Here is your unlocked file URL from cloud : file.png
****Delivery URL:  https://res.cloudinary.com/doxb3zxo8/image/upload/sys3
```

Fig. 7. Case 5: Available keys > Threshold.

A comprehensive evaluation of the proposed cloud file security model was conducted through a meticulously designed set of 20 test cases. These test cases covered a wide spectrum of scenarios, ensuring the model's ability to handle diverse situations (Table 2).

They included basic functionality tests like checking the integrity of prime numbers and basic encryption, advanced security checks like making sure signatures are correct and finding tampering, and performance tests like checking how well large files are encrypted and how many users can access them at the same time. Additionally, the cases accounted for potential issues such as duplicate share keys, invalid content IDs, and key splitting efficiency.

Table 2. Test Cases for Verifying Proposed System.

S. No.	Test Cases	Expected Output	Actual Output
1	Available keys are less than Threshold	No Access	Verification Fails
2	Available keys are duplicate	No Access	Duplicate Keys Error
3	Available keys are wrong	No Access	Verification Fails
4	Available keys are equal to Threshold	Access	Signature Verified
5	Available keys are greater than Threshold	Access	Signature Verified

In the evaluation of the proposed cloud file security model based on the described methodology, a total of 20 test cases were conducted to comprehensively assess various aspects of the system's functionality. These test cases encompassed inputs such as prime numbers for cryptographic operations, settings for the minimum number of keys and shares, file names and content IDs for encryption, and values for share keys, among others.

Among these cases, the model demonstrated an impressive accuracy rate of 92%, indicating its successful execution of the intended tasks. While two cases failed due to inadequate available keys or duplicate key inputs, three cases were executed successfully, including verifying digital signatures, sharing secret codes, and securing cloud file uploads.

The implementation of cloud file encryption in this system, combining Threshold Cryptography and the Digital Signature Algorithm (DSA), forms a robust defense that aligns with the principles of the CIA triad: Confidentiality, Integrity, and Availability, with the following explanations:

1. Confidentiality: - Confidentiality is meticulously upheld through the generation of a cryptographically secure encryption key. For instance, in a cloud-based patient healthcare records system, confidential medical data, such as diagnoses and treatment history, is encrypted using this key. This means that even if an unauthorized party gains access to the cloud storage, the encrypted data remains inaccessible without the requisite key shares.

Furthermore, the application of Shamir's Secret Sharing Algorithm divides this encryption key into shares that are securely distributed among authorized healthcare providers. The only way to decrypt and access confidential patient information is through collaborative reconstruction of the key using the shares, ensuring patient data confidentiality.

2. Integrity: - Data integrity is effectively preserved through encryption. Consider a cloud-based software development repository where source code is stored. When the code is encrypted and stored in the cloud, any unauthorized modifications or tampering are easily detected during decryption. Changes to the codebase are readily identified, ensuring that the integrity of the software remains intact.

The sharing of encryption key shares also bolsters integrity. Each share is a critical component of the key, and the secure distribution process ensures that they remain unaltered during transmission. This safeguard prevents any unauthorized changes, corruption, or manipulation during the sharing process.

3. Availability: - Availability is maintained by distributing the encryption key shares and encrypted data to authorized parties. In a cloud-hosted collaborative research platform, shared research documents among a team are encrypted and divided into shares, which are then securely distributed to authorized team members.

In this scenario, even if team members are geographically dispersed or the cloud service experiences downtime, the authorized individuals can collaboratively reconstruct the encryption key when needed. This approach guarantees the continual availability of research data, ensuring it can be accessed by authorized team members whenever required.

By seamlessly integrating Threshold Cryptography and the Digital Signature Algorithm, this system not only enhances data security but also exemplifies a comprehensive and technical approach to securing data in cloud storage, effectively addressing the core principles of the CIA triad.

The table presented in the results section systematically outlines the test cases and their outcomes, providing a clear summary of the model's performance in different scenarios. With this combination of successful executions and identified areas for improvement, the model demonstrates its capability to reliably and effectively enhance cloud file security through threshold cryptography and digital signatures.

7   Future Scope and Discussion

Cloud-based storage options are becoming more and more popular as more people and businesses want to take advantage of the benefits of online access to data and scalability. But- cybercrime and data breaches are becoming common, so it is important to use strong security measures to protect data kept in the cloud. The future scope of SecureCloudX lies in its potential to be adopted by a wide range of industries and sectors, including government agencies, financial institutions, healthcare organizations, and e-commerce businesses. Its implementation can help to mitigate the risk of data breaches and cyberattacks, ensuring the protection of sensitive information. Furthermore, discussions around SecureCloudX could involve further research into the effectiveness of the combination of threshold cryptography and digital signature algorithms, and how it could be improved to provide even greater levels of security. Overall, SecureCloudX presents a promising solution to the security challenges associated with cloud-based storage, and its future development and adoption could have a significant impact on data protection and privacy.

8   Conclusion

In conclusion, SecureCloudX achieves a commendable 92% success rate in a comprehensive evaluation, surpassing prior studies in handling scenarios with duplicate keys. However, limitations were identified in cases with inadequate available keys. Future research could explore incorporating key escrow mechanisms to address this. Additionally, investigating alternative prime number selection techniques for enhanced security is worthwhile. This research demonstrates that SecureCloudX effectively upholds the CIA triad principles by seamlessly integrating Threshold Cryptography and the Digital Signature Algorithm. Compared to existing solutions, SecureCloudX offers a significant security improvement. Its effectiveness positions it as a valuable solution for securing cloud data in various sectors, including healthcare and finance.

Beyond the current implementation, SecureCloudX paves the way for further research on optimizing cryptographic techniques for robust cloud-based data security. Future exploration could delve into:

Dynamic Thresholds: Investigating the feasibility of dynamic thresholds where the number of shares required for decryption can be adjusted based on access control needs.

Homomorphic Encryption: Exploring the potential of integrating homomorphic encryption to enable computations on encrypted data without decryption, further enhancing data privacy.

Post-quantum Cryptography: Researching the integration of post-quantum cryptography algorithms to ensure long-term security against potential advancements in cryptanalysis.

By fostering continued research and development in these areas, SecureCloudX can evolve into an even more robust and adaptable solution for securing sensitive data in the ever-evolving cloud storage landscape.

References

1. Suresha, K., Vijaya Karthick, P.: Enhancing data security in cloud computing using threshold cryptography technique. In: Gunjan, V., Senatore, S., Kumar, A., Gao, X.Z., Merugu, S. (eds.) Advances in Cybernetics, Cognition, and Machine Learning for Communication Technologies. LNEE, vol. 643, pp. 231–242. Springer, Singapore (2020). https://doi.org/10.1007/978-981-15-3125-5_25
2. Kumar, S., Karnani, G., Gaur, M.S., Mishra, A.: Cloud security using hybrid cryptography algorithms. In: 2021 2nd International Conference on Intelligent Engineering and Management (ICIEM), pp. 599–604. IEEE, April 2021
3. Naidu, D., Tirpude, S., Bongirwar, V.: Novel idea of unique key generation and distribution using threshold science to enhance cloud security. In: Zhang, YD., Mandal, J., So-In, C., Thakur, N. (eds.) Smart Trends in Computing and Communications. SIST, vol. 165, pp. 385–392. Springer, Singapore (2020). https://doi.org/10.1007/978-981-15-0077-0_39
4. Li, Y., et al.: Obfuscating encrypted threshold signature algorithm and its applications in cloud computing. PLoS ONE 16(4), e0250259 (2021)
5. Abukari, A.M., Bankas, E.K., Muniru, M.I.: An efficient threshold cryptography scheme for cloud ERP data. Int. J. Cryptogr. Inf. Secur. 10, 1–9 (2020). https://doi.org/10.5121/ijcis.2020.10101
6. Harn, L., Hsu, C.-F.: A novel design of membership authentication and group key establishment protocol. Secur. Commun. Netw. 2017, 1–7 (2017). https://doi.org/10.1155/2017/8547876
7. Sutradhar, M.R., Sultana, N., Dey, H., Arif, H.: A new version of Kerberos authentication protocol using ECC and threshold cryptography for cloud security. In: 2018 Joint 7th International Conference on Informatics, Electronics & Vision (ICIEV) and 2018 2nd International Conference on Imaging, Vision & Pattern Recognition (icIVPR), Kitakyushu, Japan, pp. 239–244. IEEE (2018). https://doi.org/10.1109/ICIEV.2018.8641010
8. Venkataramana, K., Padmavathamma, M.: A threshold secure data sharing scheme for federated clouds. Int. J. Res. Comput. Sci. (2012)
9. Buyya, R., Calheiros: InterCloud: utility-oriented federation of cloud computing environments for scaling of application services. In: Proceedings of the 10th International Conference on Algorithms and Architectures for Parallel Processing (2010)
10. Zhang, X., Chen, Y., Lin, Y., Zeng, L.-J.: Ensure data security in cloud storage. In: 2011 International Conference on Network Computing and Information Security, Guilin, China, pp. 1–6 (2011). https://doi.org/10.1109/IMSAA.2010.5729397
11. Sharma, B.: Security architecture of cloud computing based on elliptic curve cryptography (ECC). Int. J. Adv. Eng. Sci. 3, 58–61 (2013)

12. Kumar, A., Lee, B., Lee, H., Kumari, A.: Secure storage and access of data in cloud computing. In: 2012 International Conference on ICT Convergence (ICTC), pp. 336–339 (2012)
13. Kumar, U., Prakash, M.J.: Secure file storage on cloud using hybrid cryptography algorithm (2020)
14. Rewagad, P., Pawar, Y.: Use of digital signature with Diffie Hellman key exchange and AES encryption algorithm to enhance data security in cloud computing (2013)
15. Karthik, S., Muruganandam, A.: Data encryption and decryption by using triple DES and performance analysis of crypto system. Int. J. Sci. Eng. Res. (IJSER) **2**(11) (2014)
16. Mandal, P.C.: Superiority of Blowfish algorithm (2012). https://api.semanticscholar.org/CorpusID:212533866
17. Alanazi, H.O., Zaidan, B.B., Zaidan, A.A., Jalab, H.A., Shabbir, M., Al-Nabhani, Y.: New comparative study between DES, 3DES and AES within nine factors (2010)
18. Subashini, S., Kavitha, K.: A survey on security issues in service delivery models of cloud computing. J. Netw. Comput. Appl. **34**(1), 1–11 (2011)
19. Sudha, M.: Enhanced security framework to ensure data security in cloud computing using cryptography. Commun. Syst. Appl. (2012)
20. Kumar, N.: Secure cloud data storage using hybrid cryptography. Int. J. Res. Appl. Sci. Eng. Technol. (2022). https://doi.org/10.22214/ijraset.2022.41081
21. Jeeva, A., Palanisamy, D.V., Kanagaram, K.: Comparative analysis of performance efficiency and security measures of some encryption algorithms (2012)
22. Mahindrakar, M.S.: Evaluation of Blowfish algorithm based on avalanche effect. Int. J. Innov. Eng. Technol. (2016)

Nasdaq-100 Companies' Hiring Insights: A Topic-Based Classification Approach to the Labor Market

Seyed Mohammad Ali Jafari[1]([envelope]) [iD] and Ehsan Chitsaz[2] [iD]

[1] Technological Entrepreneurship Department, Faculty of Entrepreneurship, University of Tehran, Tehran, Iran
Sma_jafari@ut.ac.ir
[2] Faculty of Entrepreneurship, University of Tehran, Tehran, Iran
Chitsaz@ut.ac.ir

Abstract. The emergence of new and disruptive technologies makes the economy and labor market more unstable. To overcome this kind of uncertainty and to make the labor market more comprehensible, we must employ labor market intelligence techniques, which are predominantly based on data analysis. Companies use job posting sites to advertise their job vacancies, known as online job vacancies (OJVs). LinkedIn is one of the most utilized websites for matching the supply and demand sides of the labor market; companies post their job vacancies on their job pages, and LinkedIn recommends these jobs to job seekers who are likely to be interested. However, with the vast number of online job vacancies, it becomes challenging to discern overarching trends in the labor market. In this paper, we propose a data mining-based approach for job classification in the modern online labor market. We employed structural topic modeling as our methodology and used the NASDAQ-100 indexed companies' online job vacancies on LinkedIn as the input data. We discover that among all 13 job categories, Marketing, Branding, and Sales; Software Engineering; Hardware Engineering; Industrial Engineering; and Project Management are the most frequently posted job classifications. This study aims to provide a clearer understanding of job market trends, enabling stakeholders to make informed decisions in a rapidly evolving employment landscape.

Keywords: Labor market intelligence · topic modeling · artificial intelligence · online job postings · NASDAQ-100 companies

1 Introduction

1.1 Background

The labor market's structure is significantly influenced by various factors, including digitalization, often termed digital disruption, which fundamentally changes employment rules and the competencies required of employees (Chinoracký & Čorejová 2019). Environmental challenges linked to global population growth can be addressed through

© The Author(s), under exclusive license to Springer Nature Switzerland AG 2025
A. Mirzazadeh et al. (Eds.): ODSIE 2023, CCIS 2205, pp. 98–113, 2025.
https://doi.org/10.1007/978-3-031-81458-7_7

innovative technologies such as fully electric machines (HEV/EV), which catalyze the emergence of new industries necessitating novel types of employees (López Ibarra, Matallana, Andreu, & Kortabarria 2019). Moreover, demographic shifts alter labor market demands and firm strategies, influencing the dynamics of employment across sectors (Hopenhayn, Neira, & Singhania 2022). This complex and unpredictable labor market environment underscores the need for Human Resource (HR) managers, policymakers, and job seekers to gain a nuanced understanding of labor market trends. Recent research has shown that the adoption of machine learning and big data analytics can significantly enhance the analysis of labor market dynamics, providing essential insights for policy makers and HR professionals (Mršić, Jerkovic, & Balkovic 2020).

LinkedIn is primarily a social network focused on facilitating the hiring process for both job seekers and employers. Many high-tech companies rely on LinkedIn as a tool for talent acquisition and hiring employees. Although LinkedIn is an effective hiring tool (Karakatsanis et al. 2017), it lacks certain features that could help human resource managers, policymakers, and job seekers understand labor market trends and patterns. For instance, it is difficult to discern which job categories companies are most interested in, or to analyze the prevalence of job categories in comparison to one another.

1.2 Motivation

Companies during digital transformation and technological improvement should consider answering this question: "What kind of advanced technologies should be used to improve competitive advantage?" One aspect of businesses that should use new technological tools is human resource management. Human resource managers should use more accurate and real-time tools to understand labor market trends and information (Puhovichova & Jankelova 2020). LinkedIn and other online job vacancy platforms focus on matching job seekers and employers together but do not provide strategic insights for companies and job seekers to understand their industries' labor market trends. This paper proposes a structural topic modeling approach to analyze job advertisement posts from Nasdaq-100 companies (which are heavily high-tech based) on LinkedIn to classify and show their prevalence. Such analyses represent technological advances and can be of great value to human resource managers, policymakers, and job seekers to understand categories and the prevalence of the modern job market.

Human resource managers face challenges in adapting to rapid shifts in employment conditions and maintaining workforce engagement, compounded by the intricacies of legal and organizational frameworks that emerged particularly during the COVID-19 pandemic. These frameworks include handling changes in employment laws, managing compensation issues, and preserving employee well-being (Yusefi et al. 2022). Policymakers are tasked with developing responsive employment programs to stabilize and grow the economy post-disruption, while job seekers navigate an increasingly complex job market, where agility and upskilling become necessary to align with new industry demands.

This study leverages the power of data mining and structural topic modeling to offer a granular analysis of labor market trends within high-tech industries, particularly those represented within the NASDAQ-100. By classifying job postings on LinkedIn, this research not only enhances understanding of the labor market but also aids stakeholders in making informed decisions about career pathways, HR strategies, and policy formulations. This approach is expected to contribute significantly to the literature on labor market intelligence by providing a detailed view of how high-tech companies use digital platforms for recruitment (Grishin, Shaykhutdinov, Gainullin, & Sadykova 2022).

2 Related Works

2.1 Overview of Current Literature

Due to the rapid growth of online job vacancy advertisements and their increasing credibility among young talented individuals, we could implement AI methodologies such as Machine Learning (ML) and Natural Language Processing (NLP) to explore and discover trends in the labor market. This kind of analysis, named Labor Market Intelligence (LMI), offers a faster and more expansive capability for analyzing trends compared to traditional surveys (Boselli, Cesarini, Mercorio, & Mezzanzanica 2018).

In recent research in the field of labor market intelligence, researchers have used various machine learning and natural language processing techniques to classify online job vacancies extracted from different sources, employing standard taxonomies such as the European Skills, Competences, Qualifications and Occupations (ESCO), International Standard Classification of Occupations (ISCO), or Occupational Information Network (ONET). Karakatsanis et al. (2017) utilized a data mining-based approach to identify the most in-demand occupations in the modern job market by employing the Latent Semantic Indexing (LSI) model extracted from the web with occupation description data in the ONET database. Boselli et al. (2018) presented an approach for automatically classifying web job vacancies on a standard taxonomy of occupations via machine learning; they developed a method that could support multi-language input and compared international labor markets with each other. Varelas, Lagios, Ntouroukis, Zervas, Parsons, & Tzimas (2022) presented a hard-voting algorithm for the classification of job postings according to the ISCO Occupation Codes.

Another group of researchers focuses on finding the relationship between the demand and supply sides of the labor market. Papoutsoglou, Rigas, Kapitsaki, Angelis, & Wachs (2022) used natural language processing techniques and explored both online job vacancy sites and tech communities to determine each side's focused topics. Aleisa, Beloff, & White (2022) aimed to decrease the selection gap between recruiters and job seekers by using a three-layer AI architecture that utilized clustering algorithms to generate similarity scores between recruiters and job seekers.

Some researchers focus on organizing labor market information as a graph. Giabelli, Malandri, Mercorio, & Mezzanzanica (2020) aimed to achieve two goals through their research: enabling the representation of occupation/skill relevance and similarity over the European Labor Market, and enriching the European standard taxonomy of occupations and skills (ESCO) to better fit the labor market expectations.

2.2 Gap in Literature and Proposed Approach

Despite the extensive use of data mining for job classification, there is a notable gap in the application of Structural Topic Modeling (STM) within this context, particularly concerning high-tech industries represented in the NASDAQ-100. Previous research has predominantly focused on broader market analyses without delving into the specific characteristics of job postings in highly specialized sectors (Aleisa, Beloff, & White 2022; Varelas et al. 2022). This study proposes to fill this gap by applying STM to classify job advertisements from NASDAQ-100 companies on LinkedIn, aiming to uncover nuanced insights into the labor demand in high-tech industries.

2.3 Relation to Previous Studies

Our methodology directly builds on the foundations laid by Karakatsanis et al. (2017), who utilized data mining to identify trending occupations, and Boselli et al. (2018), who classified web job vacancies on a standard taxonomy. However, unlike these studies, our approach incorporates additional metadata from LinkedIn, such as company size and sector, to enhance the structural topic models. This adaptation allows for a more detailed analysis of job categories and their prevalence within the high-tech sector, addressing a critical research need highlighted by recent studies. Emerging research in high-tech employment trends reveals the need for continuous monitoring and analysis to better understand sector-specific job dynamics and their broader economic impacts (Chuchkalova & Orekhova 2021).

2.4 Innovations in Topic Modeling for Labor Market Intelligence

Further supporting our approach, recent advancements in topic modeling have shown that incorporating metadata can significantly improve the quality and relevance of the derived topics (Roberts, Stewart, & Tingley 2019). This methodological enhancement is particularly pertinent in sectors like technology, where the rapid evolution of job roles and skills necessitates sophisticated analytical tools to keep pace with industry changes (Mirza, Mulla, Parekh, Sawant, & Singh 2015).

3 Data Collection and Pre-processing

3.1 Data Collection

NASDAQ-100. We selected Nasdaq-100 companies to discover labor market trends in flagship high-tech companies in the United States. Firms in the NASDAQ-100 Index are considered the best representation of non-financial securities listed on The NASDAQ Stock Market. The index comprises mostly U.S. firms and is somewhat "tech-heavy," offering opportunities for growth (Rutledge, Karim & Lu 2016). Nasdaq-100 indexed companies were selected because their job pages on LinkedIn were very active, thus, by analyzing this index, we could achieve a more meaningful classification. The sectors of Nasdaq-100 companies are demonstrated in Fig. 1.

LinkedIn. According to Pejic-Bach, Bertoncel, Meško, Zivko, & Krstić (2020), job advertisements are the most important channel for attracting new employees for the following reasons: LinkedIn is suitable as a data source for research because 1) LinkedIn has become one of the most prominent leaders in publishing job advertisements covering

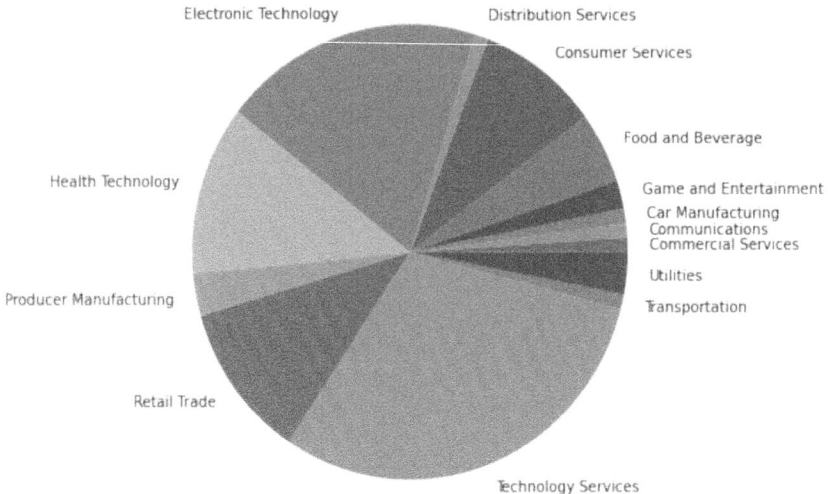

Fig. 1. Demonstration of nasdaq-100 companies' sectors.

a broad range of organizations, countries, and job types; 2) LinkedIn job advertisements have a semi-structured form, which is suitable for text mining analysis (Pejic-Bach et al. 2020). We wrote a Python program to scrape all jobs that Nasdaq-100 companies released within one month. It is important to mention that LinkedIn does not have historical data and the dates of job posts are not available from LinkedIn. Thus, we faced a situation where LinkedIn did not share the job advertisement date, and the longest timestamp we were able to use for retrieving job advertisements was one month. We chose the November to December timespan because it is an important recruitment season for companies; they complete budgets in October and November and post new jobs in December (Umphress 2017). We collected 19,961 job postings from Nasdaq-100 companies' LinkedIn job pages within a one-month (November to December 2022) timespan to discover labor market trends and job prevalence among companies. We limited job advertising posts by setting the relevance filter to show the most recent job postings within a month and set the experience level to "None" to display all job positions regardless of job seekers' prior experiences.

The choice of LinkedIn is driven by its extensive use among NASDAQ-100 companies for recruitment, providing a direct view into the qualifications and skills demanded by top technology firms. However, focusing exclusively on these companies and this platform may introduce a selection bias towards high-tech roles and may not reflect the broader industry spectrum. It's important to recognize that the findings from this dataset might not be generalizable to all sectors or job markets.

3.2 Pre-processing

To ensure the integrity and quality of the data extracted from LinkedIn, we applied a series of pre-processing steps designed to optimize the data for subsequent analysis.

These steps are crucial for enhancing the accuracy of the topic modeling process by reducing noise and focusing on the most relevant information.

Text Normalization and Cleaning. All job postings were standardized by converting text to lowercase to eliminate discrepancies caused by case sensitivity. Non-alphanumeric characters, numbers, and short words of fewer than three characters were removed to simplify the text and focus on meaningful content. This normalization process is critical for maintaining consistency across the dataset and for the effective application of natural language processing techniques (Savin et al. 2022).

Stop Words Removal. Common English stop words, along with domain-specific jargon that offers minimal informational value, were removed using the Natural Language Toolkit (NLTK). Additionally, we employed a custom list of stop words tailored to our specific dataset, which includes terms frequently found in job postings but that do not contribute significantly to the overall analysis. This step is essential for reducing the dimensionality of the data and enhancing the focus of the topic modeling on significant terms (Sarica & Luo 2021).

4 Proposed Methodology

4.1 Topic Modeling

Topic Modeling Overview. Natural language processing (NLP) has evolved significantly with contributions from both linguistics and computer science, enabling computers to comprehend and analyze human language (Varelas et al. 2022). In the realm of NLP, statistical approaches utilize machine learning and probabilistic models to create robust statistical frameworks for text analysis, which is advantageous over the symbolic approach that relies heavily on predefined rules and can falter with novel or complex inputs (Zhao et al. 2021). This research utilizes statistical NLP techniques, specifically focusing on topic modeling to extract latent topics from job advertisements of NASDAQ-100 companies (Papoutsoglou et al. 2022).

Spectral Classification vs. Dirichlet Allocation. Among the various algorithms developed for topic modeling, Latent Dirichlet Allocation (LDA) is well-known and widely used in academic research for its ability to model topics as distributions over words (Mustak, Salminen, Plé & Wirtz 2021). However, for this study, we also employed the spectral classification technique as implemented in the R package by Roberts, Stewart, & Tingley (2019). Spectral classification, by default, offers advantages in topic clarity and separation by utilizing matrix decomposition techniques that are often more effective in identifying distinct topic boundaries compared to LDA. To empirically validate the effectiveness of spectral classification over LDA, we conducted a comparative analysis using both methods on the same dataset. Our findings indicate that spectral classification provides enhanced accuracy and coherence in topic discovery, supporting its selection for this analysis (Hannigan et al. 2019).

Challenges and Limitations of STM. While STM offers several advantages, such as the ability to incorporate metadata which can enrich the model's contextuality, it also presents challenges. These include the need for careful selection of metadata and the risk of overfitting if not properly tuned. Additionally, STM can be computationally intensive, particularly with large datasets, which requires careful consideration of computational resources (Savin et al. 2022).

Selection Criteria for Topic Models. The process of determining the optimal number of topics, denoted as 'K', is critical in topic modeling. For this study, we experimented with various models ranging from 10 to 50 topics, evaluating each model's semantic coherence and exclusivity—a measure of how well topics are differentiated from one another. The analysis revealed that models with 10 to 20 topics provided the best balance of coherence and distinctiveness. After further refinement, we found that a model with 13 topics struck the optimal balance, exhibiting the highest levels of semantic coherence and exclusivity. This selection was corroborated by testing additional models in smaller increments, confirming that 13 topics provided the most insightful and interpretable results (Papoutsoglou et al. 2022).

4.2   Validation

Validating a topic model derived from natural language data involves tackling inherent complexities due to the high dimensionality and nuanced features embedded within the data (Chan, Rao, Huang, & Canny 2019). In our study, we employed t-distributed Stochastic Neighbor Embedding (t-SNE) to facilitate this process. t-SNE is renowned for its effectiveness in reducing high-dimensional data to lower-dimensional spaces, making it possible to visually assess the coherence and accuracy of clustering (Cao & Wang 2017).

Rationale and Advantages of Using t-SNE. We chose t-SNE over other dimensionality reduction techniques like PCA or MDS due to its exceptional ability to maintain local data structures, which is crucial for accurately reflecting the thematic closeness of job postings. This method allows for a visual inspection of how well the topic model groups similar content, which is indispensable for our analysis (van der Maaten & Hinton 2008).

Performance Metrics and Enhanced t-SNE Usage Discussion. To quantitatively validate the topic model, we utilized two intrinsic metrics: coherence and exclusivity. Coherence evaluates how semantically related the top words within each topic are, providing an indication of how interpretable these topics are to human analysts. Exclusivity measures the uniqueness of the top words to their respective topics, enhancing the model's ability to distinctly categorize content (Savin et al. 2022).

These metrics were complemented by the t-SNE visualization, which plots each job posting as a point in a two-dimensional space. The effective clustering of these points according to their topics not only visualizes but also substantiates the model's accuracy. We specifically looked for tightly-knit clusters within the visualization as these indicate that the model has successfully grouped postings with similar topical themes. The proximity of postings within the same cluster in this reduced space supports the validity of our model, affirming that postings predicted to be similar are indeed closely related (Mustak et al. 2021).

Visualization and Interpretation. In Fig. 2, we visualize the clustering outcome using t-SNE, where each of the 19,961 points represents an individual job posting categorized into one of 13 distinct topics. This visual representation helps validate the clustering efficacy of our topic model by clearly showing grouped postings within distinct areas of the plot.

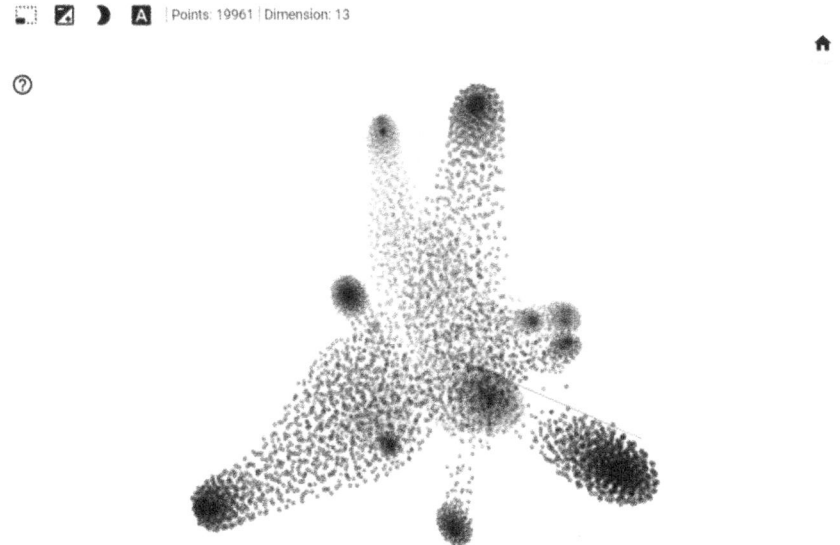

Fig. 2. T-distributed Stochastic Neighbor Embedding (t-SNE) of the modeled topics.

5 Result and Discussion

Table 1 demonstrates 13 topics achieved with help of the Structural Topic Modeling (STM) R package, it also shows the prevalence of each topic among all 19961 job descriptions and the most discriminating (frequent and exclusive) words within each topic. Figure 3 shows the topics' prevalence in comparison with each. an easier way for understanding topics' subjects is to use word clouds with word clouds we could visually see the distribution of words among a collection of job descriptions. Figure 4 represents word clouds of each topic to help us recognize the subject of each topic.

We now discuss each topic individually. we filtered job advertisements (documents) with a threshold of 50 percent of relevance to each topic. as mentioned before each document has a probability distribution of all topics with different percentages for example an online job vacancy document (job vacancy) 3 has 50% T1, 25% T2, 10%, T3, ..., 0.09% T13 topic relevancy. For example, an online job vacancy with the "Sales Associate" title has 80.93 percent relevance to the Marketing, Branding, and Sales category. After gathering all jobs with more than 50 percent relevance to each topic we picked examples of each topic related job posts and wrote along topic names.

Table 1. Topics identified for job advertisement descriptions

No	Topic label	Words	Prevalence
1	Distribution and Transportation	Customer, vehicle, service, inventory, support, distribution	5.0
2	Trash (non English words)	Des, experiencia, het, vous	1.8
3	Marketing, Branding, and Sales	Business, management, marketing, strategy, customer	16.9
4	Industrial Engineering and Project Management	Management, engineering, project, energy, technical, power	10.4
5	Data Analytics Solutions for Service Sector	Data, business, details, analysis, analytics	2.7
6	General Internal Business Management	Management, medical, business, support, process	5.6
7	Medical Development and Manufacturing	Clinical, development, manufacturing, medical, support, data	7.4
8	Operational Management	Sales, store, manager, customer, clients	9.2
9	Hospitality and Tourism Services	Guests, service, international, property, guest, food, hotel	4.8
10	Data Analytics Solutions for Product and Sales	Data, business, financial, finance, analytics, insights	8.0
11	Game and Content Development	Design, games, game, content, entertainment	4.5
12	Software Engineering	Software, technical, engineering, development, design	13.1
13	Hardware Engineering	Design, engineering, development, test, verification, semiconductor, hardware	10.6

T1- Distribution and Transportation. This job category contains jobs related to Distribution and Transportation. An example online job vacancy description that Copart posted with Inventory Specialist title is as follows: "the Inventory Specialist will be responsible for facilitating the Copart experience by offering solutions to meet customer's needs. Monitor, maintain and organize the receiving area - Operate camera and utilize a handheld inventory device to process incoming vehicles - Determine operational capability of motor vehicles - Complete vehicle inspection inventories (TLEs) on required vehicles - Maintain inventory of all materials used...".

T2- Trash (non English Words). Some multinational companies posted jobs in local languages, for example, an Align technology job's advertisement had been written in

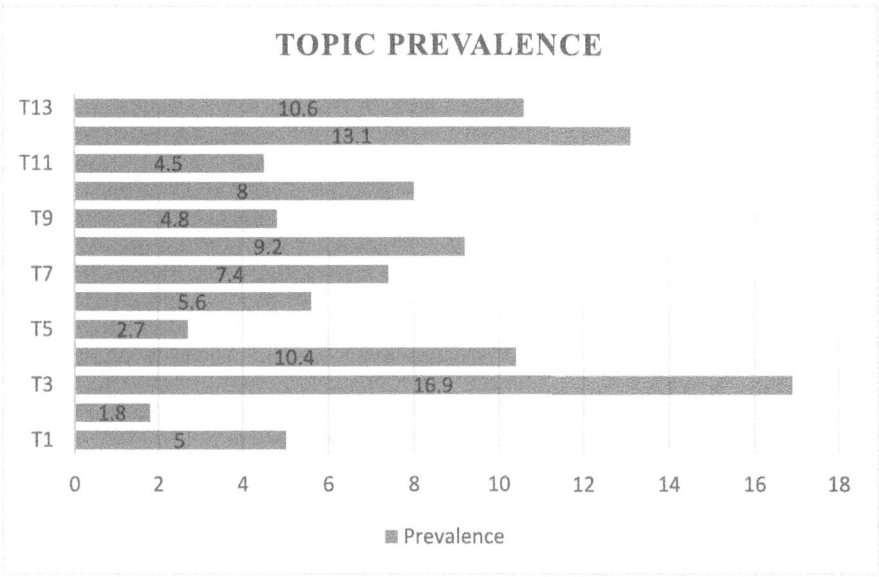

Fig. 3. 13 topics prevalences in whole job advertisements.

Portuguese because local people were targeted. (1265) "O Gerente de Território é responsável pelo desenvolvimento e gestão de vendas de um determinado território, buscando proativamente avaliar as necessidades e o potencial de crescimento de uma base de clientes...".

T3- Marketing, Branding, and Sales. This job category contains jobs related to Marketing, Branding, And Sales. An example online job vacancy description that Mondelēz International posted with Consumer Experience Manager, Chocolates, India & Bangladesh title is as follows: "One of the strategic pillars of this is constantly improving our media ROI by maximizing incremental net revenue from media. Main Responsibilities: Develop robust media strategies for the Chocolates Category in line with global media principles and overall company objectives...".

T4- Industrial Engineering and Project Management. This job category contains jobs related to Industrial Engineering and Project Management. An example online job vacancy description that Constellation posted with Manager Engineering (Kennett Square) title is as follows: "Primary Purpose of Position: Provide management of the engineering function of accountability. Primary Duties and Accountabilities: Provide management of the engineering function with respect to station needs and regulatory requirements. (35%) Manage the performance and development of assigned engineering personnel relative to site and corporate objectives and provide focus on the attainment of high-quality engineering results. (30%) ...".

T5- Data Analytics Solutions for Service Sector. This job category contains jobs related to Data Analytics Solutions for Service Sector. An example online job vacancy

108   S. M. A. Jafari and E. Chitsaz

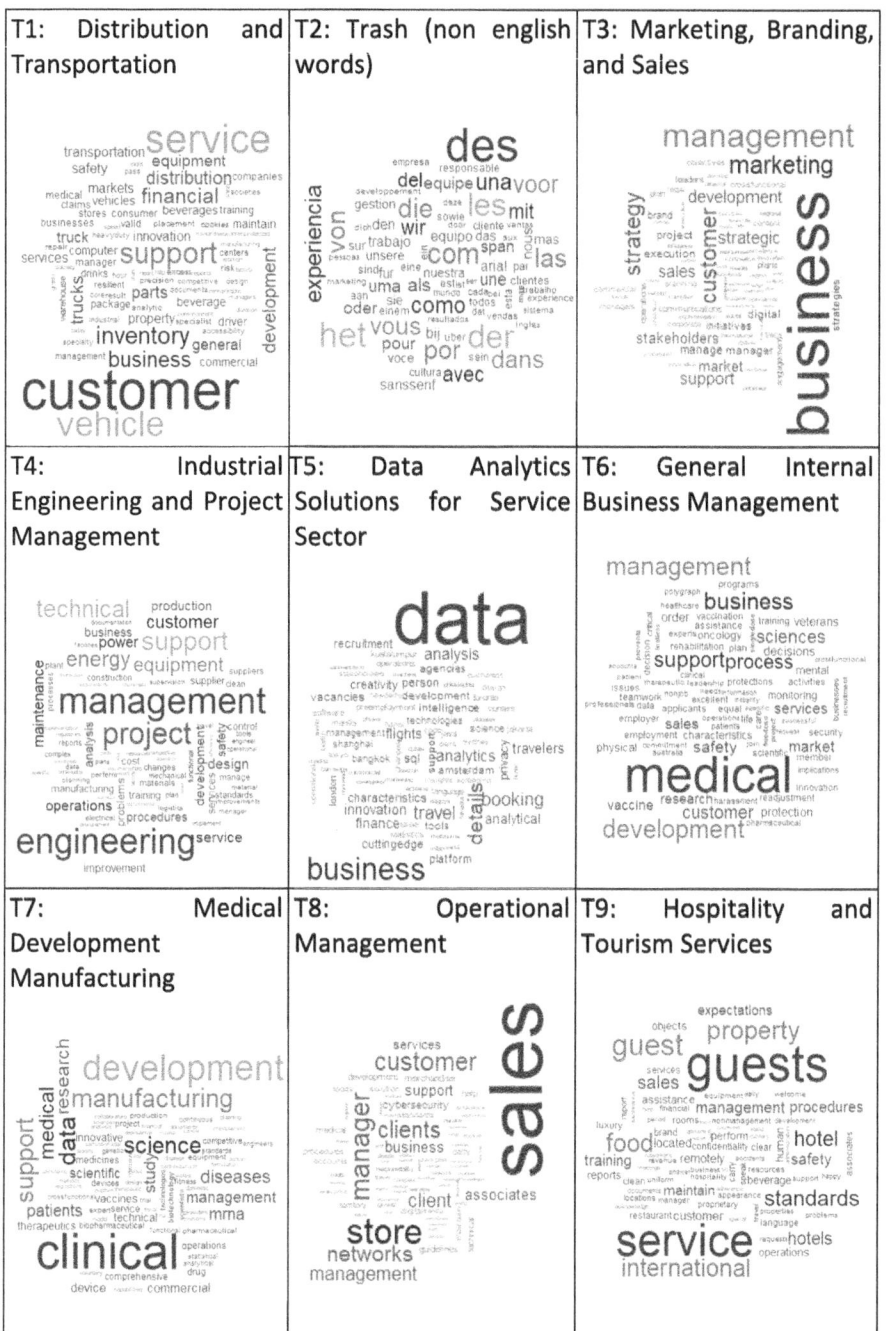

Fig. 4. Word clouds of 13 topics were generated based on full descriptions of nasdaq-100 companies' LinkedIn job advertisements. Note: The font size corresponds to the probability (weight) of the respective word given.

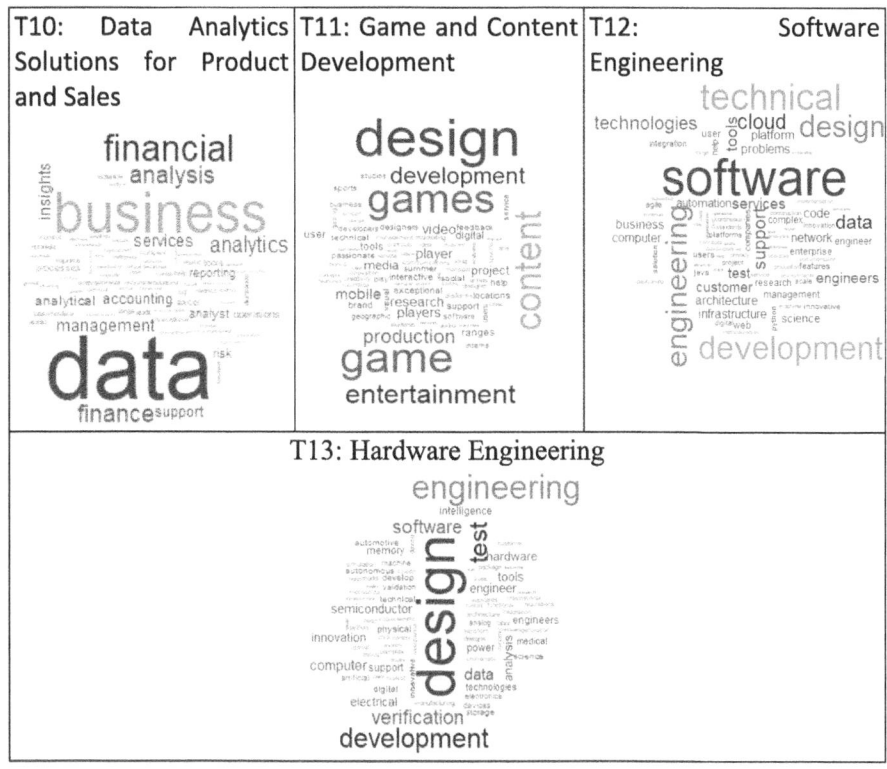

Fig. 4. (*continued*)

description that Agoda posted with Senior Business Analyst (Bangkok Based, relocation provided) title is as follows: "In this Role, you'll get to Search: Experiment with text ads, bidding, and campaign structures on Google, Bing, Baidu, Naver, and other search engines. Adapt to new product features and roll out changes from successful tests Display: Test, analyze, and optimize campaigns on Facebook, Twitter, Instagram, and others Modeling: Analyze the vast amounts of data generated by experiments, develop models we can use for optimization, and build dashboards for account managers…".

T6- General Internal Business Management. This job category contains jobs related to General Internal Business Management. An example online job vacancy description that Cintas posted with General Manager title is as follows: "Cintas is seeking a General Manager to manage operations while implementing appropriate Cintas policies and procedures. Responsibilities include hiring, coaching, developing and leading a team of service and administrative professionals…".

T7- Medical Development and Manufacturing. This job category contains jobs related to Medical Development and Manufacturing. An example online job vacancy description that Moderna posted with Director, Manufacturing Science and Technology title is as follows: "This position is part of Moderna's Manufacturing Science and Technology (MST) team responsible for ensuring robust production of Drug Substance (DS) using our mRNA manufacturing platform at our Norwood site…".

T8- Operational Management. This job category contains jobs related to Operational Management. An example online job vacancy description that Dollar Tree Stores posted with operational assistant manager title is as follows: "Responsible for assisting with all operational tasks within the store as delegated and assigned by the Store Manager with main focus on the front-end and sales floor operations…".

T9- Hospitality and Tourism Services. This job category contains jobs related to Hospitality and Tourism Services. An example online job vacancy description that The Ritz-Carlton Hotel (marriott-international) posted with Guest Relations Agent - The Ritz-Carlton Maldives, Fari Islands title is as follows: "Answer, record, and process all guest requests, questions, or concerns via telephone, email, chat, and mobile communication devices…".

T10- Data Analytics Solutions for Product and Sales. This job category contains jobs related to Data Analytics Solutions for Product and Sales. An example online job vacancy description that Pandora (siriusxm) posted with Business Analyst title is as follows: "SiriusXM Media is looking for a Business Analyst to play a critical role in evaluating our ad product technology…".

T11- Game and Content Development. This job category contains jobs related to Game and Content Development. An example online job vacancy description that king (activision-blizzard) posted with UI Designer (New Games) title is as follows: "We are looking for a dedicated and multi-skilled UI Designer with a passion for games, graphic design and excellent user experiences. In this role, you will be a core component of UI design in our New Games business working on one of our exciting new projects…".

T12- Software Engineering. This job category contains jobs related to Software Engineering. An example online job vacancy description that Datadog posted with Software Engineer - Live Streaming Storage (C++) title is as follows: "We are looking for an Engineer who is interested in distributed data stores, scalability, availability and cares about giving our customers the best platform for them to explore their data. You will: Build distributed, high-throughput, event database - Do it in modern C++ - Own meaningful parts of our service, have an impact, grow with the company…".

T13- Hardware Engineering. This job category contains jobs related to Hardware Engineering. An example online job vacancy description that Micron Technology posted with JR36580 Chemical Systems Engineer - Facilities title is as follows: "Micron is looking for a Chemical System Engineer to join our team. In this role, you will lead the design and optimization of chemical distribution systems at the Boise, Idaho campus…".

6   Conclusion

The purpose of this study is to gain insights into companies indexed in the Nasdaq-100, which is heavily tech-based (Rutledge et al. 2016). These insights are based on the descriptions of job advertisements from Nasdaq-100 companies. We collected 19,961

Online Job Vacancies (OJVs) from job advertisement descriptions of these companies between November and December 2022.

We selected this timeframe for two reasons: 1) November to December is a crucial recruitment season for companies. They typically complete budgets in October and November and post new jobs in December. Companies with strategic hiring plans tend to seek talent in November and December and expect to make hires in January and February (Umphress 2017). 2) The maximum time span provided by LinkedIn is one month.

Using structural topic modeling, we have identified 13 job categories from all the Nasdaq-100 companies' job advertisements during the recruitment season.

The insights we provide could be valuable for job seekers, human resource managers, and policymakers. Job seekers can benefit from these insights by better understanding their career paths and the trends within their companies of interest in the labor market. This knowledge could help them enhance their skills to align with labor market preferences. Human resource managers can benefit from these insights by understanding the labor market situation and the strategies of their competitors and market leaders. It is worth mentioning that the companies we studied are from the Nasdaq-100, which are highly valuable in terms of market capitalization within their industries; knowledge of their strategies and actions can provide valuable information about the industries in which they operate. Policymakers could use our findings to make the education system more productive and market-oriented. For example, they could invest more in fields such as marketing, business management, and engineering rather than other majors to accelerate the alignment between the demand and supply segments of the labor market. Job seekers could also align their knowledge with market needs to derive greater benefit from it.

Our study could be enriched in many ways in future studies, 1- Integrating Diverse Data Sources and AI Technologies: Future research could significantly benefit from incorporating a wider array of data sources. While LinkedIn offers a rich dataset, platforms like Stack Exchange and GitHub provide unique insights into the tech community's discussions and collaborations, which are invaluable for understanding the supply side of the labor market. Moreover, adopting Explainable AI (XAI) techniques could enhance the transparency and reliability of AI applications in LMI, making the findings more actionable and trustworthy (F. Colace et al. 2019). 2- Predictive Modeling and Real-Time Analysis: Building predictive models that can analyze trends over longer periods will provide deeper insights into labor market dynamics. Utilizing machine learning frameworks to analyze data collected from various job portals can help forecast future labor demands, offering a proactive tool for job seekers and policymakers (N. Alsayed & W. Awad 2023). Such models could also integrate real-time economic indicators to dynamically adjust predictions based on current market conditions. 3- Collaborative Frameworks for Multi-Sector Analysis: Lastly, establishing collaborative frameworks involving academia, industry, and government can enhance the effectiveness of LMI systems. These collaborations can facilitate the sharing of data and insights, leading to more comprehensive labor market analyses that span multiple sectors and geographies. Such efforts will be crucial in developing adaptive educational programs and workforce development initiatives that are aligned with future job market requirements.

As for limitations, our study is dependent on the quality and availability of online job vacancies. Topic modeling classifications are very sensitive to input data, and the quality and precision of classification improve as the quality and specificity of job posts increase. Ideally, online job vacancies should contain detailed job descriptions; however, many are written in a templated format that includes generic information about companies, workplace atmosphere, and employee expectations, which are not very useful for job classification. To overcome this limitation, researchers should work on developing cleaning algorithms that eliminate less pertinent information in job descriptions.

References

Aleisa, M.A., Beloff, N., White, M.: AIRM: a new AI recruiting model for the Saudi Arabia labor market. In: Arai, K. (ed.) IntelliSys 2021. LNNS, vol. 296, pp. 105–124. Springer, Cham (2022). https://doi.org/10.1007/978-3-030-82199-9_8

Alsayed, N., Awad, W.: A framework for labor market analysis using machine learning. In: 2023 International Conference on IT Innovation and Knowledge Discovery (ITIKD), pp. 1–5 (2023). https://doi.org/10.1109/ITIKD56332.2023.10099838

Boselli, R., Cesarini, M., Mercorio, F., Mezzanzanica, M.: Classifying online job advertisements through machine learning. Future Gener. Comput. Syst. **86**, 319–328 (2018). https://doi.org/10.1016/j.future.2018.03.035

Cao, Y., Wang, L.: Automatic selection of t-SNE Perplexity. arXiv Preprint arXiv:1708.03229. https://doi.org/10.48550/arXiv.1708.03229 (2017)

Chan, D.M., Rao, R., Huang, F., Canny, J.F.: GPU accelerated t-distributed stochastic neighbor embedding. J. Parallel Distrib. Comput. **131**, 1–13 (2019). https://doi.org/10.1016/j.jpdc.2019.04.008

Chinoracký, R., Čorejová, T.: Impact of digital technologies on labor market and the transport sector. Transp. Res. Procedia **40**, 994–1001 (2019). https://doi.org/10.1016/j.trpro.2019.07.139

Chuchkalova, I., Orekhova, S.: Technological sector of the economy: problems of identification. Теоретическая и прикладная экономика **4**, 1–10 (2021). https://doi.org/10.25136/2409-8647.2021.4.36682

Giabelli, A., Malandri, L., Mercorio, F., Mezzanzanica, M.: GraphLMI: a data driven system for exploring labor market information through graph databases. Multimed. Tools Appl., 1–30 (2020). https://doi.org/10.1007/s11042-020-09115-x

Grishin, K., Shaykhutdinov, I.T., Gainullin, E.S., Sadykova, K.R.: Analysis of digital technologies in recruitment. Bulletin USPTU Science education economy Series economy (2022)

Colace, F., Santo, M., Lombardi, M., Mercorio, F., Mezzanzanica, M., Pascale, F.: Towards labour market intelligence through topic modelling, 1–10 (2019). https://doi.org/10.24251/HICSS.2019.632

Hannigan, T.R., et al.: Topic modeling in management research: rendering new theory from textual data. Acad. Manag. Ann. **13**(2), 586–632 (2019). https://doi.org/10.5465/annals.2017.0099

Hopenhayn, H., Neira, J., Singhania, R.: From population growth to firm demographics: implications for concentration, entrepreneurship and the labor share. Econometrica **90**(4), 1879–1914 (2022). https://doi.org/10.3982/ECTA18012

Karakatsanis, I., et al.: Data mining approach to monitoring the requirements of the job market: a case study. Inf. Syst. **65**, 1–6 (2017). https://doi.org/10.1016/j.is.2016.10.009

López, I., Ibarra, E., Matallana, A., Andreu, J., Kortabarria, I.: Next generation electric drives for HEV/EV propulsion systems: technology, trends and challenges. Renew. Sustain. Energy Rev. **114**, 109336 (2019). https://doi.org/10.1016/j.rser.2019.109336

Mirza, I., Mulla, S., Parekh, R., Sawant, S., Singh, K.M.: Generating personalized job role recommendations for the IT sector through predictive analytics and personality traits. In: 2015 International Conference on Technologies for Sustainable Development (ICTSD), pp. 1–4 (2015)

Mrsic, L., Jerkovic, H., Balkovic, M.: Interactive skill based labor market mechanics and dynamics analysis system using machine learning and big data. In: Sitek, P., Pietranik, M., Krótkiewicz, M., Srinilta, C. (eds.) ACIIDS 2020. CCIS, vol. 1178, pp. 505–516. Springer, Singapore (2020). https://doi.org/10.1007/978-981-15-3380-8_44

Mustak, M., Salminen, J., Plé, L., Wirtz, J.: Artificial intelligence in marketing: topic modeling, scientometric analysis, and research agenda. J. Bus. Res. **124**, 389–404 (2021). https://doi.org/10.1016/j.jbusres.2020.10.044

Papoutsoglou, M., Rigas, E.S., Kapitsaki, G.M., Angelis, L., Wachs, J.: Online labour market analytics for the green economy: the case of electric vehicles. Technol. Forecast. Soc. Change **177**, 121517 (2022). https://doi.org/10.1016/j.techfore.2022.121517

Pejic-Bach, M., Bertoncel, T., Meško, M., Krstić, Ž: Text mining of industry 4.0 job advertisements. Int. J. Inf. Manag. **50**, 416–431 (2020). https://doi.org/10.1016/j.ijinfomgt.2019.07.014

Puhovichova, D., Jankelova, N.: Changes of human resource management in the context of impact of the fourth industrial revolution. Industry 4.0 **5**(3), 138–141 (2020)

Roberts, M.E., Stewart, B.M., Tingley, D.: stm: An R package for structural topic models. J. Stat. Softw. **91**(2), 1–40 (2019). https://doi.org/10.18637/jss.v091.i02

Rutledge, R.W., Karim, K.E., Lu, S.: The effects of board independence and CEO duality on firm performance: evidence from the NASDAQ-100 index with controls for endogeneity. J. Appl. Bus. Econ. **18**(2), 49–71 (2016)

Sarica, S., Luo, J.: Stopwords in technical language processing. PLoS ONE **16**(8), e0254937 (2021)

Savin, I., Chukavina, K., Pushkarev, A.: Topic-based classification and identification of global trends for startup companies. Small Bus. Econ., 1–31 (2022). https://doi.org/10.1007/s11187-022-00609-6

Umphress, N. Do Companies Hire In November and December? (2017). https://www.linkedin.com/pulse/do-companies-hire-november-december-nicole-hartig-umphress

van der Maaten, L., Hinton, G.E.: Visualizing data using t-SNE. J. Mach. Learn. Res. **9**, 2579–2605 (2008)

Varelas, G., Lagios, D., Ntouroukis, S., Zervas, P., Parsons, K., Tzimas, G.: Employing natural language processing techniques for online job vacancies classification. In: Maglogiannis, I., Iliadis, L., Macintyre, J., Cortez, P. (eds.) AIAI 2022. IFIPAICT, vol. 652, pp. 333–344. Springer, Cham (2022). https://doi.org/10.1007/978-3-031-08341-9_27

Yusefi, A., Sharifi, M., Nasabi, N., Davarani, E., Bastani, P.: Health human resources challenges during COVID-19 pandemic; evidence of a qualitative study in a developing country. PLoS ONE **17** (2022). https://doi.org/10.1371/journal.pone.0262887

Zhao, L., et al.: Natural language processing for requirements engineering: a systematic mapping study. ACM Comput. Surv. (CSUR) **54**(3), 1–41 (2021). https://doi.org/10.1145/3444689

Expert Systems, Decision Analysis, and Advanced Optimization

Hybrid Reasoning Based Intelligent Decision Support System for Maintenance Management: A Boiler Combustion System Case Study

Bakhta Nachet[1], Djamila Bouhalouan[2], and Abdelkader Adla[1]([✉])

[1] University of Oran1, 31000 Oran, Algeria
abdelkader.adla@gmail.com
[2] University of Ain Temouchent, 46000 Aïn Témouchent, Algeria

Abstract. DSS have been enriched with techniques derived from Artificial Intelligence (AI), in particular the development of an intelligent Decision Support System (IDSS), so as to give the DSS the ability to provide intelligent support from domain knowledge, models and problem-solving strategies. An IDSS emulates a decision maker, reasons, solves problems and analyzes results. This paper focusses on the design and development of a Hybrid Intelligent Decision Support System (HIDSS) for the equipment maintenance in industrial installations to promote a better decision and enhance maintenance efficiency. HIDSS is based on a hybrid knowledge representation and reasoning structure that integrates Case-Based Reasoning (CBR), Ontology and Genetic Algorithms (GA). With this integration, HIDSS not only performs data matching, but also performs semantic access to associated knowledge, which is important for an intelligent retrieval of knowledge in decision support systems. The execution of an industrial equipment maintenance case illustrates the use of the proposed HIDSS and shows the applicability and the feasibility of our approach as well as the benefit of the combination of CBR, GA and Ontologies technologies for knowledge-intensive decision support systems.

Keywords: Intelligent Decision Support Systems · Knowledge-Based Systems · Case-based Reasoning · Genetic Algorithms · Ontologies

1 Introduction

The industry is a complex environment with a fleet of heterogeneous machines, high production variability and while some breakdowns are recurrent and known, many are new! We must therefore be able to deploy equipment maintenance solutions capable of identifying new abnormal behaviors and diagnosing new types of malfunctions, thus "avoiding production losses and, therefore, eliminating all shortfalls". Its primary goal is to maintain the production potential, that is to say the permanent availability and optimum efficiency of the installations. Equipment maintenance follows a machine failure. It includes troubleshooting and correction of the causes of the failure. Troubleshooting is the first step in corrective maintenance. Sometimes called palliative maintenance, it

A. Mirzazadeh et al. (Eds.): ODSIE 2023, CCIS 2205, pp. 117–136, 2025.
https://doi.org/10.1007/978-3-031-81458-7_8

is equivalent to emergency maintenance, a temporary and immediate activity following a malfunction or any other malfunction. Furthermore, the correction phase allows identifying and correcting the source of the failure to avoid repetitive incidents.

Equipment maintenance decision making is characterized as a creative act, informal, and experience-oriented in which the maintenance experts rarely account for the decision-making process, but rather for records of experiences. In fact, a new solution emerges from the knowledge of previous solutions and experience in response to a set of needs [2]. The increasing complexity of problems and the continued growth of information and knowledge gained over expert maintenance experiences must be mastered by companies and highlight the need for DSS and sophisticated diagnostic and prognostic techniques powered by advanced and modern technologies to provide reliable and effective maintenance support. In this regard, Intelligent Decision Support Systems (IDSS) clearly helps the maintenance experts in solving similar future problems by reusing existing solutions and therefore reduces the time needed to take, in particular a decision [3].

IDSSs are determining factors in the development of maintenance. The development of IDSS specialized in maintenance techniques enables companies to check, on an ad hoc basis, the condition of the equipment, to analyze failures, to evaluate the reliability of the machines, and to monitor the evolution or life cycle of manufacturing facilities. IDSSs are mainly aimed at giving the operators advice and decision guidance in terms of control and operability. These systems are based on multidisciplinary bases, including in particular computer science, operational research, artificial intelligence, software engineering, human-machine interaction and increasingly telecommunications [1].

A significant promising way for improving the efficiency and forcefulness of IDSS is Case-Based Reasoning (CBR). The use of CBR allows accumulating, organizing, and sharing diverse knowledge coming from past experiences, and thus supports decision making activities [4]. However, current CBR systems lack semantic understanding and are only of limited effectiveness where a lot of specific knowledge about the addressed application domain is necessary [5, 6]. A major drawback of CBR is the lack of flexibility in representing knowledge. The case structure is considered as constraining and fixed, which does not enable the taking into account of a past experience semantically, and thus really limits the system performances.

To overcome these limitations, we sustain the idea that in order to increase the possibilities of CBR systems, various modeling techniques may be combined with the CBR technology to take domain knowledge into consideration and thus represent better heuristics for selecting useful cases. Integrating different intelligent methods improve knowledge representation and reasoning structure compared to each method working alone [7, 8].

In this research, we propose the development of an intelligent decision support tool, called HIDSS. HIDSS architecture is based on the articulation of three technologies, namely: Case-Based Reasoning (CBR), Genetic Algorithms (GA) and Ontologies. CBR is augmented by GA to improve case retrieval by selecting the most significant characteristics (feature weighting). Discovering or learning case characteristics weights will indicate the importance of the attributes of a case compared to the solution ones. This weighting information improves the recovery accuracy of CBR systems. In addition,

CBR is extended by ontologies to enable semantic search for solutions [9, 10]. The general domain knowledge allows the CBR system reasoning with semantic and pragmatic criteria in addition to syntactic ones and enables the user understanding the relations among data and concepts behind the stored knowledge. To validate the usefulness of our approach, we apply it to the real-world case of boiler fault diagnosis and repair and review the results produced by our approach.

The remainder of the paper is organized as follows: In Sect. 2, we present a literature review on intelligent decision support systems and some reasoning modes. Section 3 describes de case study. In Sect. 4, we present our hybrid approach to implement intelligent decision support systems. Implementation and experimentation of our system in the boiler fault diagnosis and repairing domain is presented in Sect. 5. Finally, some clues of future work are given before concluding.

2 Literature Review

The use of tools from Artificial Intelligence has strengthened DSSs by eliminating irrelevant information and by presenting facts deduced from the data and has given rise to another generation of so-called intelligent DSSs [11, 12]. IDSS is different from a simple conventional DSS; it must provide active assistance in the resolution of decision problems ideally during all the decision-making phases in the search for information, the construction and the use of models, their execution and evaluation [13].

Several works have shown the dominating role of IDSS in equipment maintenance. Guo et al. [14] develop an intelligent CBR system in injection mould design integrating domain ontology. In [15] a decision support system for industrial equipment diagnosis and repair is proposed which combined CBR approach with the knowledge engineering techniques. Zhang et al. [16] design and develop IDSS_EER (Intelligent Decision Support System of Engineering Equipment Repair) embodying both qualitative and quantitative analyses. Rodrigues et al. [17] propose a decision model built on a Simple Multi-Attribute Rating Technique using Swings (SMARTS) method to evaluate Computerized Maintenance Management Systems (CMMS). The model integrated perspective within Integrated Maintenance System (IMS). In Xu [18], a DT-driven intelligent maintenance decision-making system is developed. Ni and Jin [19] develop decision support tools for effective maintenance operations. Vafaei et al. [20] used Fuzzy inference system for composing rule set and generating maintenance scenarios. Accorsi et al. [21] propose a condition based maintenance framework. In [22] a Bayesian approach to maintenance action recommendation is proposed, which utilizes past event data to classify maintenance problems and estimate their probability. Nunez and Borsato [23] propose a semantic framework in prognostic maintenance practice. Boral et al. [24] develop a CBR system to detect and isolate equipment faults.

Different reasoning or inference technologies have been used in the development of these systems, the most widespread of which are rule-based reasoning (RBR), model-based reasoning (MBR) and case-based reasoning (CBR). The RBR systems relied on a knowledge base containing rules [25, 26]. However, systems such as diagnostic tools developed under the compiled rules paradigm tend to lack coverage, clarity, flexibility, and cost-effective extensibility. The model-based reasoning paradigm was developed to overcome these shortcomings [27–29]. But in many such systems, whether the knowledge is surface or deep, an explicit model of the domain has yet to be obtained and implemented often in the form of rules or object models. Despite their clear success in many areas, the model-based reasoning approaches are often reasoning algorithms with combinatorial complexity, leading to excessive or unacceptable computation times; the creation of the required model may include knowledge and effort that is not available, or the model may in fact be unknown (e.g. certain physiological processes). To deal with these situations, a CBR approach is then proposed as an alternative paradigm to RBR and MBR approaches. It is intended to be close to the way people solve problems and which overcomes the fragility of rule-based and model-based reasoning systems. However, CBR systems are poor at semantics understanding ability [30], and are further optimized by other intelligent methods [13, 31] such ontology, genetic algorithms, fuzzy sets, etc.

3   Case Study

One of the tasks of maintenance is to repair breakdowns. A breakdown is the unforeseen cessation of the operation of a device or the deviation of a mechanism from its normal state. It will therefore have the effect of interrupting or altering the service provided. The breakdown does not necessarily imply the complete shutdown of the installation; it can also result in a slowdown in production.

We applied the IDSS Concepts to a fault diagnosis of the boiler combustion management system. The latter is one of the most critical systems for the proper operating of an installation as it has a significant impact on the methods to apprehend the various problems in equipment maintenance. Operating personnel often find themselves faced with situations that require a rapid decision-making reaction. Therefore, it is essential to have material and human resources adapted to the required qualifications.

Different sensors are installed and connected to the boiler combustion management system to allow monitoring and continuous data collection on boilers. This data will be used by IDSS detect and identify anomalies and trends that may indicate an impending breakdown at the different stages of operating of the boiler. The Company uses this information to plan preventive maintenance interventions, reducing downtime and maintenance costs.

Usually, in a crisis situation (if there is a boiler fault), the operator strives to identify, analyze and diagnose the failure. First the failure is detected; this may occur in two ways:

- It may be automatically signaled by an alarm to the operator: the alarm flag is indicated on the panel in the control room. The operator locates and identifies the fault following a tedious search based on voluminous technical documentation.
- It may be detected directly on site by the operator. This case occurs when the sensor is defective (failure not automatically reported where the boiler works abnormally but no alarm is triggered)). The operator must explore a larger fault space through a series of tests to identify the breakdown.

At time even the technician of the manufacturer company (generally located abroad) is solicited when the problem cannot be solved locally. This situation condemns the installation to operate in degraded mode if not to shutdown pending resolution of the problem. To remedy these situations, the IDSS that we offer will support operators to solve the problems and keep persistently the installations in working order.

4   The Proposed HR-IDSS System

The structure of the HIDSS system refers to the DSS architecture proposed by G. Marakas [32]. In this architecture, emerges a technological module stemming from AI, known as a knowledge base manager system, integrating the problem model and emphasizing reasoning in decision-making.

4.1   Case Base Representation and Indexing

We consider a case consisting of two parts: a situation description that represents a "Problem" and a "Solution" which recovers this situation.

Let P be the problem space and S the solution space, there is a binary relation r on P $\times$ S (*Problems X Solutions*) meaning "*has solution*". To solve a problem p means finding a solution s belonging to *Solutions* such that: p "*has solutions*"s. Therefore, a case c of the set of cases C can be considered as a point in the problem space P and the solution space S. The case c is therefore a Cartesian product of a problem and its solution $c = (p, s)$ where:

p: represents the problem and corresponds to an element of the problem space P.

s: represents the solution and corresponds to an element of the solution space S.

Two kinds of cases are distinguished:

1) A source case: referenced as a past case, a previous case, a recorded case or a retained case. This case represents specific knowledge of previously experienced experiences stored in a case base. A source case (cs) is written as: $cs = (source, solution(source))$
2) A target case: Along with the source case, the target case is a new case or an unresolved case. A target case (cc) is written as: $cc = (target, solution(target))$.

A descriptor d is defined by a pair $d = (a,v)$ where a is an attribute and v the value associated with it relative to this case. In accordance with this vocabulary, source and target are defined as follows:

- source $= \{ds_1..ds_n\}$ where ds_i is a descriptor of the source problem.
- solution(source) $= \{Ds_1..Ds_m\}$ where Ds_i is a descriptor of the source solution.
- target $= \{dc_1..dc_n\}$ where dc_i is a descriptor of the target problem.
- solution(target) $= \{Dc_1..Dc_n\}$ where Dc_i is a descriptor of the target solution.

A case base is therefore constituted of a finite set of source cases, generally, structured in two parts each of which is described by a set of descriptors which can be simple or complex, and defined in a dedicated ontology. (Fig. 1) presents the UML class diagram modeling the case base in the diagnostic domain.

For a quick access and effective case retrieval, cases must be indexed. Case indexing means assigning indexes to cases. The choice of these clues is important to find the right case at the right time. For that, we implement the group indexing mode consisting of

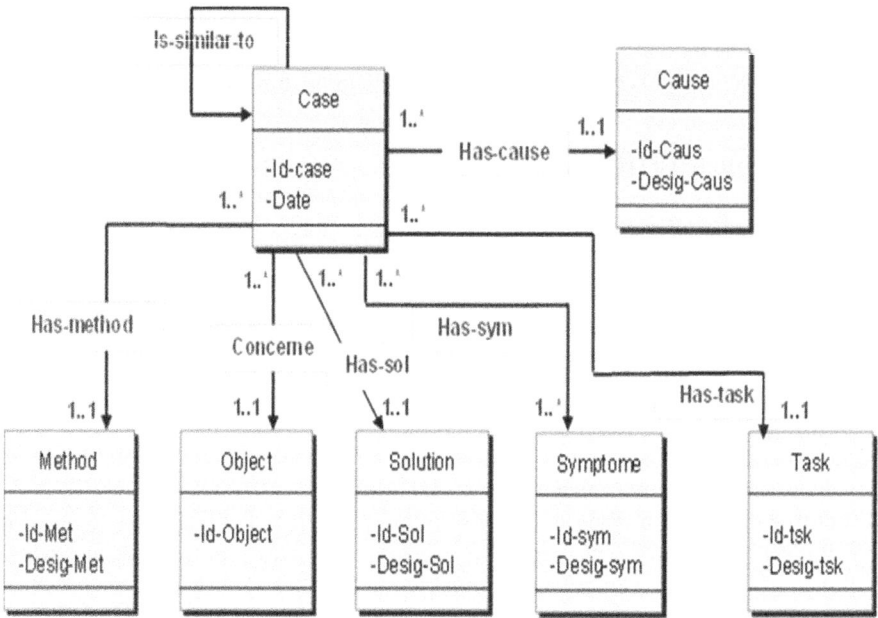

Fig. 1. Excerpt of Case Base Class Diagram.

grouping the cases by category in order to reduce the number of available cases and then calculating the measure of similarity and comparing the resemblance. This mode of indexing allows reliable and fast extraction in a subset of relevant cases. The most famous and widely applied hierarchical structures for this type of organization are index trees with their various evolutions. They allow partitioning at each node, in a recursive way, all the cases into subsets, by structuring the case base in the form of a decision tree.

4.2 Ontology Modeling

We use three ontologies to improve the case search in the case base:

1) Domain ontology concerns the vocabulary used to express components of equipment. This ontology specifies explicitly the concepts relating to the material to be maintained and the relations between concepts. The latter are mainly aggregation and composition relationships. Figure 2 represents the conceptual model of the domain ontology.

2) Task ontology describes the problems of solving breakdowns of equipment. For this, we considered the concepts task, cause, symptom, solution and object, as well as the relationships between them. Figure 3 represents the conceptual model of the task ontology.

3) Application ontology describes (Fig. 4):
   • Concepts referring to the decisions as well as the objects concerned by the decisions.

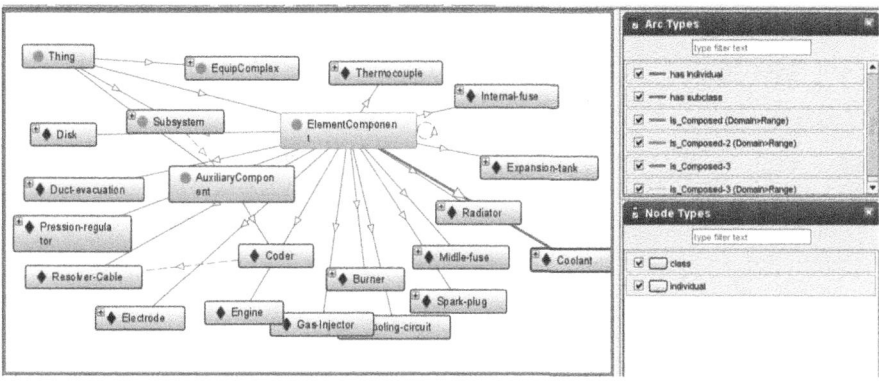

Fig. 2. Excerpt of the Domain Ontology.

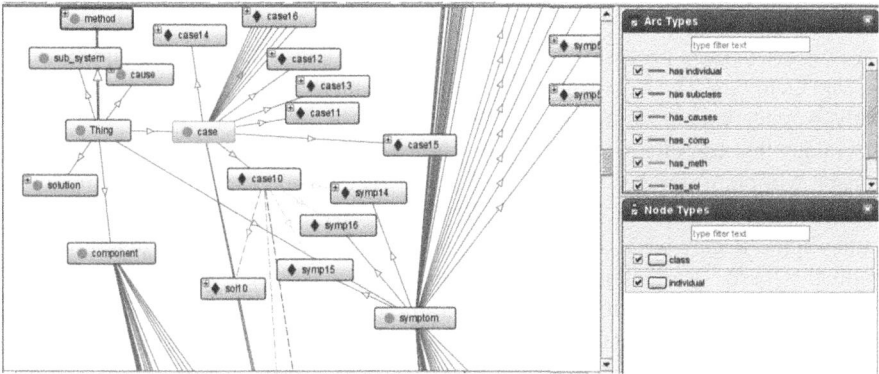

Fig. 3. Excerpt of the Task Ontology.

• Semantic relationships between the concepts related to the application.

To formalize the ontologies and represent the case base, we used the OWL language [33]. The ontologies were created using Protégé, and then generated in the OWL format. This would allow having certain compatibility in the formalization languages of the different knowledge handled by the system, as well as in its operating by tools such as Simple Protocol And RDF Query Language (SPARQL) [34].

The ontologies are further expressed in an operational language. The computer specification of the operations is then applied to the concepts in an operational language, offering reasoning mechanisms adapted to the knowledge manipulations. To operationalize the ontologies we used the NetBeans development environment [35] with the Java language [36]. We also used the Jena framework [37] which provides a programming environment for RDF [38], OWL and a query engine which allows processing SPARQL queries.

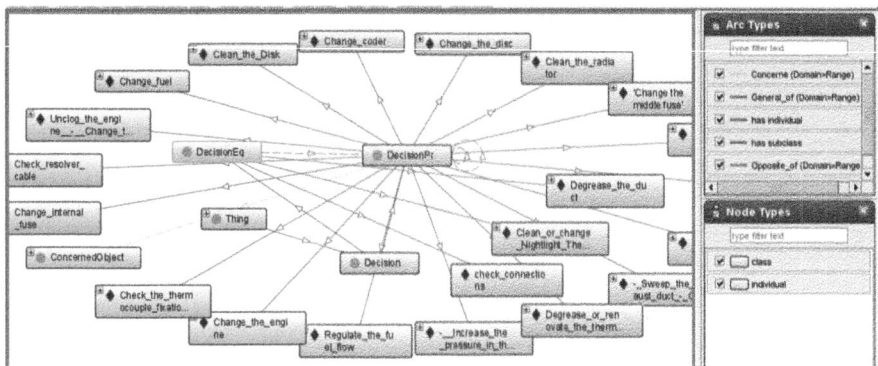

Fig. 4. Excerpt of Application Ontology.

4.3 HIDSS Process

The problem-solving process lies on the hybridization of three techniques, namely CBR, GA and Ontology. The key to CBR is to retrieve the most similar case quickly and accurately. Application of GA to CBR is performed to learn case feature weights and extract knowledge that can effectively guide the retrieval of relevant cases. Discovering these weights will point out the importance of the characteristics of a case compared to the characteristics of the solution and improve the recovery accuracy of CBR mechanisms. Furthermore, it may be necessary to infer features from a case description, involving ontologies, based on domain knowledge represented by ontologies. By explicating general domain knowledge, the system can reason and deal with a current situation in a more flexible and contextual way. Figure 5 depicts the problem-solving process of the HIDSS system is thus articulated around the following steps:

Problem Identification. In the IDSS system, each case is structured in two parts:

1) Problem section: describing the problem to be solved.
2) Solution section: describing the problem solution.

To express his request for solving a problem, the participant introduces a description of the problem by giving its characteristics. Two types of data formats are specified to describe the characteristics of a problem: nominal (textual) or numeric. As for the solution, associated with the problem, it is described by a single textual data field representing the decision taken by the user of the system.

Problem Solving. To reduce the search time and improve the case retrieval efficiency, the search space of the most similar source cases in the case base must be narrowed. To this end, this case search and retrieval operation is carried out in two reasoning cycles: search for similarities and ontological reasoning. This operation follows a preliminary step which consists of indexing the case base.

Step 1: GA-Based Attributes Selection. In order to improve the case-based reasoning process and ensure the effective case retrieval process, we propose to carry out the matching and the calculation of similarities using only the attributes deemed important.

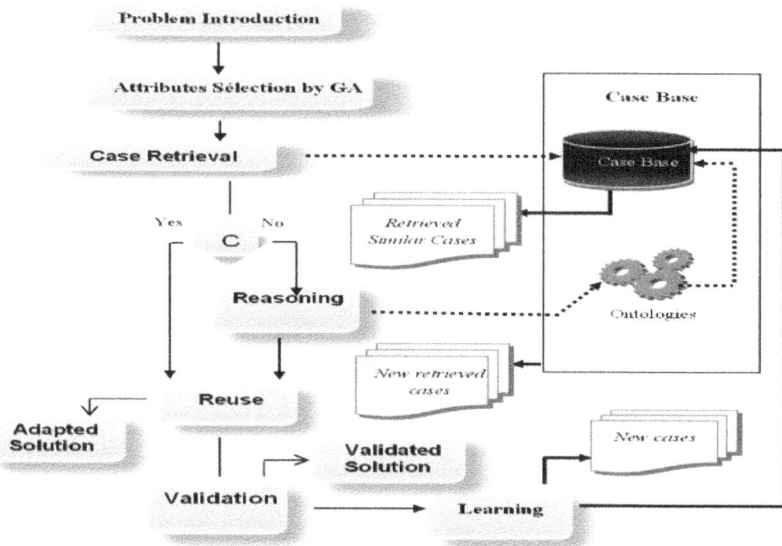

Fig. 5. Problem Solving Process.

The problem attributes are then weighted and assigned weights reflecting their respective importance. We assign importance values the optimal feature weights to the different attributes by applying genetic algorithms (Fig. 6).

Step 2: Case-Based Reasoning Cycle. Case retrieval is strongly related to the matching mechanism used to measure the similarity between cases. Indeed, the retrieval of the most relevant case and the most closely similar to the new case, among the group or subset, is based on the similarity measure. The objective is to establish similarities between the cases. More precisely, it is a question of calculating the degree of similarity between the source cases and the target case. This operation is carried out in two steps:

1) **Local Similarities Measure:** The measurement of partial similarities between the attributes of a source case and those of the target case is carried out on the basis of their relevant descriptive attributes. From these different types of data, numerical or textual, the similarity is calculated for each attribute of a case, called a partial similarity.

a) *Numerical attributes Similarity:* For numeric data, the similarity between two values is determined based on a distance between two values. The distance between two cases X and Y is given by the equation (Eq. 1):

$$d(X, Y) = \left(\sum_{i=1}^{L} w_i |x_i - y_i|^p \right)^{1/p} \tag{1}$$

In equation (Eq. 1), x_i and y_i represent, respectively, the *ith* attribute of a source case X and a target case Y where w_i is the weight associated with this attribute. When the attribute is of numerical type, the partial distance is given by the equation (Eq. 2),

Input : T : List of attributes (symptoms), **P** : List of GA parameters
Output : Ranking of individuals (weighting)
Initialisation : GenNb=50 ; PopSize=200.
Randomly generate the N individuals of the initial population P_0.
SolList, InitialPop : List of chromosomes.
S_1, S_2, S_3, S_4 : Chromosomes.
For each generation
Do

      Evaluate each individual in the population with the objective function.
      SoList=**ObjFunction**(InitialPop)
      Sort the evaluated set.
      Store the best individuals (until the population size is reached) from the sorted population in an archive.
        If one of the stopping criteria is true
          Stop the GA process
          Return individuals sorted by weighting (solution).
        Else
          For j =1 à PopSize
           / Selection: selection of parent individuals */*
           (S_1, S_2) = Selection(InitialPop)
           If "Crossover" condition is satisfied
             Carry out the "crossover" to generate Children
             (S_3, S_4) = CrossingOnePoint(S_1, S_2)
          End
          If Condition of the "Mutation" is satisfied
            Carry out the mutation to generate the Children individuals
          End
          Insert the individuals (S_1, S_2, S_3, S_4) in the new population P_{I+1}
          InitialPop = **Replace** (NewPopulation)
        End
      End
While Stopping criteria are false

Fig. 6. Genetic Algorithm

derived from the equation (Eq. 1):

$$d(x_i, y_i) = |x_i - y_i| \qquad (2)$$

However, this way of measuring can distort the results when the attributes have different sizes. So, the distance calculation must be standardized. To do this, the domain of definition has to be explicitly expressed and this expression implemented in the calculation by using a function Inti (Eq. 3), which expresses the difference between the minimum and maximum values for an attribute, i:

$$d(x_i, y_i) = \frac{|x_i - y_i|}{int_i} \qquad (3)$$

Finally, from the distance given by the equation (Eq. 4), the partial similarity on a numerical attribute expressing that two problems are closer and the most similar, is given by the equation (Eq. 4):

$$sim(x_i, y_i) = 1 - \frac{|x_i - y_i|}{int_i} \qquad (4)$$

The equation (Eq. 5) is used when the value of the ith attribute of the target problem is known in an exact way. But generally, to compare the similarity of a numerical attribute of a target case (problem) with the same numerical attribute of a source case (solution), the knowledge of an exact value is not required to decide their similarity. In the majority of cases, only a range of possible values or an approximate value is known.

b) *Nominal Attributes Similarity:* For the nominal attributes (text values), the partial similarity is classical and is determined by two values of equation (Eq. 5):

$$sim(x_i, y_i) = \begin{cases} 1 \; if \; x_i = y_i \\ 0 \; if \; x_i \neq y_i \end{cases} \qquad (5)$$

So, if two features (x_i, y_i) have the same textual value, then the similarity is equal to 1, otherwise it is equal to 0.

2) **Global Similarity Measure:** This can be calculated from all the partial, numerical or nominal similarities, calculated beforehand. This similarity makes it possible to establish, more precisely, the degree of similarity between the source problem and the target problem. The overall similarity is obtained by comparing the two cases, target (Y) and source (X); it is equal to the weighted sum (wi) of the partial similarities: Sim(xi, yi), given by the equation (Eq. 6):

$$sim(X, Y) = \frac{\sum_i^n w_i sim(x_i, y_i)}{\sum_i^n w_i} \qquad (6)$$

where wi is the weight of a feature calculated by equation (Eq. 1) above.

If a perfectly similar source case is not found, the second cycle of reasoning is triggered.

Step 3: Ontology-based Reasoning Cycle. The system aims to find cases similar to the new problem from the source cases. But, when the case retrieval process is unsuccessful or when the retrieved case is not satisfactory, then the ontological reasoning is invoked. It exploits the semantic relationships between the concepts of the ontologies to generate other solutions to the new problem.

By exploiting the application ontology, solutions that are more general or more specific to the new solution are generated. For example, we can situate the solution relative to the material by displaying the relevant component and using the domain ontology to locate the component concerned relative to other neighboring components or relative to the component that contains it. Similarly, other case attributes can be used as inputs to ontologies in order to expand the solutions or conversely restrict them. When the extension or the restriction concerns a task or an attribute, the Task ontology is used.

When it concerns a faulty component, the domain ontology is used, and if it concerns the solution of the problem, the application ontology (solutions) is used.

The usefulness of the ontological reasoning resides in bringing out from the ontologies the knowledge that exists between the different parameters of the new problem. This would allow to converge towards the semantically closest case or even, to start searching a structurally close case from the case base. Then, according to the attributes of this case, the ontologies are exploited to generate other solutions to the new problem. The decision-maker will be able to choose, among the alternative solutions, the one he deems appropriate for the new problem.

Step 4: Validation and Learning. Once a suitable solution has been generated, it is evaluated. To test a solution, some of the criteria must be set to evaluate the proposed solution. The knowledge base can then be updated considering any new information discovered while processing the new solution.

5   System Implementation

The first step in the resolution process consists of launching Genetic Algorithms (GA) to weight the problem attributes (symptoms) (Fig. 7). After generating the initial population, the GA module is executed. Then a second step of the process is initiated; it concerns the exploitation of the case base initially to find similar cases.

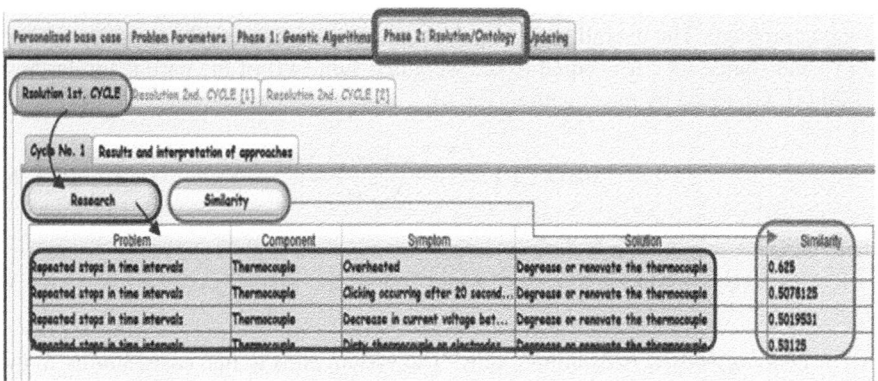

Fig. 7. 1st resolution cycle: Retrieve solutions.

The retrieved cases by GA weighting of the problem attributes and the cases found by similarity search are compared. The results are then presented to the decision maker. We calculated some measures of reconciliation between the results (Fig. 8).

This stage is completed by the launch of a second cycle in ontology resolution (Fig. 9). In this case, the most general solutions (decisions) of those found in the first cycle are thus generated.

Some solutions are then suggested to the decision-maker. The latter choose the appropriate solution. For example, the solution chosed is «Desgrease or renovate the

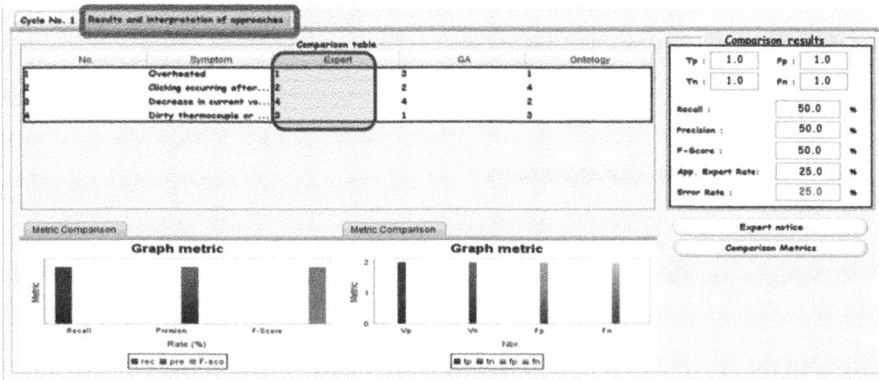

Fig. 8. Results nterpretting and comparison.

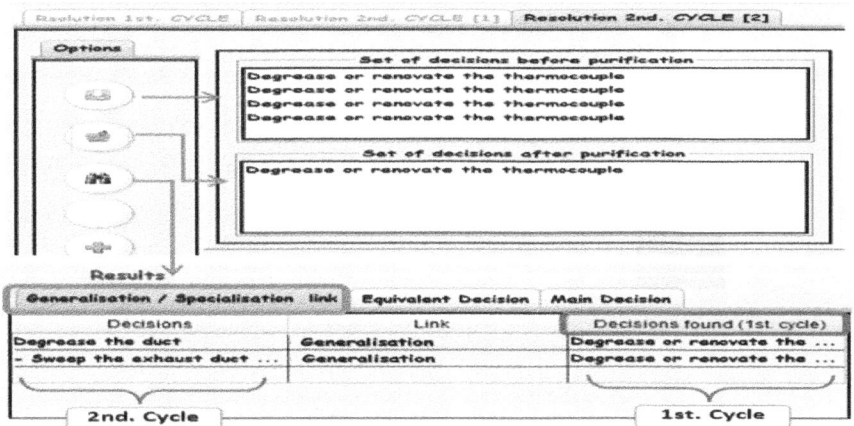

Fig. 9. 2nd resolution cycle: Retrieve most general solutions.

thermocouple". The chosen solution is therefore validated and saved in the case base (Fig. 10).

Figure 11 depicts the resolution graph showing the solved problems rate before and after IDSS operating.

6   Experimentation Results and Discussion

The performance of our system depends mainly on the case retrieval process and the relevance of the cases retrieved to provide the decision-makers with effective help and respond to their needs. Thus, to evaluate the system performance we measure the extent to which a search, carried out with respect to certain parameters, can provide us with satisfactory results by comparing it with other systems.

We collected a sample of 122 breakdowns from technical documentation on the boiler management system. Based on this data, we carried out a simulation of the boiler

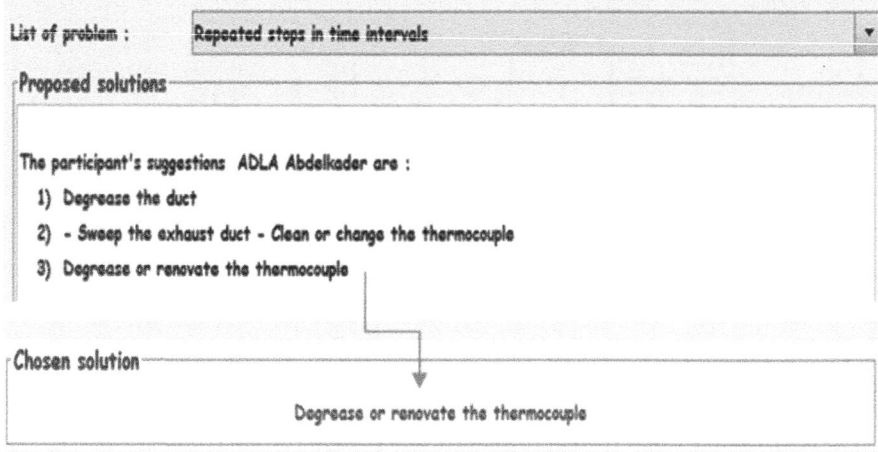

Fig. 10. Solution Choice.

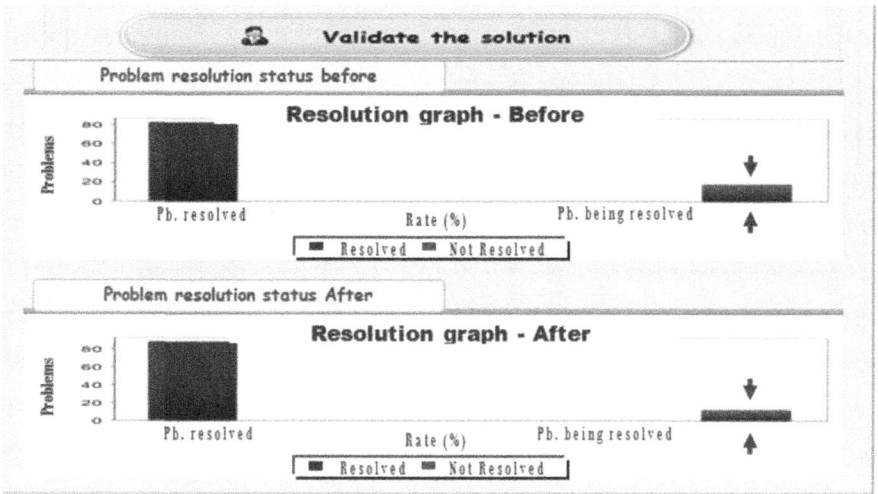

Fig. 11. Case base evolution.

maintenance actions. More precisely, we performed the fault diagnosis and repair process on two systems: our proposed HIDSS system and the FreeCBR system. FreeCBR is a conventional, open source system implemented in Java and based on case-based reasoning (CBR) [39].

For effective assessment, we executed five fault diagnosis requests for five boiler breakdowns on the two systems: HIDSS and FreeCBR. The evaluation and comparison of these two systems will be based on statistical parameters such as Precision, Recall, Accuracy and F-measure. Both Precision and Recall parameters are biased by individual perception. Recall is difficult to calculate because it is often difficult to know how many relevant cases exist in a case base. Often Recall is estimated by identifying a pool of relevant cases against which the proportion of relevant cases retrieved is determined. A pool of relevant cases (retrieved (A) + non retrieved (C)) can be created in different ways. For the purpose of our study, we manually performed a multiple scan to find all relevant cases from different searches and create a pool of relevant cases (Table 1).

Table 1. Relevant cases.

Queries	Q1	Q2	Q3	Q4	Q5
Relevant Cases	17	17	10	12	11

Table 2 summarizes the results of the test relating to the first query, comparing the cases retrieved by the two systems, HIDSS and FreeCBR.

Table 2. Cases retrieved by HIDSS for the 1st query compared to FreeCBR.

Systems	Retrieved		Non-retrieved		Total
	Relevant	Irrelevant	Relevant	Irrelevant	
HIDSS	9	19	8	86	122
FreeCBR	14	51	3	54	122

From these results, we calculated the different statistical parameters for the two systems (Table 3):

Table 3. Statistical parameters for HIDSS and FreeCBR systems.

	Precision	Recall	Accuracy	F-measure
HIDSS	0.321	0.529	0.779	0,4
FreeCBR	0.215	0.824	0.557	0.341

In order to generalize these results and better evaluate the performance of the two systems, we repeated the test with the four remaining queries. The execution results of the five queries provide the following performance thresholds for the two systems (Table 4).

From these results, a performance comparison between the two systems HIDSS and FreeCBR is carried out in relation to statistical values (Table 5).

The FreeCBR system retrieves more relevant cases (A) than HIDSS. But HIDSS recovers much fewer irrelevant cases (D) than FreeCBR. That said, HIDSS retrieves fewer relevant cases compared to FreeCBR, but eliminates many more unnecessary cases.

Table 4. Comparison of relevant cases retrieved by HIDSS and FreeCBR for all five queries.

Queries	Q1	Q2	Q3	Q4	Q5
HIDSS	9	3	3	7	5
FreeCBR	14	17	10	10	10

Table 5. Performance comparison based on statistical values.

Queries	Q.1		Q2		Q3		Q.4		Q5	
	HIDSS	FreeCBR	HIDSS	FreeCBR	HIDSS	FreeCBR	HIDSS	FreeCBR	HIDSS	FreeCBR
Accuracy	0.779	0.557	0.852	0.623	0.910	0.631	0.820	0.582	0.934	0.918
F-Measure	0.400	0.301	0.250	0.425	0.353	0.308	0.389	0.282	0.556	0.667

- F-measure

F-measure refers to the inverse relationship between precision and recall. It seeks a compromise between these two parameters. Some tasks particularly require good precision while others need good recall. Of the five tests, HIDSS has, on three occasions, a higher F-measure than that of FreeCBR, that is, HIDSS has a better trade-off between precision and recall than FreeCBR (Fig. 12).

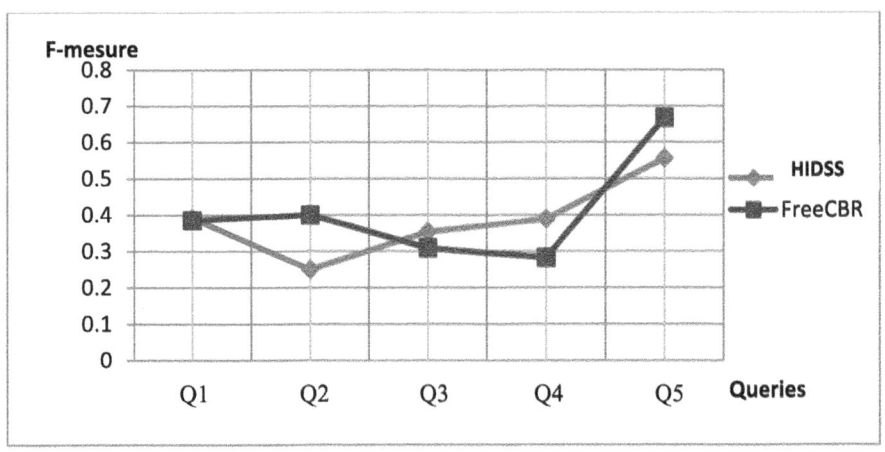

Fig. 12. F-measure.

- Accuracy

According to the figure (Fig. 13), we notice that, out of the five tests, HIDSS has a higher Accuracy than that of FreeCBR. This means that HIDSS has a higher proportion of cases correctly classified, relevant or not.

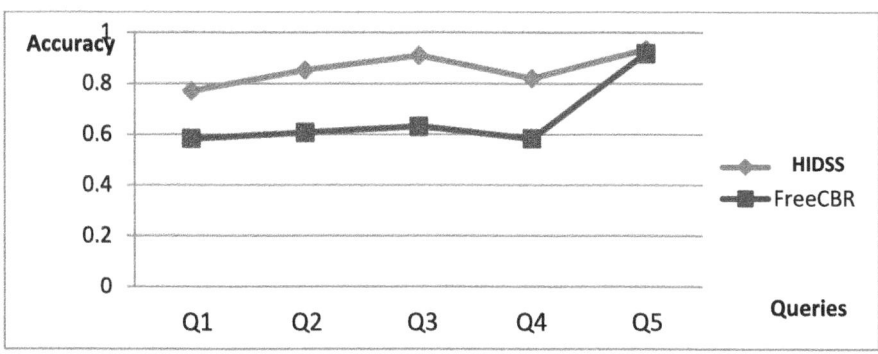

Fig. 13. Accuracy.

A further investigation of the results shows that FreeCBR with similarity function and equal weights provided an inferior result. By adopting similarity functions with weights, HIDSS has a superior result compared to FreeCBR with equal weights. Several similarity measures were used, depending on the type and value domain of the attributes. Also, attribute weighting was performed which is useful in situations where searching for cases similar to the target problem has not yielded any satisfactory results. Thus, the weighting of the attributes will be able to specify the descriptors most relevant to the problem to be solved.

In fault diagnosis activity, maintenance experts must make a decision within a short period of time. This is a critical decision task where retrieving a single relevant case is sufficient. This means that a complete recall of all relevant cases is not necessary. In fact, experts need a quick search for relevant cases to make the decision quickly; which requires better precision than higher recall, i.e. a higher F-measure.

Considering the previous results, we can conclude that the HIDSS system outperforms FreeCBR overall. However, we need more experiments and comparisons with other systems to set up in the performance of our system. Furthermore, by exploiting the boiler equipment-ontology to the automated fault diagnosis provides a uniform understanding for domain knowledge; removes the ambiguity of heterogeneous concept, ensures the substance of knowledge, expresses various fault diagnosis in different granularity, and improve the maintenance quality. For this reason, one important feature of the proposed approach is its high flexibility, expandability potential applicability to all kinds of equipment.

7 Conclusion

The development of an intelligent decision support system requires advances in emulating behaviors that humans take for granted. Building such a system endowed with human capabilities likely requires a hybrid approach that combines various AI techniques.

In this article, the HIDSS prototype is developed and experimented in industrial equipment maintenance domain which demonstrates that the three key-enabling reasoning technologies CBR, GA and Ontology are feasible and practical. The interest of our approach lies first of all in the use of GA to weight the characteristics of a problem and to select only the most relevant ones to reduce the search space and ensure the recovery of similar relevant cases. Moreover, by exploiting the semantic relations, it is possible to deduce new knowledge from those introduced. Also, by exploring the ontologies new knowledge can be inferred which will be used for a new search cycle. In addition to reason with purely syntactic criteria, the use of ontologies allows the system to use semantic and pragmatic ones,

To assess the effectiveness and performance of HIDSS as a support tool to boiler maintenance experts, we conducted a comparative study with the conventional FreeCBR system based on statistical parameters. The results obtained are generally encouraging. The result of all the tests carried out is that the FreeCBR system reports more similar cases (a high recall rate) than the HIDSS system, but the latter remains more precise and more exact in retrieving relevant cases.

We believe that the proposed approach integrates, shares, and expands maintenance knowledge, and finally achieves a flexible and cost-saving tool to determine the actions of fault diagnosis and repair for boiler equipment. The HIDSS system indeed provides substantial assistance to the various actors in the process of maintenance. However, we are fully aware of the limitations of the proposed system. It is not universal and in no way claims to be. Also, we used a sample of 122 cases. It would be interesting to see how the system performs against a larger number. Further experimentation would undoubtedly improve its effectiveness.

References

1. Gachet, A., Haettensschwiler, P.: Developing intelligent decision support system: a bipartite approach. In: Palade, V., Howlerr, R.J., Jain, L. (eds.) Knowledge-Based Intelligent Information and Engineering System. KES 2003. Lecture in Notes Computer Science, Vol 2774. Springer, Heidelberg (2003)
2. Zaraté, P.: Des Systèmes Interactifs d'Aide à la Décision Aux Systèmes Coopératifs d'Aide à la Décision: Contributions conceptuelles et fonctionnelles. HDR dissertation, Polytechnic National Institute of Toulouse (INPT) (2005)
3. Adam, F.: 20 years of decision making and decision support research published by the Journal of Decision Systems. J. Decis. Syst. **21**(2), 93–99 (2012)
4. Cobo, M.G., Martínez, M., Gutiérrez-Salcedo, M., Fujita, H., Herrera-Viedma, E.: 25 years at knowledge-based systems: a bibliometric analysis. Knowl.-Based Syst. **80**, 3–13 (2015)
5. Richter, M.R., Weber, R.O.: Case-Based Reasoning: A Textbook. Springer, Heidelberg (2013)
6. Lamontagne, L., Plaza, E.: Case-Based Reasoning Research and Development. LNCS 8765. Springer, Switzerland (2014)
7. Bumblauskas, D., Gemmill, D., Igou, A., Anzengruber, J.: Smart Maintenance Decision Support Systems (SMDSS) based on corporate big data analytics. Expert Syst. Appl. **90**, 303–317 (2017)
8. Shana, W., Dongbob, L., Gaoc, J.: Jinga, L: A knowledge based machine tool maintenance planning system using case-based reasoning techniques. Robotics Comput. Integrated Manuf. **58**, 80–96 (2019)
9. Ming, Z., Sharma, G., Allen, J.K., Mistree, F.: An ontology for representing knowledge of decision interactions in decision-based design. Comput. Ind. **114**, 103145 (2020)
10. Kendall, E.F.: Deborah L : McGuinness. Springer, Ontology Engineering (2022)

11. Phillips-Wren, G., Lakhmi, J.: Artificial Intelligence for Decision Making, Lecture Notes in Computer Science (2006)
12. Colson, E.: What AI-Driven Decision Making Looks Like. Harvard Business review, July 08 (2019)
13. Adla, A.: Aide à la facilitation pour une prise de décision collective : Proposition d' un model et d'un outil. Université Toulouse III Paul Sabatier (2010)
14. Guo, Y., Peng, Y., Hu, J.: Research on high creative application of case-based reasoning system on engineering design. Comput. Ind. **64**(1), 90–103 (2013)
15. Rasovska, I., Chebel-Morello, B., Zerhouni, N.: A mix method of knowledge capitalization in maintenance. J. Intell. Manuf. **19**(3), 347–359 (2008)
16. Zhang, M., Yu, J., Wang, M.H., Qi, C.Y.: Study on design of intelligent decision support system of engineering equipment repair. In: Applied Mechanics and Materials (Vol. 127, pp. 36–41). Trans Tech Publications, Ltd. (2011)
17. Rodrigues, R.C., Sousa, H., Gondim, I.A.: SMARTS-based decision support model for CMMS selection in integrated building maintenance management. Buildings **13**(10), 2521 (2023)
18. Xu, Q., Zhou, G, Zhang, C., et al.: Digital twin-driven intelligent maintenance decision-making system and key-enabling technologies for nuclear power equipment, Digital Twin **2,** 14 (2022)
19. Ni, J., Jin, X.: Decision support systems for effective maintenance operations. CIRP Ann. **61**(1), 411–414 (2012)
20. Vafaei, N., Ribeiro, R.A., Camarinha-Matos, L.M.: Fuzzy early warning systems for condition based maintenance. Comput. Ind. Eng. **128**, 736–746 (2019)
21. Accorsi, R., Manzini, R., Pascarella, P., Patella, M., Sassi, S.: Data mining and machine learning for condition-based maintenance. Procedia Manuf. **11**, 1153–1161 (2017)
22. Katsouros, V., Papavassiliou, V., Emmanouilidis, C.: A Bayesian approach for maintenance action recommendation. Int. J. Prognostics Health Manage. **4**(2)
23. Nuñez, D.L., Borsato, M.: An ontology-based model for prognostics and health management of machines. J. Ind. Inf. Integr. **6**, 33–46 (2017)
24. Boral, S., Chaturvedi, S.K., Naikan, V.N.: A case-based reasoning system for fault detection and isolation: a case study on complex gearboxes. J. Quality Maintenance Eng. (2019)
25. Russell, S., Norvig, P.: Artificial Intelligence: A Modern Approach, 2nd edn. Prentice Hall, Upper Saddle River, NJ (2003)
26. Shang, Y.: Expert Systems in The Electrical Engineering Handbook (2005)
27. Wolfgang E.: Introduction to Artificial Intelligence, second edition, Springer (2017)
28. Stefik, M.: Introduction to Knowledge Systems. Elsevier, Amsterdam (2014)
29. Witten, I.H., Fran, E., Hall, M.A., Pal, C.J.: Data Mining: Practical Machine Learning Tools and Techniques. Morgan Kaufmann Publishers Inc., San Francisco (2016)
30. Benmessaoud, N., Adla, A.: Intelligent semantic case based reasoning system for fault diagnosis. J. Digital Inf. Manage. (JDIM) **17**(2), 75–86 (2019)
31. Liu, S., Zaraté, P.: Knowledge based decision support systems: a survey on technologies and application domains. In: Joint International Conference between INFORMS and EWG-DSS, Lecture Notes in Business Information Processing, Toulouse (2014)
32. Marakas, G.: Decision Support Systems In the 21st Century. 2nd edn. Prentice Hall (2003)
33. OWL 2 Web Ontology Language New Features and Rationale (Second Edition) W3C Recommendation 11 December 2012. http://www.w3.org/TR/owl-new-features/
34. SPARQL Query Language for RDF W3C Recommendation 15 January 2008. http://www.w3.org/TR/rdf-sparql-query/
35. Apache NetBeans (2018). https://netbeans.apache.org/
36. Java SE (2012). http://java.sun.com
37. Apache Jena (2012). http://jena.sourceforge.net/

38. RDF (Resource Description Framework): Model and Syntax Specification W3C Proposed Recommendation, 05 January 1999. http://www.w3.org/TR/PR-rdf-syntax
39. ElKafrawy, P., Mohamed, R.A.: Comparative study of case based reasoning software. IJSRMS vol. 1, Issue 6, pp. 224–233 (2014)

The Mediating Role of Artificial Intelligence Strategy on Strategic Thinking Skills and Digital Leadership at Baghdad's General Company for Electrical and Electronic Industries

Saba Noori Alhamdany[1(✉)], Mohammed Noori Alhamdany[2], and Saad Noori Alhamdany[3]

[1] Management and Economy, University of Fallujah, Fallujah, Iraq
saba.noori@uofallujah.edu.iq
[2] Technical Institute – Anbar, Middle Technical University, Fallujah, Iraq
mohammed.noori@mtu.edu.iq
[3] Management and Economy, University of Fallujah, Fallujah, Iraq
saad.noori@uofallujah.edu.iq

Abstract. Public attention is paid to digital leadership as an up-to-date issue that is only now beginning to gain traction, however still very little is known about its nature, definition, and structures in Iraq. This paper aims to explore the impact of strategic thinking skills and its three dimensions (System thinking skills, Reframing thinking skills, and reflective thinking skills) on artificial intelligence strategy and the mediating role of these skills in digital leadership. We contacted the managers of Baghdad's General Company for Electrical and Electronic Industries to test our hypotheses utilizing the upper echelons theory. The data have been collected based on a survey questionnaire, 100 questionnaires out of 120 have been returned successfully, The analysis of our data was conducted using AMOS and SPSS software, The study used "reliability", "correlation", and "multiple regression", analysis techniques to examine the relationship between study variables. It's resulted that an artificial intelligence strategy partially mediates the association between strategic thinking skills and digital leadership.

Keywords: Artificial intelligence strategy · System thinking skills · Digital leadership · thought leader

1 Introduction

Organizations are struggling mightily to adjust to a digital world that is constantly evolving. Therefore, Scholars have suggested that digital leadership (DL) plays a crucial role in enabling organizations to navigate the challenges and leverage the opportunities presented by the digital era [11] According to [45], DL is defined "as an approach suitable for the digital age", which is similar to [9] "Doing the right things for the strategic success of digitalization for the enterprise and its business ecosystem", it's a complex construct aiming for a customer-centered, digitally enabled, leading-edge business model by (1)

A. Mirzazadeh et al. (Eds.): ODSIE 2023, CCIS 2205, pp. 137–153, 2025.
https://doi.org/10.1007/978-3-031-81458-7_9

transforming the role, skills, and style of the digital leader, (2) realizing a digital organization, including governance, vision, values, structure, culture, and decision processes following the Covid-19 pandemic, businesses have veteran a sizeable digital variation, and it has been realized that digital leadership is straight mandatory for businesses. The study of digital leadership is a subset of leadership theory based on [16], has become a catalyst for examining how the characteristics and experiences of leaders shape their perceptions, choices, and actions in ways that ultimately affect various corporate outcomes The perspective of strategic thinking skills (STS) has been gaining more attention in the strategic management field, In line with [8] research, we used the definition of STS proposed by [35], "Three processes support leaders in (a) understanding the situation through the process of reframing; (b) formulating theories of practice to guide actions through the process of reflecting; and (c) using systems thinking in a holistic manner", Some research has focused on the function of STS generation in the formation of a corporation, Thus the literature suggests there will be positive effects of STS on a firm's DL depending on the STS dimension analyzed (systems thinking, reframing, and reflection), We looked at new explanatory parameters that help us understand how a company grows DL through their STS, The AIS approach, in particular, aids in highlighting and comprehending the relationship between STS and DL, Thus [36] It has been discovered that DL is determined by its AIS, We contend that Artificial Intelligence (AI) will have a significant impact on strategy execution and decision-making in the very near future, As a result, strategy should be a component of the overall digital strategy [6]; [37], The involvement of AI strategy can help to shed light on the relationship between the three dimensions of STS and DL, resolving any remaining uncertainties. This study addresses a gap in the current literature by providing a conclusion to the debate on the varying impacts of each type of STS on DL. Our objective is to find out how the AI approach affects the interaction between STS and DL. We contend that the development of DL will only occur if STS is concentrated on creating and improving AI strategy. The main focus of this research is to determine whether the use of artificial intelligence strategies can help bridge the gap between strategic thinking skills and digital leadership in Baghdad's General Company for Electrical and Electronic Industries In the AL-Waziria region. Three significant contributions are made in this paper. First, we note that the link between STS and DL is driven by AI strategy. As a result, the AI technique enables us to comprehend and dispel any concerns we may have about this outcome. Second, in keeping with those works that need the independent examination of each component, we offer a thorough investigation of the heterogeneous effect of each STS dimension. Third, to meet the demands of earlier publications, this study links three theoretical approaches—STS, DL, and AI strategy—that have attracted increasing interest in the business administration literature over the past two decades. It does this by looking at the main theoretical and empirical antecedents of DL [23], We first, therefore, explain the theoretical basis of our work and the hypotheses. Second, we describe the methodology and the results obtained. Third, we present the discussion of these results, the main conclusions that can be drawn from them, and the wider implications that follow.

WHAT IS A STRATEGY?

In general, strategy offers solutions to problems like where to compete and how to compete to outperform rivals. A good and practical way to define strategy or competitive advantage is by creating and capturing economic value (EV), which is the difference between a customer's willingness to pay (WTP) and a supplier's opportunity cost (SOC). It is reasonable to think of competitive advantage as the straightforward price difference between a firm's goods and services and their cost because WTP and SOC are challenging to calculate. According to publicly accessible financial data, competitive advantage, or EV, is roughly the profit and is based on a company's overall sales and expenses. CEOs and other senior management personnel must reconsider their business, competition, and functional digital and business strategies to take advantage of the potential that Artificial intelligence presents, which differs depending on the organization.

2 Theoretical Framework

2.1 Strategic Thinking Skills and Digital Leadership

Strategic thinking is a skill that must be studied, developed, practised, and utilized because the majority of us are static thinkers who tend to make decisions for only a short period. To develop planning actions that will have the maximum possible beneficial influence toward a pre-defined goal, leaders must integrate elements such as analysis, exploration, understanding, and describing a complex scenario [3], According to [34], in chaotic, complicated, and unstable circumstances, leaders frequently fail. This frequently occurred as a result of managers' training in linear systematic thinking, which does not allow for optimal performance in all circumstances. Other research findings that show that linear thinking does not fit with the current stage of human history, where enormous changes are occurring, have been used to corroborate these conclusions [27], This leads to the conclusion that a new class of strategic CEOs with advanced cognitive and analytical abilities is needed in the current external environment [32], conducted a significant multi-cultural study in which they defined cognition as the process by which humans use mental processing to learn, manipulate ideas, and process knowledge and beliefs. In this study, we used the Strategic Thinking Questionnaire which evaluates three of these mental processes: systems thinking, reframing, and reflection. In this regard, the digital leader must develop a digital vision that is clearly stated and, as a result, is acceptable to staff members. Empty statements like "we have to digitalize all of our processes" are too vague and inappropriate to persuade staff to adopt digitization [18], The upper echelon theory developed by [16] is the foundation for the study of digital leadership [43], The upper echelon theory's main claim is that leaders' decisions are influenced by their experience, values, and personalities [15], which in turn affect organizational performance [16], A digital leader has both the right mindset (i.e., a digital skillset) and the necessary skills (i.e., a digital skillset) to successfully implement the vision in a company (i.e., digital implementation), They are able to articulate a meaningful strategic vision for a digital future, a leader must concurrently meet all three criteria (i.e., have a digital attitude and skill set and implement digitally), Strategic thinking is impacted by a leader's behavior and judgments when they exhibit unique qualities and performance that come from having a digital leadership style, Digital leaders may effectively manage change inside the company by fostering cooperation, establishing agreement, and using strategic thinking. Leaders may create strategies that encourage buy-in and the successful implementation of digital initiatives by taking into account the demands and viewpoints of various stakeholders. Innovation and constant learning: Strategic thinking promotes an attitude of constant innovation and learning. To spur digital innovation, digital leaders must maintain their curiosity, accept new technology, and promote experimentation. Leaders may foster a culture of learning, flexibility, and creativity within their organization by exercising strategic thinking [18, 20, 31, 44], We developed DL that offers strategic thinking abilities, A summary of the factors that determine the idea (Fig. 1).

Fig. 1. Mechanism for building digital leadership (Source: Researchers' work).

From these arguments, we propose the following hypothesis: H1: Strategic thinking skills have a positive effect on digital leadership.

2.2 Strategic Thinking Skills and Artificial Intelligence Strategy

If a business wants to use AI to gain a competitive advantage in the AI era, it must have a digital strategy that accepts and supports its AI strategy [3], It necessitates continual assessment, modification, and synchronization with changing business needs and technological advances [13, 29, 31], It is critical to create clear objectives, assess preparedness and capabilities, identify use cases, design a robust data strategy, build or buy choices, invest in AI expertise, and address ethical concerns while developing AI plans. Adopt an iterative and experimental culture, fostering cooperation and partnerships with AI professionals, academic institutions, startups, and industry peers, Use metrics and KPIs to monitor and assess performance, and plan for scalability and integration, staying up to current on breakthroughs in AI technologies and market trends [26], Organizations may maximize their AI implementation and achieve their strategic goals by concentrating on five critical issues. Managers will need to connect the AI strategy with the competitive strategy to do this, A differentiation competitive strategy and a cost leadership competitive strategy will have various effects on the AI strategy, To focus on uniqueness, it may be necessary to prioritize revenue growth in AI strategy and invest in technologies for customer relationship management and personalization. Instead of focusing on cost leadership, sophisticated AI solutions that give complete cost awareness and help effective business process execution may be necessary [6]; ztürk, 2022; [28], As a result, we propose the following hypothesis: H2: Strategic thinking skills has a positive effect on Artificial intelligence strategy.

2.3 Artificial Intelligence Strategy and Digital Leadership

In the digital age, effective leadership requires assisting subordinates in maximizing the use of the organization's online resources for the good of all [41], Because of the recent rapid development of digital technology, many firms are undergoing significant changes to their organizational structures and the responsibilities that employees play, To adapt to the new situation, the corporation will need to adjust a lot of things, These include the available job categories, the workplace culture, and the technology, In addition to setting the basis for an uncertain future, transformational efforts promote shifts to better fulfil immediate requirements [40], Digital leaders require a special set of talents in order to effectively address these issues and support the transformation [10], Leaders wield significant power because they design their organizations to face an increasingly uncertain and volatile future, For example, because of the inherent uncertainty of the future of digital technology, it is challenging for digital leaders to encourage their teams to work with a new set of technologies that may or may not be embraced in the future [39], This is a typical problem for digital leaders, which is aggravated by the fact that many leaders lack the skills necessary to be effective digital leaders. According to [23], Artificial intelligence (AI) strategy and digital leadership go hand in hand, as organizations leverage AI technologies to drive digital transformation and achieve strategic objectives In essence, a "digital vision" involves having a clear and meaningful strategy for the digital future of a company, along with the right mindset "a digital mindset" and skills a "digital skillset" to bring this vision to life through implementation known as "digital implementation", To be considered a digital leader, an individual must possess all three of these characteristics (i.e. "a digital attitude and skillset, as well as the ability to undertake digital implementation"), It's important for digital leaders to understand the potential of AI and how it can improve corporate processes, enhance customer experiences, and drive innovation, These leaders provide strategic direction and ensure that AI initiatives are aligned with the organization's wider digital transformation objectives, As a result, traditional leadership paradigms such as transformational or entrepreneurial leadership are not outdated; instead, they are becoming increasingly relevant [36]; [22], We suggest the following hypothesis based on these arguments: H3: Artificial intelligence strategy has a positive effect on digital leadership.

2.4 Artificial Intelligence Strategy as a Mediate

When implementing AI, organizations may encounter several risks and challenges that need to be addressed. Here are some potential risks and challenges associated with AI implementation: 1- Ethical Concerns: AI systems can raise ethical concerns related to privacy, bias, fairness, and transparency. Organizations need to ensure that AI algorithms and models are developed and deployed ethically, taking measures to address biases, protect user privacy, and provide transparency in decision-making processes. 2-Data Quality and Availability: AI relies heavily on data, and organizations may face challenges related to data quality, availability, and relevance, Insufficient or poor-quality data can lead to inaccurate or biased AI outcomes, Organizations must invest in data management practices, data cleansing, and data governance to ensure the availability of high-quality data for AI applications. 3- Lack of Skilled Talent: AI implementation requires expertise in areas such as data science, machine learning, and AI engineering, The shortage of skilled professionals in these domains can pose a challenge for organizations, It is essential to invest in talent acquisition, training, and upskilling programs to build a capable AI workforce [17], To do that AI strategy can operate as a bridge builder by assisting leaders and stakeholders with decision-making, Organizations can obtain important insights from massive amounts of data by employing AI technologies such as machine learning, natural language processing, and predictive analytics. These insights can help leaders make educated decisions and navigate complicated company challenges by informing strategic decision-making, It helps organizations leverage the power of AI to achieve their objectives and drive meaningful transformation [19]; [25], We suggest the following hypothesis based on these arguments: H4: Artificial intelligence strategy mediate the relationships between strategic thinking skills (System thinking Reframing thinking Reflecting thinking) and digital leadership.

3 Methodology

3.1 Research Model

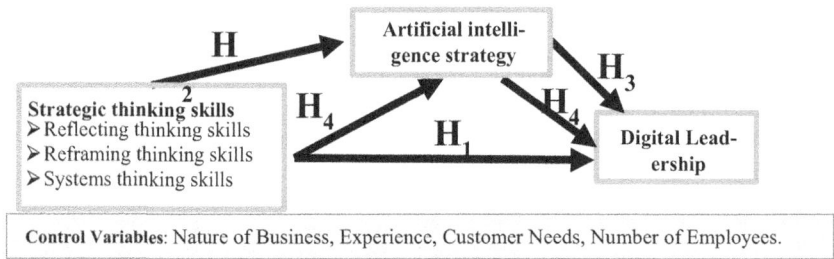

Fig. 2. Research Model.

H1: Dimensions of strategic thinking skills have a positive effect on digital leadership (Fig. 2).
H2: Dimensions of strategic thinking skills have a positive effect on artificial intelligence strategy.
H3: artificial intelligence strategy has a positive effect on digital leadership.
H4: Dimensions of strategic thinking skills have a positive effect on digital leadership through artificial intelligence strategy.

3.2 Sample and Data Collection

This research aims to analyze how artificial intelligence strategy influences the relationship between strategic thinking skills and digital leadership in Baghdad's General Company for Electrical and Electronic Industries, There are numerous justifications for selecting this particular company as the research setting, Primarily, The company likely has extensive knowledge and experience in the electrical and electronic industries, Over time, they have likely gained specialized expertise and skills in the manufacturing of electrical and electronic products, This expertise enables them to deliver high-quality products and services, Additionally, the company may offer a wide range of products and services in these sectors, allowing customers to find multiple solutions in one place, This comprehensive product range simplifies procurement and improves operational efficiency. Being closely connected to the market, the company is well-suited for developing an artificial intelligence strategy. This flexibility is particularly valuable for businesses with specific needs or those looking for customized solutions. Additionally, the company can tailor products and solutions to meet individual customer needs. This adaptability is advantageous and can result in improved results. As a mediator between these two aspects, they play a vital role in steering the economy. Our study, conducted from July 20 to September 16, 2023, used a quantitative approach. The questionnaire was initially developed in Arabic and then translated into English using the back-to-back method [7], After eliminating questionnaires with missing values in more than 6% of the variables, only 100 out of the 120 managerial employees (i.e., "executives, section managers, section heads, accountants, supervisors, departmental managers, general manager, director, chief executive officer…etc.") who worked in manufacturing enterprises of various sizes (e.g., Air conditioner factory, High and Low Voltage Motors Factory, Transformer and generator factory, Feeding Industries Factory, Engines and Home Appliances Factory, Safety, fire and health protection equipment factory) as its target population are shown in Table 1, The questionnaire was created based on the researchers' prior experience working for companies and their own judgment, The researchers purposefully chose a sample and asked them to respond, The AL-Waziria area was the subject of the study in particular since it is thought to be Baghdad's most industrialized district, We used a Likert-type scale with five points (ranging from 1 = Almost never to 5 = Frequently) to assess all the items, except for Control Variables, "Nature of Business", "Experience", "Customer Needs Number of Employees", Face-to-face interviews and self-administered questionnaires were used to collect data, The questionnaire consists of 18 items in three sets of questions related to strategic thinking skills (reflecting thinking skills, reframing thinking skills, and systems thinking skills)developed by [33], A scale consisting of seven items, created by [42], has been utilized to evaluate the strategy of artificial intelligence, whereas [44], created a scale consisting of ten items to assess digital leadership performance, "Statistical Package for Social Sciences" (SPSS) Version 22.0" was used to analyze the data collected from "validity of the constructs" [24].

Table 1. Findings from the distribution of the survey questionnaire.

Number of questionnaires distributed	Number of questionnaires retrieved	Number of damaged surveys	Number of questionnaires valid for analysis	recovery ratio	Valid response rate
120	108	8	100	90%	83%

4 Analysis

4.1 Factor Analysis

The scales underwent an "exploratory factor analysis", and the most accurate results were achieved through a principal component analysis using "a Promax rotation" [14]. There are six items that pertain to Reflecting thinking skills, another six items that pertain to Reframing thinking skills, a set of System thinking skills, seven items for Artificial intelligence strategy, and a total of ten items for digital leadership, After the analysis was conducted, one paragraph from each dimension was removed. Table 2 shows the factor loadings for Reflecting thinking skills, Reframing thinking skills, System thinking skills, artificial intelligence strategy, and digital leadership.

The information presented in the table shows indicators of the research model for each dimension, as well as for the research variables as a whole, that the KMO = 93%, GFI = 93%, TLI = 94%, CFI = 95%, RMSEA = .058, CR = .891, This means that the factors obtained from the factorial analysis are more dependable. Furthermore, the Bartlett = .000.

4.2 Correlations Statistics

All variables are positively correlated with one another [38]; [12] as indicated in Table 3.

4.3 Hypothesis Testing

Multiple regression analysis of Strategic thinking skills on (Digital Leadership and Artificial Intelligence Strategy) (Table 4).

By examining the table, it is evident that hypothesis H1 is supported. This is evidenced by the $R2$ value of .652 and the F value of 64.01, which is statistically significant at the .05 level, Therefore, hypothesis H1: can be accepted, as it states Strategic thinking skills has a positive effect on digital leadership", Given $R2 = .586$ and $F = 47.18$, with a statistically significant value of 0.05, hypothesis H2: can be accepted, which states: "Strategic thinking skills has a positive effect on Artificial intelligence strategy", It is evident from the table that both $R2$ (.727) and F (119.71) values are statistically significant ($p < 0.05$), Therefore hypothesis H3″ can be accepted, which states: "Artificial intelligence strategy has a positive effect on digital leadership", Hence, AI opens avenues for innovation and can provide a competitive edge to organizations, By leveraging AI technologies, leaders can explore new business models, develop innovative products

Table 2. The results of the factorial analysis of the research dimensions

ITEMS	EFA (loading >.05)					SFA			quality indicators
	1	2	3	4	5	SD.	t	P	
Ask why questions to develop an understanding of problems.	.721					.557			
Try to apply your experience and knowledge to any problem.	.977					.719	12.691	.000	GFI=.911 TLI=.921
Try to find a common goal that will allow two parties within your organization who are competing or in conflict to succeed.	.433					Deleted			CFI=.936 IFI=.964 RMSEA=.066 KMO=.678
Seek coaching from colleagues or professionals when thinking about past decisions that you have made.	.655					.821	17.481	.000	
Seek to frame problems from different perspectives.	.585					.741	16.221	.000	
Discover how you could have handled a situation better when thinking about a past decision you have made.	.511					.653	16.712	.000	
Try to create and evaluate a larger number of possible solutions and perceptions when the problem is more complex.		.713				.771			
Engage in discussions with those who hold a different worldview and different beliefs.		.812				.708	16.251	.000	
Try first to examine the problem at its face value and create plans to solve it before seeking other people's opinions.		.636				.819	15.746	.000	GFI=.914 TLI=.915 CFI=.924 IFI=.929 RMSEA=.053 KMO=.726 CR=.879
Examine a problem by using one viewpoint.		.418				Deleted			
Track trends by asking everyone around you what is changing or what is new.		.711				.732	17.574	.000	
Try first to examine the problem at its face value and create plans to solve it before seeking other people's opinions.		.546				.844	17.451	.000	
Find that in most cases external environmental changes require changes internally.			.612			.692			
Try to look for changes in the company's structure that lead to significant enduring improvements.			.681			.742	16.443	.000	

(continued)

Table 2. (*continued*)

Focus on searching for the cause before taking any action.	.581		.741	17.748	.000
Search for specific feedback on your company's performance.	.564		.672	15.448	.000
Search for specific feedback on your company's performance.	.413		Deleted		
Concentrate on developing the capabilities of company employees to solve the problem when they are faced with a problem needing resolution.	.628		.631	17.593	.000
Artificial Intelligence Strategy					
AI strategy failures can lead to significant financial losses for companies.	.617		.767		
Companies invest resources in implementing and maintaining AI solutions with the expectation of improved performance.	.647		.787	17.584	.000
Integrating AI strategy into a company's overall strategy involves careful planning and consideration of various factors.	.471		Deleted		
AI strategy can provide value, such as improving operational efficiency, enhancing customer experience, or enabling data-driven decision-making.	.694		.633	16.228	.000
Companies can use AI strategies to streamline operations, increase efficiency, and reduce costs.	.673		.776	16.794	.000
Companies invest in research and development to develop new AI strategy technologies or improve existing ones.	.671		.610	15.748	.000
AI strategy technologies can automate repetitive and mundane tasks, allowing employees to focus on more complex and creative work.	.701		.575	18.117	.000
Digital Leadership					
The top leader of our company is a creative business leader with thought ability.		.637	.721		

GFI=.912
TLI=.913
CFI=.919
IFI=.931
RMSEA=.63
KMO=.782
CR=.915

GFI=.934
TLI=.927
CFI=.918
IFI=.952
RMSEA=.054
KMO=.884
CR=.894

(*continued*)

Table 2. (*continued*)

The top leader of our company has a creative and innovative mindset.	.732	.744	17.591	.000	GFI=.936 TLI=.951 CFI=.973 IFI=.916 RMSEA=.67 KMO=.892 CR=.868
The top leader of our company could formulate the idea of the future into the reality of business.	.747	.553	16.681	.000	
The top leader of our company always reflects.	.754	.508	16.338	.000	
The top leader of our	.567	.691	18.511	.000	
company continues to explore problems at work.					
The top leader of our company keeps his/her thirst for knowledge to learn and adapt to change.		.577	.716	16.527	.000
The top leader of our company has the learning capability.		.471	Deleted		
The top leader of our company masters the trend of scientific and technological development.		.667	.806	17.831	.000
The top leader of our company has a global vision and vision.		734	.647	16.412	.000
The top leader of our company actively builds strong domestic and global networks.		.581	.779	17.468	.000
Explained variance ratio%	77.46	GFI	.937		
KMO	.937	TLI	.947		
Bartlett	.000	CFI	.951		
CR	.891	RMSEA	.058		

and services, and gain a market advantage, AI-driven innovation can help organizations differentiate themselves [6] adapt to changing market dynamics, and stay ahead of competitors.

4.4 Mediation Test

To investigate further the mediator effect [4] of artificial intelligence strategy on the relationship between Reflecting thinking, Reframing thinking, System thinking and digital leadership, According to the results artificial intelligence strategy is a partial mediator variable in this relationship (Table 5).

We see that Reframing thinking skills and digital leadership have a remarkable association that was demonstrated by using Artificial Intelligence Strategy as an intermediary;

Table 3. Correlations, Chronba, AVE, Standard Error, VIF, t, and the mean

		Chronba	AVE	Mean	Std.Est	Toler-ance	VIF	t	Sig.	1	2	3	4	5
1	Reflecting thinking skills	.845	.67	3.11	.65	.315	3.159	18.63	.00	1				
2	Reframing thinking skills	.781	.67	3.74	.71	.311	3.111	15.71	.00	.631**	1			
3	Systems thinking skills	.954	.71	3.84	.69	.258	3.841	17.34	.00	.572**	.712**	1		
4	Artificial intelligence strategy	.924	.68	4.6	.72	.259	4.113	18.64	.00	.576**	.681**	.597**	1	
5	Digital Leadership	.847	.69	3.27	.68	.251	4.131	16.57	.00	.671**	.621**	.591**	.671**	1

$**p<0,01.$

Table 4. The outcome of testing the three hypotheses.

	Dependent Variable								
Variables	**Digital Leadership**			**Artificial Intelligence Strategy**			**Digital Leadership**		
	B	t	Sig	B	t	Sig	B	t	Sig
Nature of Business	29.31	6.29	.00	.15	4.19	.00	.21	7.21	.00
Experience	.41	7.97	.00	.13	2.11	.20	.46	6.97	.01
Customer Needs	79.81	2.01	.17	-.08	-.62	.35	.19	3.01	.00
Number of Employees	67.88	7.24	.01	.16	4.13	.00	/	/	/
Reflecting thinking skills	.07	2.03	.00	.17	3.29	.00	/	/	/
Reframing thinking skills	.16	3.24	.00	.21	3.99	.00	/	/	/
Systems thinking skills	.15	3.27	.00	.34	4.87	.00	/	/	/
Artificial Intelligence Strategy	/	/	/	/	/	/	.48	12.29	.00
r		.667			.612			.699	
R^2		.652			.586			.727	
R^2_{adj}		.633			.574			.683	
F		64.01			47.18			119.71	
Sig.		.00			.00			.00	

as a result, there is complete mediation between these two concepts via Artificial Intelligence Strategy, The results suggest that there was only a limited amount of mediation; even though systems thinking skills had less of a direct impact on digital leadership, they still had a sizable influence when digital leadership was included as a mediator in the model. Therefore, H4 is partially supported.

5   Discussion

The major findings highlight the importance of enabling a change in organizational culture through supportive activities, addressing appropriate leadership capabilities, in terms of, knowledge, vision and communication, coherence and support, responsibility and sustainability. This research explores the importance of digital leadership and identifies key capabilities to facilitate and sustain the transition to digital transformation, The data from this study indicate that STS and DL have a positive relationships, thus strategic thinking skills can enhance a leader's ability in Baghdad's General Company for

Table 5. Mediation assessment.

DV	Path	IV	Mediation	Direct	Indirect
Artificial intelligence strategy	←	Reflecting thinking skills	.117**	/	.169
Digital Leadership	←	Artificial intelligence strategy	.470**	/	.101
Digital Leadership	←	Reflecting thinking skills	.161**	.152	/
Digital Leadership	←	Reframing thinking skills	.351**	.379	/
Artificial intelligence strategy	←	Reframing thinking skills	.262**	/	.131
Digital Leadership	←	Systems thinking skills	−.112**	−.119	/
Artificial intelligence strategy	←	Systems thinking skills	−.761**	/	−.109

Note. p** < .01, (CMIN/df = 4.56, p < .01, CFI = .934, GFI = .94, NFI = .937, TLI = .931, and RMSEA = .06)

Electrical and Electronic Industry to navigate the rapidly evolving digital landscape and leverage technology for organizational success, a deeper understanding is required in decision concerns to AI's development, regulation, uses and operations. The results align with past research indicating notable connections within this area of study [18]; [44], The findings showed that STS and AIS have a positive relationship, keep pace with AI technological innovations, and ensure the safe and secure use of AI strategy. According to a recent study by Montasari (2023) the organization's AI strategy with its overall strategic goals is important to ensure that AI strategy initiatives make a meaningful contribution to the organization's success. Such initiatives could involve improving operational efficiency, enhancing the customer experience, enabling data-driven decision-making, or promoting innovation. Additionally, this research plays a significant part in digital leadership to stay updated with the latest advancements, trends, and best practices in AI to help them acquire more knowledge of strategy AI technologies, algorithms, and methodologies. By staying well-informed, the leaders of the company can make better decisions about AI adoption, and implementation, This aligns with [36] findings that empowering digital leaders with knowledge, insights, and tools effectively leads organizations in AI strategy by supporting strategic decision-making, driving innovation, addressing challenges, and promoting responsible and ethical AI adoption, By utilizing the research, leaders can stay at the forefront of AI advancements and effectively navigate the dynamic digital landscape, certainly, all aspects of strategic thinking abilities are discovered to benefit artificial intelligence strategy. This research boosts the notion that businesses in Iraq can develop a strategy based on strategic thinking abilities rather than a profit-oriented perspective to contribute to their long-term survival since it focuses on the declared or unstated needs and wishes of the customers. One significant finding from our research is that artificial intelligence strategy plays a partial intermediary role in strategic thinking skills and digital leadership, outcomes and mechanisms of digital leadership can be explored at the leader innovation behaviour and organizational agility.

These findings state that the Company for Electrical and Electronic Industries in Baghdad can use application AI technologies to enhance operational efficiency within the company [23], So, AI-powered automation can streamline production processes, reduce errors, and optimize resource utilization. In addition, the AI algorithms can also be utilized for predictive maintenance, enabling proactive identification and prevention of equipment failures. Implementing AI solutions in areas such as supply chain management, inventory optimization, quality control cost reduction, productivity improvement, and faster time-to-market.

6   Limitations and Future Direction

This study has several limitations. The main limitation is the technological complexity involved in digital leadership. It requires a deep understanding of complex technologies and their implications, and implementing digital initiatives often involves significant changes in processes, workflows, and organizational culture. Data privacy and security are also a concern in digital leadership, as it involves leveraging data and technology. Additionally, the analysis is limited to the AL-Waziria area in Baghdad. This area covers most of the General Company for Electrical and Electronic Industries in the city. It would be beneficial for future research to include an analysis of the AL-Taji location as well. As previously stated, the variables we utilized in our analysis are not commonly utilized simultaneously in existing literature, as a result, our objective is to fill this void as extensively as possible and believe that this analysis can be applied to other companies operating in various regions around the globe. Additionally, this study offers convincing evidence for future research to build upon, More in-depth understanding can be gained regarding the abilities involved in strategic thinking, as well as strategies related to artificial intelligence and the parameters of digital leadership, To enhance future studies, it would also be beneficial to consider additional strategic thinking skills, Digital leaders need to inspire innovation, drive digital transformation, and navigate technological disruption. Parameters include fostering a culture of experimentation, fostering digital fluency across the organization, taking strategic risks, and creating agile structures.

References

1. Abbu, H.M., Gudergan, P., Hoeborn, G., Kwiatkowski, A.: Measuring the human dimensions of digital leadership for successful digital transformation. Res. Technol. Manage. **65**(3), (2022). https://doi.org/10.1080/08956308.2022.2048588
2. AlAjmi, M.K.: The impact of digital leadership on teachers' technology integration during the COVID-19 pandemic in Kuwait. Int. J. Educ. Res. **112**, 101928 (2022). https://doi.org/10.1016/j.ijer.2022.101928
3. Alireza, G., Ali, A., Mostafa, H.-K.: Efficient multi-objective meta-heuristic algorithms for energy-aware non-permutation flow-shop scheduling problem. Expert Syst. Appl. **213**(Part b) (2023)
4. Alireza, G., Ali, A., Mirjalili, S.: A robust possibilistic programming framework for designing an organ transplant supply chain under uncertainty. Ann. Oper. Res. **328**, 493–530 (2023)

5. Avolio, B.J.: Full leadership development: Building the vital forces in organizations, Thousand Oaks, CA7 Sage (1999)
6. Baron, R.M., Kenny, D.A.: The moderator-mediator variable distinction in social psychological research: conceptual, strategic, and statistical consideration. J. Pers. Soc. Psychol. **51**, 1173–1182 (1986)
7. Benitez, J., Arenas, A., Castillo, A., Esteves, J.: Impact of digital leadership capability on innovation performance: the role of platform digitization capability. Inf. Manage. **59**(2) (2022)
8. Borges, A.F., Laurindo, F.J., Spínola, M.M., Gonçalves, R.F., Mattos, C.A.: The strategic use of artificial intelligence in the digital era: Systematic literature review and future research directions. Int. J. Inf. Manage. **57**, 102225 (2021)
9. Brislin, R.W.: Translation and content analysis of oral and written material. In: Berry (1980)
10. Casey, A.J., Goldman, E.F.: Enhancing the ability to think strategically: a learning model. Manage. Learn. **41**(2), 167–185 (2010). https://doi.org/10.1177/1350507609355497
11. ElAkid, I.: The challenges of digital leadership—a critical analysis in times of disruptive changes. In: Digital Management in Covid-19 Pandemic and Post-Pandemic Times: Proceedings of the International Scientific-Practical Conference (ISPC 2021), pp. 117–129 (2023)
12. ElSawy, O.A., Kræmmergaard, P., Amsinck, H., Lerbeck Vinther, A.: How LEGO built the foundations and enterprise capabilities for digital leadership. MIS Q. Executive **15**, 141–166 (2016)
13. Erhan, T., Uzunbacak, H.H., Aydin, E.: From conventional to digital leadership: exploring digitalization of leadership and innovative work behaviour. Manag. Res. Rev. **45**(11), 1524–1543 (2022). https://doi.org/10.1108/MRR-05-2021-0338
14. Fornell, C., Larcker, D.F.: Evaluating structural equation models with unobservable variables and measurement error. J. Mark. Res. **18**(1), 39–50 (1981)
15. Haefner, N., Wincent, J., Parida, V., Gassmann, O.: Artificial intelligence and innovation management: a review, framework, and research agenda☆. Technol. Forecast. Soc. Chang. **162**, 120392 (2020)
16. Hair J.F., Black W.C., Babin, B.J., Aderson R., E.: Multivariate Data Analysis, Seventh, Prentice Hall, Englewood Cliffs (2010)
17. Hambrick, D.C.: Upper echelons theory: an update. Acad. Manag. Rev. **32**, 334–343 (2007). https://doi.org/10.5465/AMR.2007.24345254
18. Hambrick, D.C., Mason, P.A.: Upper echelons: the organization as a reflection of its top managers. Acad. Manag. Rev. **9**, 193–206 (1984). https://doi.org/10.5465/AMR.1984.4277628
19. Han, T.A., Pereira, L.M., Lenaerts, T., Santos, F.C.: Mediating artificial intelligence developments through negative and positive incentives. PLoS ONE **16**(1), e0244592 (2021). https://doi.org/10.1371/journal.pone.0244592
20. Hensellek, S.: Digital leadership: a framework for successful leadership in the digital age. J. Media Manage. Entrepreneurship **2**(1), 55–69 (2020). https://doi.org/10.4018/JMME.2020010104
21. John, A.M.: Artificial Intelligence in Information Services: Revolution or Survival? In: Proceedings of the IATUL Conferences (1991). https://docs.lib.purdue.edu/iatul/1991/papers/8
22. Kieser, H.: The influence of digital leadership, innovation and organisational learning on the digital maturity of an organisation. University of Pretoria, Pretoria (2017)
23. Klus, M.F., Müller, J.: The digital leader: what one needs to master today's organisational challenges. J. Bus. Econ. **91**(8), 1189–1223 (2021)
24. Kollmann, T.: Digital Leadership: Grundlagen der Unternehmensführung in der Digitalen Wirtschaft, 2nd edn. Wiesbaden: Springer Gabler (2022c)

25. Kollmann, T., Kollmann, K., Kollmann, N.: Artificial leadership: digital transformation as a leadershiptask between the chief digital officer and artificial intelligence. Int. J. Bus. Sci. Appl. Manage. **18**(1), 76–95 (2023)
26. Kumar, P., Kumar, A., Palvia, S., Verma, S.: Online business education research: systematic analysis and a conceptual model. Int. J. Manage. Educ. **17**(1), 26–35 (2019). https://doi.org/10.1016/j.ijme.2018.11.002
27. Li, X., Lin, B.: The development and design of artificial intelligence in cultural and creative products. Math. Problems Eng. **2021**, Article ID 9942277, 10 (2021). .1109/TEM.2020.2996175
28. Mithas, S., Murugesan, S., Seetharaman, P.: What is Your Artificial Intelligence Strategy? In: IT Professional, vol. 22, no. 2, pp. 4–9, 1 (2020). https://doi.org/10.1109/MITP.2019.2957620
29. Moldoveanu, M.: Thinking strategically about thinking strategically: the computational structure and dynamics of managerial problem selection and formulation. Strateg. Manag. J. **30**(7), 737–763 (2009). https://doi.org/10.1002/smj.757
30. Montasari, R.: National artificial intelligence strategies: a comparison of the UK, EU and US approaches with those adopted by state adversaries. In: Countering Cyberterrorism. Advances in Information Security, vol. 101. Springer, Cham (2023). https://doi.org/10.1007/978-3-031-21920-7_7
31. Öztürk, H.: Artificial intelligence strategy in archives. Current Perspectives Soc. Sci. **26**(1), 54–61 (2022)
32. Pepe, G., Pavone, P.: Conceptual basis for the definition of digital leadership. In: Hundal, S., Kostyuk, A., Govorun, D. (eds.) Corporate Governance: A Search for Emerging Trends in the Pandemic Times, pp. 48–50 (2021)
33. Petrella, S., Miller, C., Cooper, B.: Russia's artificial intelligence strategy: the role of state-owned firms. Orbis **65**(1), 75–100 (2020). https://doi.org/10.1016/j.orbis.2020.11.004
34. Pisapia, J., Morris, J., Cavanaugh, G., Ellington, L.: The strategic thinking questionnaire: Validation and confirmation of constructs. In: The 31st SMS Annual International Conference, Miami, Florida November 6–9, 2011 (2011)
35. Pisapia, J., Pang, N.S.K., Hee, T.H., Lin, Y., Morris, J.D.: A comparison of the use of strategic thinking skills of aspiring school leaders in Hong Kong, Malaysia, Shanghai, and the United States: an exploratory study. Int. J. Educ. Stud. **2**(2), 48–58 (2009)
36. Pisapia, J.R.: Mastering Change in a Globalizing World: New Directions in Leadership (Education Policy Studies Series, No. 61), The Hong Kong Institute of Educational Research, Hong Kong (2006)
37. Pisapia, J., Reyes-Guerra, D., Coukos-Semmel, E.: Developing a strategic mindset: Constructing the measures. Leadership Review. Kravis Leadership Institute, Leadership Review, 1 (5), pp. 41–68. cited in Scopus; questionnaire response as an element of a previously tested method." Journal of Applied Psychology. 59(3): 297–301 (2005)
38. Quaquebeke, N.V., Gerpott, F.H.: The Now, New, and Next of Digital Leadership: How Artificial Intelligence (AI) Will Take Over and Change Leadership as We Know It (2023). https://doi.org/10.1177/15480518231181731
39. Raed, H., Qawasmehb, E., F, Alserhanc, A., F, Ahmadd, H., Hammourie, Q., Halimf, M., Darawsheh, S.R.: Utilizing business intelligence and digital transformation and leadership to enhance employee job satisfaction and business added value in greater Amman municipality. Int. J. Data Network Sci. **7** (2023). 1
40. Roldán, J.L., Sánchez-Franco, M.J.: Variance-based structural equation modeling: guidelines for using partial least squares in information systems research. In: Mora, M., Gelman, O., Steenkamp, A., Raisinghani, M. (eds.) Research Methodologies, Innovations and Philosophies in Software Systems Engineering and Information Systems, pp.193–221. Hershey, PA: Information Science Reference (2012). https://doi.org/10.4018/978-1-4666-0179-6.ch010

41. Sheninger, E.: Digital leadership: Changing paradigms for changing times, Corwin Press (2019)
42. Shin, J., Mollah, M.A., Choi, J.: Sustainability and Organizational Performance in South Korea: The Effect of Digital Leadership on Digital Culture and Employees' Digital Capabilities. Sustainability **15**(3), 2027 (2023). https://doi.org/10.1016/j.ijinfomgt.2020.102225
43. Tigre, F.B., Curado, C., Henriques, P.L.: Digital leadership: a bibliometric analysis. J. Leadership Organizational Stud. **30**(1), 40–70.077–1084 (2023)
44. Wang, W., Siau, K.: Artificial intelligence, machine learning, automation, robotics, future of work and future of humanity: a review and research agenda. J. Database Manage. **30**(1), 61–79 (2019)
45. Wasono, L.W., Furinto, A.: The effect of digital leadership and innovation management for incumbent telecommunication company in the digital disruptive era. Int. J. Eng. Technol. Innov. **7**, 125–130 (2018). https://doi.org/10.14419/ijet.v7i2.29.13142
46. Zhu, P.: Digital Master: Debunk the Myths of Enterprise Digital Maturity. Lulu Publishing Services, Morrisville, NC (2015)
47. Zulu, S.L., Khosrowshahi, F.: A taxonomy of digital leadership in the construction industry. Constr. Manage. Econ. **39**(7), 565–578 (2021). https://doi.org/10.1080/01446193.2021.1930080

A Two-Phase Time-Based Robust Shift Genetic Algorithm for Multi-resource Flexible Project Scheduling Under Uncertainties

Do Hong Nhat[✉] and Nguyen Van Hop[✉]

School of Industrial Engineering and Management, International University, VNU-HCMC, Quarter 6, Linh Trung Ward, Thu Duc, Ho Chi Minh City, Vietnam
nhatdh31@phd.hcmiu.edu.vn, nvhop@hcmiu.edu.vn

Abstract. To effectively respond to the volatile environment, the project schedule should be adaptable in a timely manner when uncertain events occur during the project process. The paper introduces a two-phase time-based robust shift genetic algorithm, designed to efficiently generate schedules for multi-resource flexible projects in the presence of uncertainties. The first phase involves developing a critical path analysis to select not only the appropriate mode but also determine the best resource type for each activity, considering renewable and non-renewable resources. The primary objective of this phase is to minimize the total makespan (Cmax) of the project. In the second phase, the project schedule is updated in response to uncertain events, with the stochastic project's completion time constraint, minimizing the overall cost. A time-based robust measure is calculated for each task, utilizing historical data. This measure serves as the foundation for a novel time-based robust shift operation, incorporated with the genetic algorithm's crossover and mutation operations to produce flexible project schedule solutions. To validate the performance of the proposed approach, numerical experiments were conducted using datasets from the standard "Project Scheduling Problems Library" (PSPLIB), yielding promising results.

Keywords: Genetic Algorithm · Time-Based Robust Shift · Multi-Mode Resources · Flexible Project Scheduling Problem

1 Introduction

Project scheduling is a challenging task due to the project's limited but volatile resources. Chance events can occur during the project's progression. Many authors are interested in the field of project scheduling to solve the problem of determined or uncertain resources, in which there are many types of resource that can be chosen: renewable resources and non-renewable resource. And multi-mode with uncertainty event is another level when the project schedule has more than one option or one mode to choose from at each activity operating in an uncertain environment. There is a brief summary table next section at Literature review, and interesting readers could find in an interesting work of Paraskevopoulos et al. (2017).

This paper addresses the multi-mode resource-constrained project scheduling problem, considering uncertain factors. The proposed approach aims to answer the following questions:

- How to identify a robust project schedule that suggests modes of activity sequences while considering limited resources?
- How to leverage experience from project history data to apply to a project scheduling problem and find an optimal scheduling instance that minimizes the total project cost in the event of unexpected task duration changes?

To answer these questions, a two-phase approach has been developed. In Phase 1, with the objective of minimizing the makespan, Critical Path Analysis and Genetic Algorithm are employed to select appropriate task modes that satisfy resource constraints and provide optimal activity sequences. In Phase 2, the objective is to minimize the total cost while considering uncertainty conditions, such as unexpected activity duration changes. A probability is set to control the makespan within the total project time.

Following this introduction, the next section will investigate current developments in project management. We develop a mathematical model for project management in Sect. 3. In Sect. 4, we use the Genetic Algorithm to handle both strict completion times and uncertain scarce resources. In the same section, we conducts numerical experiments to validate the proposed method, using datasets from the Project Scheduling Problems Library (PSPLIB) as benchmarks for validation purposes. Finally, conclusions are summarized, and some future research directions are recommended in Sect. 5.

2   Literature Review

In the last decades, the resource constrained project scheduling problem (RCPSP) has been studied extensively in different directions such as deterministic or stochastic resource constraints, mode selection, etc. The following Table 1 provides a summary of current development of these streams of research.

In terms of resource utilization, many research works study sharing, allocating, aggregating, and prioritizing different resources for use in a project. El-Abbasy et al. (2016) developed an evolutionary algorithm to share and allocate different resources such as cash, equipment, and manpower for various projects. An automated system optimizes project scheduling to trade-off between multiple objectives, including the duration of multiple projects, total cost, financing cost, maximum required credit, profit, and resource fluctuations and peak demand. In another work, Morin et al. (2017) took average resource usage over aggregated intervals. The project is scheduled with estimated capacity over consecutive periods. Later, Morin et al. (2022) modified resource constraints by limiting average resource usage over periods. A new period-indexed mixed-integer programming formulation with stronger relaxations is introduced by disaggregating the precedence constraints. Recently, Luo et al. (2022) also proposed a genetic programming hyperheuristic to learn priority rules for resource-constrained project scheduling problems. In addition, the duplicate removal technique and a compact training data selection could also efficiently improve the evolving process of priority rules.

Table 1. Literature Summary

References	Methodology	Deterministic Resources				Uncertain Resources		Mode of Activity	
		Aggregate	Share	Prioritize	Allocate	Resources	Duration	Single	Multiple
El-Abbasy et al. (2016)	Evolutionary Algorithm		X						
Morin et al. (2017)	MILP	X							
Morin et al. (2022)	MILP	X							
Luo et al. (2022)	Genetic Programming			X					
Servranckx and Vanhoucke (2019a)	Tabu Search							X	
Tian et al. (2016)	MMEDA								X
Wang and Zheng (2018)	Fly Optimization Algorithm (FOA)								X
Abido and Elazouni (2021)	Evolutionary Algorithm								X
Brcic et al. (2019)	Robustness measure, Metaheuristics						X		
Zahid et al. (2019)	Robust measures						X		
Servranckx and Vanhoucke (2019b)	Tabu Search and Simulation	X				X			
Ma et al. (2019)	Surrogate measures					X	X		
Ning et al. (2017)	Simulated annealing (SA) and tabu search (TS)	X				X		X	X
Balouka and Cohen (2021)	Benders decomposition				X		X		X
Paraskevopoulos et al. (2017)	Review research	X	X	X	X	X	X	X	X

Dealing with uncertain resources is also a challenge in the resource-constrained project scheduling problem. Many studies have investigated various ways to adapt to the variation of the project environment, such as changes in activities' parameters, unavailable resources, new customer requirements, etc. In an interesting work, Brcic et al. (2019) addressed stochastic duration in a proactive-reactive scheduling model for coordinating the project's activities. A new robustness measure, namely, Threshold Cost-based Flexibility (TCBF), is introduced to proactively update the baseline schedule. Two metaheuristics based on rollout-based and iterative policy search are also developed to improve the overall performance of project schedule solutions. Similarly, Zahid et al. (2019) suggested surrogate measures of robustness to determine the most robust schedule to avoid large tardiness due to uncertainties. The results of this work showed that among the proposed robust measures, the Average Float Index provides the best indicator for evaluating the robust nature of schedules. Besides modeling alternative execution modes for activities, Servranckx and Vanhoucke (2019b) developed alternative schedules included in the set of proactive schedules to deal with uncertainty. This set of schedules can be switched to bring the project back on track if uncertain events happen during project execution. In another interesting work, Tian et al. (2019) designed a robust scheduling approach based on the combination of the Estimation of Distribution Algorithm and Genetic Algorithm to address time uncertainty in the resource-constrained project scheduling problem (RCPSP). Two robustness measures based on time and capacity are also suggested to evaluate the flexibility of scheduling solutions. Moreover, Ma et al. (2019) also studied both stochastic resource availabilities and stochastic activity durations. Several surrogate robustness measures are investigated to generate robust baseline schedules. Different buffering strategies are also proposed to deal with uncertainties.

In addition to addressing resource utilization, mode selection is another crucial factor that significantly impacts the performance of solutions to the Resource-Constrained Project Scheduling Problem (RCPSP). On one hand, some research works solely focus on the selection of a single activity mode. On the other hand, several studies investigate the impact of different activity modes. Regarding the selection of a single activity mode, Servranckx and Vanhoucke (2019a) opted for the activity execution mode using alternative subgraphs. The dependencies between alternatives in the project structure are characterized in a classification matrix. Subsequently, this classification matrix guides a Tabu Search algorithm in finding the optimal project schedule. Expanding into multiple mode selections, Tian et al. represented a Markov network for a multi-objective multi-mode resource-constrained project scheduling problem. This approach not only satisfies precedence constraints but also allocates appropriate resources for the activities. The Markov network-based Estimation of Distribution Algorithm (EDA) models the interrelation of mode selections that share the same resources. An enhanced EDA with mutation operations and a specific local search are employed to improve the search process for the project scheduling solution. Furthermore, Wang and Zheng (2018) investigated the multi-skill resource project scheduling problem, where a resource masters several skills, and each task requires a certain skill. The smell-based and vision-based search integrated with a multi-swarm strategy were adopted in a knowledge-guided search Fruit Fly Optimization Algorithm to enhance the exploration of project schedule solutions. Recently, Abido and Elazouni (2021) designed a Multi-Objective Evolutionary Programming (MOEP) algorithm to explore multi-mode activities in time–cost trade-off and finance-based scheduling with resource leveling in project scheduling problems.

Besides the aforementioned research directions, combinations of these aspects have also been investigated by Ning et al. (2017) and Balouka and Cohen (2021). On one hand, Ning et al. (2017) explored both multi-mode cash flow and stochastic duration of activities in the Resource-Constrained Project Scheduling Problem (RCPSP). Two metaheuristics, Simulated Annealing (SA) and Tabu Search (TS), were employed to generate a robust baseline schedule, minimizing the maximal gap of cumulative cash flows. The determination of the best activity modes, time buffers, and activity start times was simultaneous with the project schedule.

On the other hand, Balouka and Cohen (2021) addressed the multi-mode issue with uncertain activity duration. Balouka (2021) proposed a Benders decomposition approach with specialized cuts for the multi-mode resource-constrained project scheduling problem with uncertain activity durations. A robust optimization approach aimed to minimize the worst-case project duration by deciding on activity modes, resource allocations, and a schedule baseline under varying levels of uncertainty, conservatism, and different types of duration distributions. Insights about the price of robustness were obtained to guide multi-mode project scheduling implementations with incomplete distribution information of activity durations.

In summary, while various aspects of the Resource-Constrained Project Scheduling Problem have been investigated, there is a need to carefully consider many factors to improve project schedules. To the best of our knowledge, no work has addressed a flexible scheduling method to deal with uncertain resources, multi-mode, and both types of non-renewable and renewable resources together. Therefore, the primary goal of this work is to use a shift operation to manage uncertain events and maintain an optimal overall cost. Phase 1 introduces a structural project scheduling algorithm to address the traditional multi-mode resource-constraint project scheduling problem. Phase 2 introduces a measure called a novel time-based robust shift operation by utilizing historical data. In this phase, a chance constraint is added to the mathematical model to find a project schedule that minimizes the overall cost.

3   Problem Description and Solution Development

3.1   Problem Description

In the investigated multi-mode resource constrained project scheduling (SMRCPSP) problem, there are J jobs in the project. Each activity (job, task) j ($j = 1, \ldots, J$) could be executed in one of Mj modes. The activities cannot be preempted, and a mode once selected cannot be changed, e.g., once an activity started in mode m, it must be completed in mode m without interruption. Each activity j in mode m has a duration djm and consume a set R and N of renewable and nonrenewable resources, respectively.

Given an upper bound $\overline{T}$ on the project's makespan, K^p units of renewable resource $r \in R$ are available at period t, $t = 1, \ldots, \overline{T}$ The overall capacity of the nonrenewable resource $r \in N$ is given by K_r^v. If activity j is scheduled in mode m, then k_{jmr}^p units of the renewable resource $r \in R$ are used for each period of activity j is in process. Additionally, k_{jmr}^v units of the nonrenewable resource $r \in N$ are consumed. The requirement is to find the best project schedule in terms of minimal makespan such that the precedence relations and the limited resource availabilities. All parameters are assumed as integer valued. All notations are defined in the following Table 2.

Table 2. Notations.

i, j	Index of activities, $i, j = 1, 2, ..., J$,
J	Number of activities
Mj	Number of modes of activities j can be performed in, $j = 1,..., J$
m	Index of modes of activity j, $m = 1, ..., Mj$
tj	surplus completion time of activity j from its EF, e.g., $EFj \leq tj \leq LFj$
r	Index of non-renewable resources, $r = 1, 2. ..., N$
k	Index of renewable resources, $k = 1,..., K$
djm	Duration of activity j performed in mode m, $j = 1, ..., J$
$\overline{T}$	Upper bound on the project's makespan
t	Index of time period, $t = 1, ..., \overline{T}$
$K^p_k \geq 0$	Number of units of renewable resource k ($k = 1,..., K$) available in period t ($t = 1, ..., \overline{T}$)
$K^v_r \geq 0$	Total number of units of nonrenewable resource r ($r = 1,2,..., N$) available in period t ($t = 1, ..., \overline{T}$)
$k^p_{jmk} \geq 0$	Number of units of renewable resource k ($k = 1,..., K$), used by activity j being performed in mode m
$k^v_{jmr} \geq 0$	Number of units of nonrenewable resource r ($r = 1,2,..., N$), consumed by activity j being performed in mode m
$f_{jmk} \geq 0$	Unit cost of each renewable resource k ($k = 1,2,..., K$), charged by activity j being performed in mode m
$f_{jmr} \geq 0$	Unit cost of each nonrenewable resource r ($r = 1,2,..., N$), charged by activity j being performed in mode m
$Pj (Sj)$	Set of immediate predecessors (successors) of activity j
$ESj (EFj)$	Earliest start time (finish time) of activity j
$LSj (LFj)$	Latest start time (finish time) of activity j. $LFj \leq \overline{T}$
sjm	Starting time of activity j under mode m
cjm	Completion time of activity j under mode m
$NSuccj$	Number of immediate successors of activity j
sj	Slack time of activity j
$rjmu$	Amount of the resource u (sum of renewable resource and non-renewable resource) occupied by the activity j in mode m
$xjmt$	1, if activity j is performed in mode m at time t. 0, otherwise

Phase 1.

The mathematical model of the problem will be:

$$\text{Minimize E(Cmax)} = \sum_{m=1}^{Mj} \sum_{t=EFj}^{LFj} t * x_{jmt} \tag{1}$$

$$\sum_{m=1}^{Mj} \sum_{t=EFj}^{LFj} x_{jmt} = 1, j = 1, 2, \ldots, J. \tag{2}$$

$$\sum_{m=1}^{M_i} \sum_{t=EF_i}^{LF_i} t * x_{imt} \leq \sum_{m=1}^{Mj} \sum_{t=EFj}^{LFj} (t - d_{jm}) * x_{jmt}; i, j = 1, 2, .., J; i \in P_j \tag{3}$$

$$\sum_{j=1}^{J} \sum_{m=1}^{Mj} k_{jmk}^{p} \sum_{q=\max\{t,EFj\}}^{\min\{t+djm-1,LFj\}} x_{jmq} \leq K_{k}^{p}, k = 1, 2, \ldots K; \ t = 1, \ldots, \overline{T}. \tag{4}$$

$$\sum_{j=1}^{J} \sum_{m=1}^{Mj} k_{jmr}^{v} \sum_{t=EFj}^{LFj} x_{jmt} \leq K_{r}^{v}, \ r = 1, 2, .., N \tag{5}$$

In this model, two objective functions are: (1) to minimize the total project's makespan. Constraint (2) ensures that exactly one mode and one completion time is assigned to each activity. Constraint (3) considers the precedence relations. Constraint (4) guarantees that the period availabilities of the renewable resources are not exceeded. Constraint (5) secures feasibility with respect to the consumable (nonrenewable) resource.

Phase 2.

In this phase, uncertain events are taken into account. Therefore, the duration of tasks is evaluated using historical data. A stochastic project completion time constraint is formulated as follows.

$$Prob\left\{ \max_{j=1,2,\ldots,J} (C_j) \leq \left(RequiredC_{max} + \delta\right) \right\} \geq \alpha \tag{6}$$

This chance constrain is converted into the equivalent form as follows:

$$\sum_{m=1}^{Mj} \sum_{t=EF_j}^{LF_j} t \times x_{jmt} + Z_\alpha \sigma_{C_j} \leq \left(RequiredC_{max} + \delta\right) \tag{7}$$

where:

σ_{C_j} = Standard deviation of makespan is calculated from a set of historial project data.

$RequiredC_{max}$ = Required makespan is an expected makespan calculated by a model updated the standard activity duration.

δ = A satisfy adjustment coefficient that can be added to decrease strictly constraint for $RequiredC_{max}$.

The mathematical model of Phase 2 will be:

$$\text{Minimize COST} = \left(\sum_{j=1}^{J} \sum_{k=1}^{K} \sum_{m=1}^{Mj} k_{jmk}^{p} * f_{jmk} * x_{jmt} \right)$$

$$+ \left(\sum_{j=1}^{J} \sum_{r=1}^{N} \sum_{m=1}^{Mj} k_{jmr}^{v} * f_{jmr} * x_{jmt} \right) \qquad (8)$$

Subject to

$$\sum_{m=1}^{Mj} \sum_{t=EF_j}^{LF_j} x_{jmt} = 1, j = 1, 2, \ldots, J \qquad (9)$$

$$\sum_{m=1}^{Mj} \sum_{t=EF_i}^{LF_i} t \times x_{imt} \leq \sum_{m=1}^{Mj} \sum_{t=EF_j}^{LF_j} (t - d_{jm}) \times x_{jmt}; i, j = 1, 2, \ldots, J; i \in P_j \qquad (10)$$

$$\sum_{j=1}^{J} \sum_{m=1}^{Mj} k_{mjk}^{p} \sum_{q=max\{t,EF_j\}}^{min\{t+d_{jm}-1,LF_j\}} x_{jmq} \leq K_k^p; k = 1, 2 \ldots, K; t = 1, 2, \ldots, \overline{T} \qquad (11)$$

$$\sum_{j=1}^{J} \sum_{m=1}^{Mj} k_{mjr}^{v} \sum_{t=EF_j}^{LF_j} x_{jmq} \leq K_r^v; r = 1, 2 \ldots, N \qquad (12)$$

$$\sum_{m=1}^{Mj} \sum_{t=EF_j}^{LF_j} t \times x_{jmt} + Z_\alpha \sigma_{C_j} \leq \left(RequiredC_{max} + \delta \right) \qquad (7)$$

In this model, the objective is to minimize the overall cost to find out new activity sequence with appropriate modes satisfied the resource constraints. Constraint (7) is added to ensure Cmax result will be less than required makespan with the expected probability.

3.2 Solution Development

In this section, the proposed approach is described in detail. Figure 1 illustrates the solution process in two phases. The duty of phase one is to develop a critical path with the appropriate mode and best resource type for each activity. The objective of this phase is to minimize the total makespan (Cmax) of the project. The second phase will update the project schedule in response to uncertain events, with the stochastic project's completion time constraint, minimizing the overall cost.

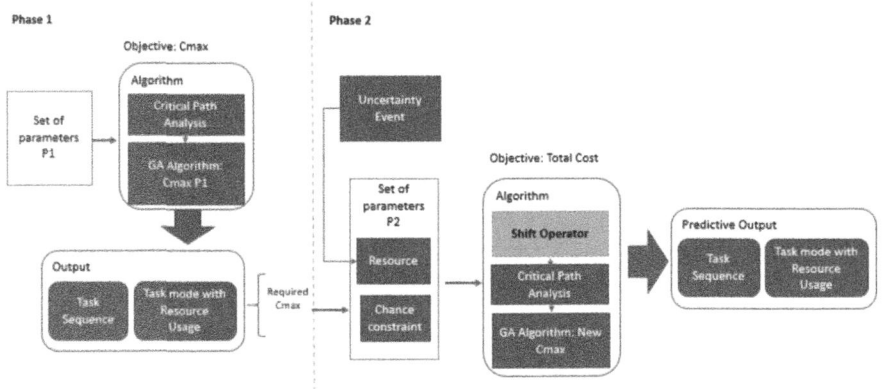

Fig. 1. The Proposed Approach

Phase 1.

The complexity of the MMRCPSP problem is not only to develop an efficient project schedule solution that minimizes the total project's makespan but also to design an effective way to propose a good project schedule with appropriate modes satisfied the limited resource. The challenge jobs are to arrange the activities in a good sequence and choose the appropriate mode for the activities of the project with two types of renewable and non-renewable resources. Therefore, the MMRCPSP problem is structurally dealt with in three critical decision-making processes: (1) critical path analysis to create the best sequence of activities with minimal makespan, (2) generating a set of solutions to deal with multi-mode activities and two types of renewable and non-renewable resources using the Genetic algorithm in which the best fitness solution will be chosen. The following algorithm of our proposed structural flexible project scheduling procedure finds the best project schedule that minimize makespan with the chosen modes of activities satisfied the limited resource.

Algorithm 1: Structural Flexible Project Scheduling (SFPS).

Input data:
- Number of tasks in the project.
- Number of modes for each task.
- Duration of each activity performed in each mode.
- Required non-renewable resource for each task in each mode.
- Required renewable resource for each task in each mode.
- Total non-renewable resource.
- Total renewable resource available in each period.
- Population size (GA's parameters) =10.
- Chromosome's size =number of genes in a chromosome (GA's parameters) = $J \times M \times \sum_{j=1}^{J} [LF_j - EF_j]$.
- Maximum number of iterations = number of generations = 20.
- Crossover rate = 0.15
- Mutation rate = 0.7

Output: The best project schedule X^*, total makespan T^*

Begin

Stage 1. Critical Path Analysis - Activity assignment sequencing (*a-seq*): determine the sequence of activities subject to precedence relationships.

Step 1. Calculate upper bound $\overline{T}$ of the project makespan:

$$\overline{T} = \sum_{j=1}^{J} max_{m=1}^{Mj} \{d_{jm}\} \tag{13}$$

Step 2. Traditional forward and backward recursions are applied to calculate time windows $[EF_j, LF_j]$ for each activity j ($j = 1,\ldots,J$) subject to precedence-constraints and modes of shortest duration:

$$EF1 = d11 \tag{13}$$

$$EFj = max \{EFi; i \in IP_j\} + d_{jt}; j = 2,\ldots,J. \tag{14}$$

$$LFj = \overline{T} \tag{15}$$

$$LFi = min \{LF_j - d_{j1}; j \in IS_i\}; i = J-1,\ldots,1. \tag{16}$$

Step 3. Sequence all activities so that the precedence relationships are satisfied.

Stage 2. Mode selection (*m-select*): Genetic Algorithm is used to generate the set of solutions based on determining the mode of activity that can be processed among the multiple mode candidates based on the most defined fitness value. In our GA, the chromosomes are designed as finite-length strings of binary values represent for a task j with its associated mode m at a time period t in time window $[EFj, LFj]$. The genes in a chromosome have three dimensions $J \times M \times \sum_{j=1}^{J} [LF_j - EF_j]$. The representation of a chromosome is illustrated in Fig. 2.

Step 4. Initialization: the initial population is generated randomly subject to the problem's constraints.

Repeat: Set iteration counter n = 0, while (n < Double Population)

Step 5. Compute the fitness value of each chromosome
Step 6. Crossover and mutation:

Crossover: two points crossover is applied.
Mutation: changes randomly some random cells to create new offspring.

Step 7. Constraint handling: all constraints are applied to each chromosome.
Step 8. New population: the chromosomes with highest fitness value are selected to create new population.

x_{111}	x_{112}	...	$x_{11(t+1)}$	...	x_{121}	x_{122}	...	$x_{jm(t+1)}$	...	$x_{JM(LF_J-EF_J)}$

Fig. 2. The structure of a chromosome (Algorithm 1)

Until (Termination condition(s) is satisfied).
End.

Phase 2.

Algorithm 2: Shift Operation Flexible Project Scheduling (SOFPS).

Input data:
- Number of tasks in the project and historical data.
- Number of modes for each task and historical data.
- Duration of each activity performed in each mode and historical data.
- Required non-renewable resource for each task in each mode and historical data.
- Required renewable resource for each task in each mode and historical data.
- Total non-renewable resource and historical data.
- Total renewable resource available in each period and historical data.
- Population size (GA's parameters) =10.
- Chromosome's size =number of genes in a chromosome (GA's parameters) = $J \times M \times \sum_{j=1}^{J} [LF_j - EF_j]$.
- Maximum number of iterations = number of generations = 20.
- Crossover rate = 0.15
- Mutation rate = 0.7

Output: The best overall project cost, updated project schedule X *, total makespan T*

Begin

Stage 1. Shift Operator

Step 1. Calculate each task duration by historical data
d_{jm} = average of each task at activity j and mode m
Step 2. Update each task d_{jm} of the project by the calculated task
Step 3. Update the affected task by event d_{jm} where event happens at activity j and mode m
Step 4. Calculate the standard deviation of makespan by historical data σ_{C_j}
Step 5. Update Required Makespan by Cmax at Phase 1 and plus a satisfy adjustment coefficient δ
$RequiredC_{max} + \delta$

Stage 2. Critical Path Analysis - Activity assignment sequencing (*a-seq*): determine the sequence of activities subject to precedence relationships.

Step 6. Calculate upper bound $\overline{T}$ of the project makespan:

$$\overline{T} = \sum_{j=1}^{J} max_{m=1}^{Mj} \{d_{jm}\} \tag{13}$$

Step 7. Traditional forward and backward recursions are applied to calculate time windows $[EFj, LFj]$ for each activity j ($j = 1, ..., J$) subject to precedence-constraints and modes of shortest duration:

$$EF1 = d11 \tag{13}$$

$$EFj = \max \{EFi; i \in IPj\} + djt; j = 2, \ldots, J. \tag{14}$$

$$LFj = \overline{T} \tag{15}$$

$$LFi = \min \{LFj - dj1; j \in ISi\}; i = J-1, \ldots, 1. \tag{16}$$

Step 8. Sequence all activities so that the precedence relationships are satisfied.

Stage 3. Mode selection (*m-select*): Genetic Algorithm is used to generate the set of solutions based on determining the mode of activity that can be processed among the multiple mode candidates based on the most defined fitness value. In our GA, the chromosomes are designed as finite-length strings of binary values represent for a task j with its associated mode m at a time period t in time window $[EFj, LFj]$. The genes in a chromosome have three dimensions $J \times M \times \sum_{j=1}^{J} [LF_j - EF_j]$. The representation of a chromosome is illustrated in Fig. 2 (Fig. 3).

Step 9. Initialization: the initial population is generated randomly subject to the problem's constraints.

Repeat: Set iteration counter n = 0, while (n < Double Population)

Step 10. Compute the fitness value of each chromosome
Step 11. Crossover and mutation:

- Crossover: two points crossover is applied.
- Mutation: changes randomly some random cells to create new offspring.

Step 12. Constraint handling: all constraints are applied to each chromosome.
Step 13. New population: the chromosomes with highest fitness value are selected to create new population.

Until (Termination condition(s) is satisfied).
End.

x_{11}	x_{11}	...	$x_{11(t+1)}$	...	x_{12}	x_{12}	...	$x_{jm(t+1)}$	...	$x_{JM(LF_J-EF_J)}$
1	2				1	2		)		

Fig. 3. The structure of a chromosome (Algorithm 2)

We are using the following numerical example to illustrate our proposed *SFPS* algorithm. We use PyCharm Community Edition 2022.1.1 for solving GA programs with default parameters. All experiments were run with 32 GB RAM, Intel Xeon CPU E5-2630 v3 @ 2.40 GHz cores running Windows Server 2012 R2 Standard.

In this example, the input parameters of the project are given in Table 3 and as follows:

Total number of units of available nonrenewable resource 1 = 20.
Total number of units of available nonrenewable resource 2 = 18.
Total number of units of available renewable resource = 8.

Table 3. Renewable and Nonrenewable Resources and Durations

Activity	Mode 1				Mode 2			
	NR1	NR 2	RR1	Duration	NR1	NR 2	RR1	Duration
Source (S)	0	0	0	0	0	0	0	0
A	2	3	3	4	3	6	4	3
B	4	3	2	1	6	6	3	3
C	3	2	1	3	2	4	3	2
D	2	4	2	3	4	2	1	3
Target (T)	0	0	0	0	0	0	0	0

Notes: NR = Non-renewable Resource; RR = Renewable Resource

Applying the SFPS algorithm, the calculation process is as follows (Tables 4 and 5).

Stage 1. Critical Path Analysis

Step 1. Calculate upper bound $\overline{T}$ of the project makespan
Using the result of column Max djm, we calculate $\overline{T} = 0 + 4 + 3 + 3 + 3 + 0 = 13$.
Step 2. Calculate time windows
Step 3. Sequence all activities (Fig. 4).

Table 4. Duration Time of Activities

Activity	Duration Mode 1	Duration Mode 2	Min djm	Max djm	ESj	EFj	LSj	LFj
Source (S)	0	0	0	0	0	0	5	5
A	4	3	3	4	0	3	5	8
B	1	3	1	3	3	4	9	10
C	3	2	2	3	3	5	8	10
D	3	3	3	3	5	8	10	13
Target (T)	0	0	0	0	8	8	13	13

Table 5. Time Windows [*EFj, LFj*] of Activities

Task	*EFj*	*LFj*	Time Windows [*LFj - EFj*]
Source (S)	0	5	5
A	3	8	5
B	4	10	6
C	5	10	5
D	8	13	5
Target (T)	8	13	5

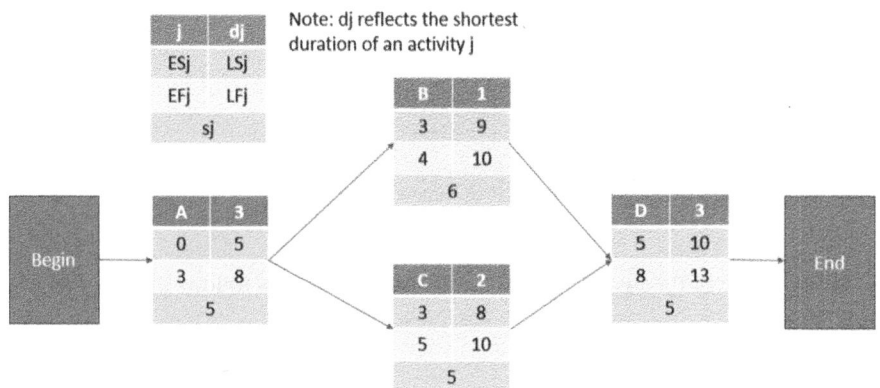

Fig. 4. Critical Path Evaluation

Time Windows = [[*LF0 − EF0*], [*LF1 - EF1*], [*LF2 − EF2*], [*LF3 − EF3*], [*LF4 − EF4*], [*LF5-EF5*]] = [5, 5, 6, 5, 5, 5]

Stage 2. Mode selection (*m-select*)

Step 4. Initialization

Generate initial solution for Step 4 based on the structure of a chromosome. With population size = 10, we generate 10 solutions and present in form [[Sol 1], [Sol 2], [Sol 3], [Sol 4], [Sol 5], [Sol 6], [Sol 7], [Sol 8], [Sol 9], [Sol 10]]

[[[0, 0, 0, 0, 0, 0, 0, 1, 0, 0, 0, 0], [0, 1, 0, 0, 0, 0, 0, 0, 0, 0, 0, 0], [0, 1, 0, 0, 0, 0, 0, 0, 0, 0, 0, 0, 0], [1, 0, 0, 0, 0, 0, 0, 0, 0, 0, 0, 0], [0, 0, 0, 0, 0, 0, 0, 1, 0, 0, 0, 0, 0], [1, 0, 0, 0, 0, 0, 0, 0, 0, 0, 0]], [[0, 1, 0, 0, 0, 0, 0, 0, 0, 0, 0, 0], [1, 0, 0, 0, 0, 0, 0, 0, 0, 0, 0, 0, 0], [0, 1, 0, 0, 0, 0, 0, 0, 0, 0, 0, 0, 0], [1, 0, 0, 0, 0, 0, 0, 0, 0, 0, 0, 0, 0], [0, 0, 0, 0, 0, 0, 1, 0, 0, 0, 0, 0], [1, 0, 0, 0, 0, 0, 0, 0, 0, 0, 0, 0]], [[0, 0, 0, 0, 0, 0, 0, 0, 1, 0, 0, 0], [0, 1, 0, 0, 0, 0, 0, 0, 0, 0, 0, 0], [0, 0, 1, 0, 0, 0, 0, 0, 0, 0, 0, 0, 0], [1, 0, 0, 0, 0, 0, 0, 0, 0, 0, 0, 0, 0], [1, 0, 0, 0, 0, 0, 0, 0, 0, 0, 0, 0], [0, 0, 0, 0, 0, 0, 1, 0, 0, 0, 0, 0]], [[0, 0, 0, 0, 0, 0, 0, 0, 1, 0, 0, 0], [0, 0, 0, 0, 0, 0, 0, 1, 0, 0, 0], [0, 0, 1, 0, 0, 0, 0, 0, 0, 0, 0, 0, 0], [1,

0, 0, 0, 0, 0, 0, 0, 0, 0, 0, 0], [1, 0, 0, 0, 0, 0, 0, 0, 0, 0, 0, 0, 0, 0], [0, 1, 0, 0, 0, 0, 0, 0, 0, 0, 0, 0]], [[0, 0, 0, 0, 0, 0, 0, 1, 0, 0, 0, 0, 0], [1, 0, 0, 0, 0, 0, 0, 0, 0, 0, 0, 0, 0, 0], [0, 0, 0, 0, 0, 0, 0, 0, 0, 1, 0, 0, 0, 0], [1, 0, 0, 0, 0, 0, 0, 0, 0, 0, 0, 0, 0], [0, 0, 0, 0, 0, 0, 1, 0, 0, 0, 0, 0], [1, 0, 0, 0, 0, 0, 0, 0, 0, 0, 0, 0]], [[0, 0, 0, 0, 0, 0, 0, 0, 0, 0, 0, 0, 1], [0, 0, 0, 0, 0, 0, 0, 0, 0, 1, 0, 0, 0], [0, 0, 1, 0, 0, 0, 0, 0, 0, 0, 0, 0, 0, 0], [0, 0, 0, 0, 0, 0, 0, 0, 1, 0, 0, 0, 0, 0], [0, 0, 0, 0, 0, 0, 0, 1, 0, 0, 0, 0, 0], [1, 0, 0, 0, 0, 0, 0, 0, 0, 0, 0, 0, 0]], [[0, 0, 0, 0, 0, 0, 0, 1, 0, 0, 0, 0], [0, 0, 0, 0, 0, 0, 0, 0, 1, 0, 0], [0, 0, 0, 1, 0, 0, 0, 0, 0, 0, 0, 0, 0, 0], [0, 0, 0, 0, 0, 0, 0, 0, 1, 0, 0, 0], [0, 0, 0, 0, 0, 0, 1, 0, 0, 0, 0, 0], [1, 0, 0, 0, 0, 0, 0, 0, 0, 0, 0, 0, 0]], [[0, 0, 0, 0, 0, 0, 1, 0, 0, 0, 0, 0], [0, 0, 0, 0, 0, 0, 0, 0, 1, 0, 0, 0], [0, 0, 1, 0, 0, 0, 0, 0, 0, 0, 0, 0, 0, 0], [0, 0, 0, 0, 0, 0, 0, 1, 0, 0, 0, 0], [0, 0, 0, 0, 0, 0, 0, 1, 0, 0, 0, 0], [1, 0, 0, 0, 0, 0, 0, 0, 0, 0, 0, 0, 0]], [[0, 0, 0, 0, 0, 0, 0, 1, 0, 0, 0, 0, 0], [0, 0, 0, 0, 0, 0, 1, 0, 0, 0, 0, 0], [0, 0, 1, 0, 0, 0, 0, 0, 0, 0, 0, 0, 0, 0], [0, 0, 0, 0, 0, 0, 0, 0, 1, 0, 0, 0], [0, 0, 0, 0, 0, 0, 1, 0, 0, 0, 0, 0], [1, 0, 0, 0, 0, 0, 0, 0, 0, 0, 0, 0]], [[0, 0, 0, 0, 0, 0, 0, 1, 0, 0, 0, 0], [0, 0, 0, 0, 0, 0, 0, 0, 0, 0, 0, 0, 0], [0, 0, 0, 1, 0, 0, 0, 0, 0, 0, 0, 0, 0, 0], [0, 0, 0, 0, 0, 0, 0, 0, 1, 0, 0, 0], [0, 0, 0, 0, 0, 0, 0, 1, 0, 0, 0, 0], [0, 0, 0, 0, 0, 0, 1, 0, 0, 0, 0, 0], [0, 0, 0, 0, 0, 0, 0, 0, 0, 1, 0, 0]]]]

The 10 objectives of these solutions are presented in form [[53], [53], [63], [72], [76], [82], [82], [82], [82], [82]], which corresponds to the order of the 10 solutions presented.

Now, we understand that one solution or one chromosome in form [[0, 0, 0, 0, 0, 0, 0, 1, 0, 0, 0, 0], [0, 1, 0, 0, 0, 0, 0, 0, 0, 0, 0, 0], [0, 1, 0, 0, 0, 0, 0, 0, 0, 0, 0, 0, 0, 0], [1, 0, 0, 0, 0, 0, 0, 0, 0, 0, 0, 0], [0, 0, 0, 0, 0, 0, 1, 0, 0, 0, 0, 0], [1, 0, 0, 0, 0, 0, 0, 0, 0, 0, 0, 0, 0]] get one objective in form [53]

Run Repeat procedure of GA for **Step 5. Step 6. Step 7. Step 8.**

After completing the GA running procedure, the obtained 10 solutions with makespan objective values are summarized in Table 6.

Table 6. A set of solutions

Solution	Sol 1	Sol 2	Sol 3	Sol 4	Sol 5	Sol 6	Sol 7	Sol 8	Sol 9	Sol 10
Makespan	8	8	8	8	8	8	8	8	8	8

Take the following chromosome to illustrate an obtained solution in the chosen solutions: [[0, 0, 0, 0, 0, 0, 1, 0, 0, 0, 0, 0], [0, 0, 0, 0, 0, 0, 0, 1, 0, 0, 0, 0, 0], [0, 1, 0, 0, 0, 0, 0, 0, 0, 0, 0, 0, 0], [1, 0, 0, 0, 0, 0, 0, 0, 0, 0, 0, 0, 0], [0, 0, 0, 0, 0, 0, 1, 0, 0, 0, 0, 0], [1, 0, 0, 0, 0, 0, 0, 0, 0, 0, 0, 0]], This chromosome has 6 activities (including source and target activities). Each activity in a chromosome is a combination of two modes and the length of each mode in binary sequence generated by time windows (LF-EF) to form its associated surplus completion time. For example, consider the activity 1: [0, 0, 0, 0, 0, 1, 0, 0, 0, 0], we have 5 cells (0, 0, 0, 0, 0) in mode 1 and 5 cells (1, 0, 0, 0, 0) in mode 2. The number of cells in each mode are determined by (LF1-EF1)+1. The detailed completion time for each activity and other results are also summarized in Tables 8 and 9 (Table 7).

Phase 2. A change happens to a task that makes its duration longer than the expected plan. This phase predicts new project schedule with the change in considering an optimal overall cost.

Table 7. Finish Time of Each Activity in a Chromosome.

Activity	EFj in time windows	Result	Finish time
Source (S)	0	[0, 0, 0, 0, 0, 0, 1, 0, 0, 0, 0, 0] Activity 0 starts at position 1 (Index 0) in mode 2	Finish time = EFj + Index in mode = 0 + 0 = 0
A	3	[0, 0, 0, 0, 0, 0, 1, 0, 0, 0, 0, 0] Activity 1 starts at position 1 (Index 0) in mode 2	Finish time = EFj + Index in mode = 3 + 0 = 3
B	4	[0, 1, 0, 0, 0, 0, 0, 0, 0, 0, 0, 0, 0, 0] Activity 2 starts at position 2 (Index 1) in mode 1	Finish time = EFj + Index in mode = 4 + 1 = 5
C	5	[1, 0, 0, 0, 0, 0, 0, 0, 0, 0, 0, 0] Activity 3 starts at position 1 (Index 0) in mode 1	Finish time = EFj + Index in mode = 5 + 0 = 5
D	8	[0, 0, 0, 0, 0, 0, 1, 0, 0, 0, 0, 0] Activity 4 starts at position 1 (Index 0) in mode 2	Finish time = EFj + Index in mode = 8 + 0 = 8
Target (T)	8	[1, 0, 0, 0, 0, 0, 0, 0, 0, 0, 0, 0] Activity 5 starts at position 1 (Index 0) in mode 1	Finish time = EFj + Index in mode = 8 + 0 = 8

Table 8. Detailed Results of a Chromosome in the solution set.

Task	Start time	End time	Chosen Mode	Non-renewable resource 1	Non-renewable resource 2	Renewable resource
Source (S)	0	0	Mode 2	0	0	0
A	0	3	Mode 2	3	6	4
B	4	5	Mode 1	6	6	3
C	2	5	Mode 1	2	4	3
D	5	8	Mode 2	4	2	1
Target (T)	6	8	Mode 1	0	0	0

We get a set of historical data for parameter input. Assumption that each task in instance requires the same resource non-renewable resource and renewable resource requirement at Table 3.

Stage 1: Shift Operator

> **Step 1**. Calculate each task duration by historical data (Table 10)
> **Step 2.** Update each task d_{jm} of the project by the calculated task

Table 9. Historical data

Instance	Mode	Target (T)	Task A	Task B	Task C	Task D	Source (S)
1	Mode 1	0	3	1	1	2	0
	Mode 2	0	1	1	1	2	0
2	Mode 1	0	3	1	2	1	0
	Mode 2	0	3	3	3	1	0
3	Mode 1	0	3	1	2	3	0
	Mode 2	0	3	1	3	2	0
4	Mode 1	0	2	2	3	1	0
	Mode 2	0	2	1	3	2	0
5	Mode 1	0	2	2	2	2	0
	Mode 2	0	2	1	1	1	0
6	Mode 1	0	1	2	2	2	0
	Mode 2	0	3	3	3	3	0
7	Mode 1	0	3	3	1	2	0
	Mode 2	0	3	2	3	3	0
8	Mode 1	0	3	2	3	3	0
	Mode 2	0	1	2	3	2	0
9	Mode 1	0	2	3	1	1	0
	Mode 2	0	2	1	3	1	0
10	Mode 1	0	2	1	1	3	0
	Mode 2	0	2	1	2	3	0

Based on historical data, we update each activity with an average duration (Table 11).

Step 3. Update the affected task by event

Table 10. Historical data task duration

Activity	Source (S)	A	B	C	D	Target (T)
Duration Mode 1	0	2	2	2	2	0
Duration Mode 2	0	2	2	3	2	0

Here, we analyze the change for activity B. Duration is increasing from 1 to 3 if mode 1 is chosen.

Table 11. Task duration updated historical data

Activity	Mode 1				Mode 2			
	NR1	NR 2	RR1	Duration	NR1	NR 2	RR1	Duration
Source (S)	0	0	0	0	0	0	0	0
A	2	3	3	2	3	6	4	2
B	4	3	2	2	6	6	3	2
C	3	2	1	2	2	4	3	3
D	2	4	2	2	4	2	1	2
Target (T)	0	0	0	0	0	0	0	0

We have the original activity duration at each mode in the table below and highlight the change point (Table 12).

Table 12. Activity duration updated the changed event

Activity	Source (S)	A	B	C	D	Target (T)
Duration Mode 1	0	2	1 → 3	2	2	0
Duration Mode 2	0	3	2	3	2	0

Step 4. Calculate the standard deviation of makespan by historical data

Calculate standard deviation of makespan

Instance	1	2	3	4	5	6	7	8	9	10
Makespan	6	7	8	7	6	9	9	9	6	7

Standard deviation = 1.1442

Step 5. Update Required Makespan by Cmax at Phase 1 and plus a satisfy adjustment coefficient δ.

$$RequiredC_{max} + \delta = 8 + 2 = 10$$

We set the satisfy adjustment coefficient $\delta = 2$ and mean that even with a required probability applied to an expected Cmax ($RequiredC_{max} + \delta$) that is still less than $RequiredC_{max}$ and event duration. This expected Cmax in this case can be understood as the optimal Cmax reached for the original plan starts to run after this event duration.

We follow **Stage 2.** Critical Path Analysis and **Stage 3.** Mode selection to calculate the Required Cmax. The result is Required Cmax = 8 with parameters in Table 13. Note that we update the activities with historical data but still respect the changed event happens at activity B.

Table 13. Duration Time of updated activities with historical duration data and the changed event

Activity	Duration Mode 1	Duration Mode 2	Min d_{jm}	Max d_{jm}	ES_j	EF_j	LS_j	LF_j
Source (S)	0	0	0	0	0	0	4	4
A	2	2	2	2	0	2	4	6
B	3	2	2	3	2	4	6	8
C	2	3	2	3	2	4	6	8
D	2	2	2	2	4	6	8	10
Target (T)	0	0	0	0	8	8	12	12

To find new project schedule with minimize Overall Cost objective under this uncertain change, we establish the constraint (7) below:

- Required probability $= 0.95 \rightarrow$ Zalpha $= 1.645$
- Standard deviation of Cj $= 1,1442$
- Required makespan = Required Cmax + Delta $= 8 + 2 = 10$

The parameters are providing for cost of nonrenewable resource in each mode in Table 14 and Table 15.

Table 14. Cost of nonrenewable resource in mode 1

Activity	Resource Type	Resource 1	Unit Cost (r1)	Resource 2	Unit Cost (r2)	Total cost
Source (S)	Nonrenewable	0	0	0	0	0
A	Nonrenewable	2	1	3	2	8
B	Nonrenewable	4	2	3	3	17
C	Nonrenewable	3	2	2	1	8
D	Nonrenewable	2	3	4	2	14
Target (T)	Nonrenewable	0	0	0	0	0

Table 15. Cost of nonrenewable resource in mode 2

Activity	Resource Type	Resource 1	Unit Cost (r1)	Resource 2	Unit Cost (r2)	Total cost
Source (S)	Nonrenewable	0	0	0	0	0
A	Nonrenewable	3	1	6	2	15
B	Nonrenewable	6	2	6	3	30
C	Nonrenewable	2	2	4	2	12
D	Nonrenewable	4	1	2	1	6
Target (T)	Nonrenewable	0	0	0	0	0

Each activity in each mode needs a number of renewable resource quantity with unit cost respectively (Table 16).

Table 16. Cost of renewable resource in mode 1 and mode 2

Activity	Resource Type	Mode 1			Mode 2		
		Resource 1	Unit Cost	Total cost	Resource 1	Unit Cost	Total cost
Source (S)	Renewable	0	0	0	0	0	0
A	Renewable	3	2	6	4	2	8
B	Renewable	2	1	2	3	4	12
C	Renewable	1	2	2	3	6	18
D	Renewable	2	3	6	1	4	4
Target (T)	Renewable	0	0	0	0	0	0

We have the result of optimal overall cost = 53, a new makespan = 8 and chosen activities by sequence below (Table 17).

After applying the shift operation to predict the new makespan, we find that the project can be completed at 8 and maintain the optimal overall cost at 53.

4 Validation

In this section, numerical experiments are conducted. We run the mathematical model at phase 2 by SOFPS algorithm formed by GA combined the novel shift operation with the original genetic algorithm. The results are observed at the same limited iteration. Unexpected event causes duration longer at 2-time units than the original plan makespan. The case study's parameters have been presented will be used to run SOFPS and the original genetic algorithm to validate the results.

Table 17. Result of Activity mode

Task	Start time	End time	Chosen Mode	Non- renewable resource 1	Non- renewable resource 2	Renewable resource
Source (S)	0	0	Mode 2	0	0	0
A	0	3	Mode 1	3	6	4
B	3	6	Mode 1	6	6	3
C	3	6	Mode 1	2	4	3
D	6	8	Mode 2	4	2	1
Target (T)	8	8	Mode 1	0	0	0

Case 1: Shift Operation with Historical data has higher upper bound than the current data.

We observe the results while increasing the iteration. Below charts are the results in shorter iteration between GA applied Shift Operation and Original GA which show positively the average percentage of Cmax Gap of −3.22% and Cost Gap of 5.09% (Figs. 5 and 6).

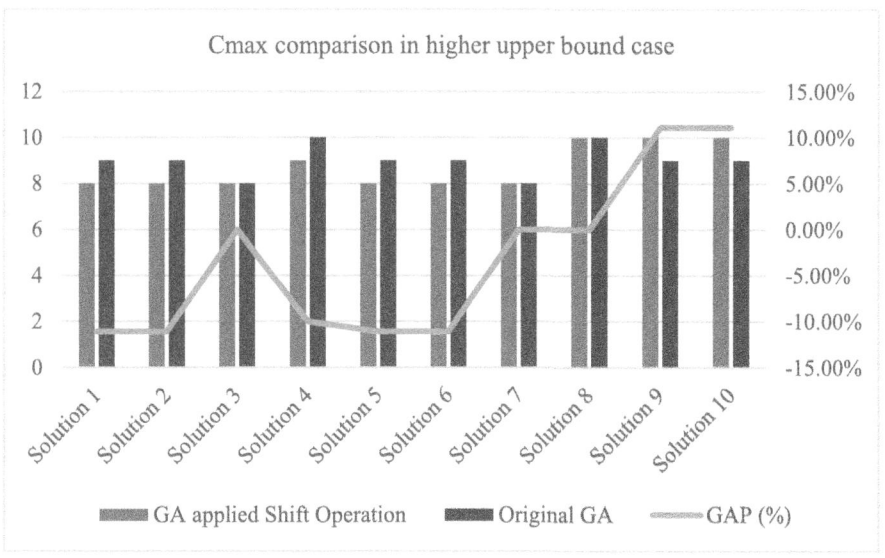

Fig. 5. Comparison Results of Cmax in higher upper bound case (Low Iteration)

We increase number of iteration and observe the result in longer Iteration. Below charts are the results in longer iteration between GA applied Shift Operation and Original GA which show positively the average percentage of Cmax Gap of −8.44% and Cost Gap of 0.00% (Figs. 7 and 8).

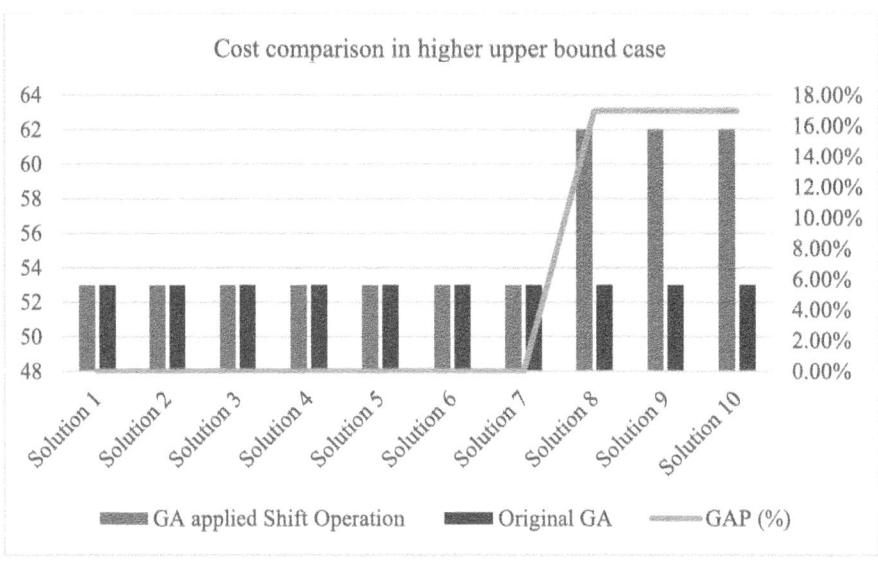

Fig. 6. Comparison Results of Cost in higher upper bound case (Low Iteration)

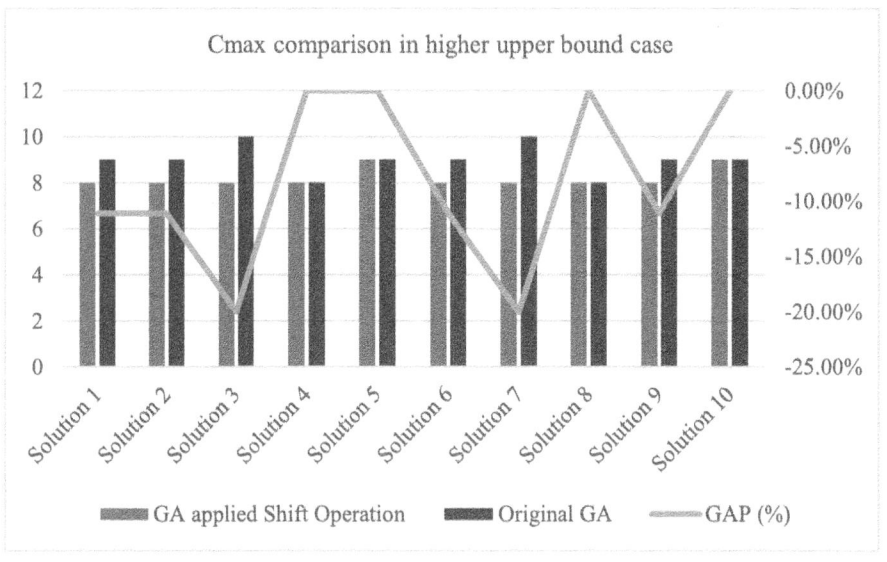

Fig. 7. Comparison Results of Cmax in higher upper bound case (High Iteration)

We observe that more iteration will shorten GAP between 2 algorithms. The GAP of makespan is decreased when COST is optimized by increasing iteration (Table 18).

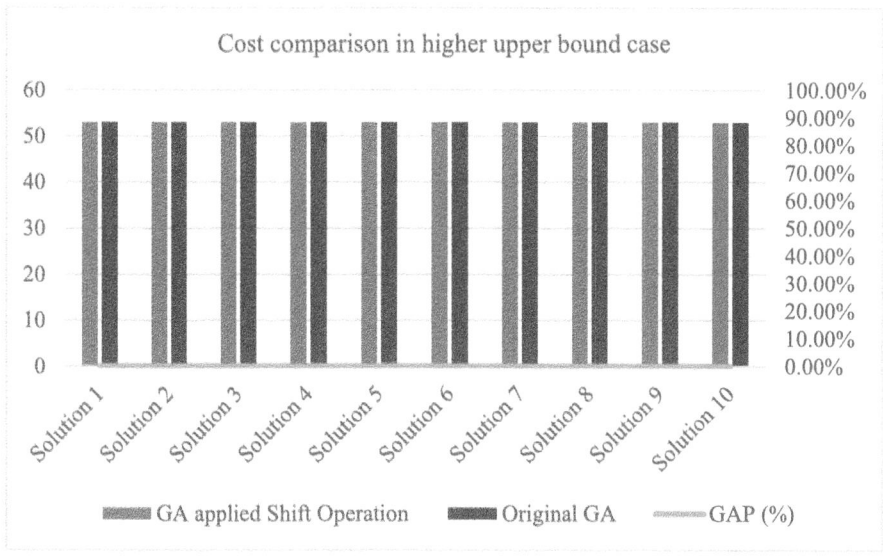

Fig. 8. Comparison Results of Cost in higher upper bound case (High Iteration)

Table 18. Delta observation

Running time (second)			
	Delta = 3	Delta = 4	GAP (%)
SOFPS	585.8538616	50.23558736	1066%

We observe that when delta is changed, either increasing or decreasing it the speed of running time will be affected. The more strictly delta or when delta decreases to 0, running time will be longer. So, we can get better solutions by decreasing delta but need to consider running time.

Case 2: Shift Operation with Historical data has lower upper bound than the current data

We observe the results while increasing the iteration. Below charts are the results in shorter iteration between GA applied Shift Operation and Original GA which show positively the average percentage of Cmax Gap of –24.58% and Cost Gap of 0.00% (Figs. 9 and 10).

We increase number of iteration and observe the results in longer Iteration. Below charts are the results in longer iteration between GA applied Shift Operation and Original GA which show positively the average percentage of Cmax Gap of –23.75% and Cost Gap of −1.45% (Figs. 11 and 12).

We observe that when historical data has lower upper bound than current data, the result is better than the higher case. The more iteration, the faster convergence to the optimal cost will be (Table 19).

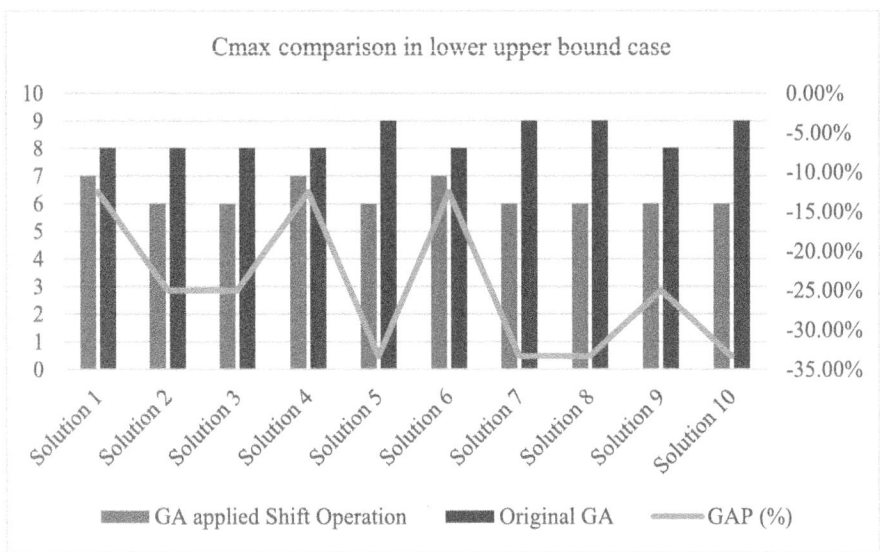

Fig. 9. Comparison Results of Cmax in lower upper bound case (Low Iteration)

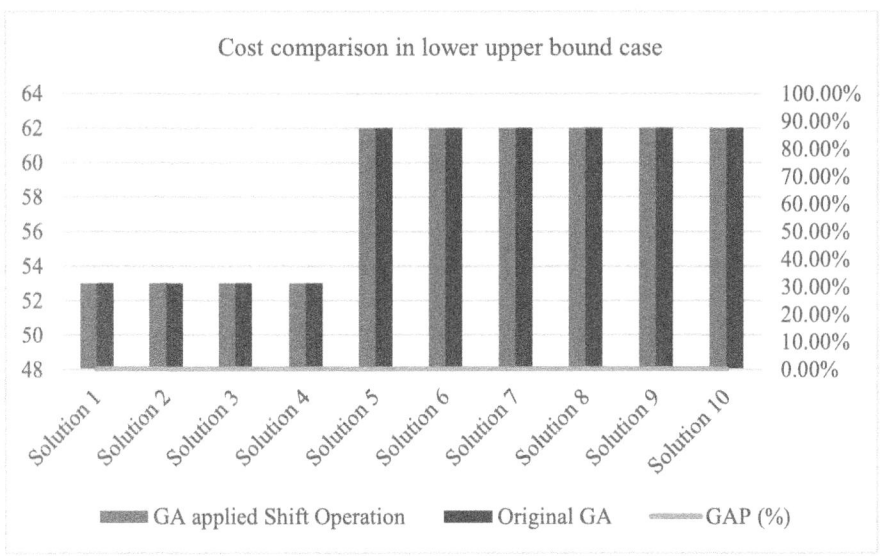

Fig. 10. Comparison Results of Cost in lower upper bound case (Low Iteration)

Running time by SOFPS is still better than the original GA. The same observation is applied for Delta. We still can increase Delta to quickly get the results.

When there is an update event on resource, here we refer to the task duration. For the current project schedule, all the tasks followed by the updated task will be affected and the makespan will be longer than the original one.

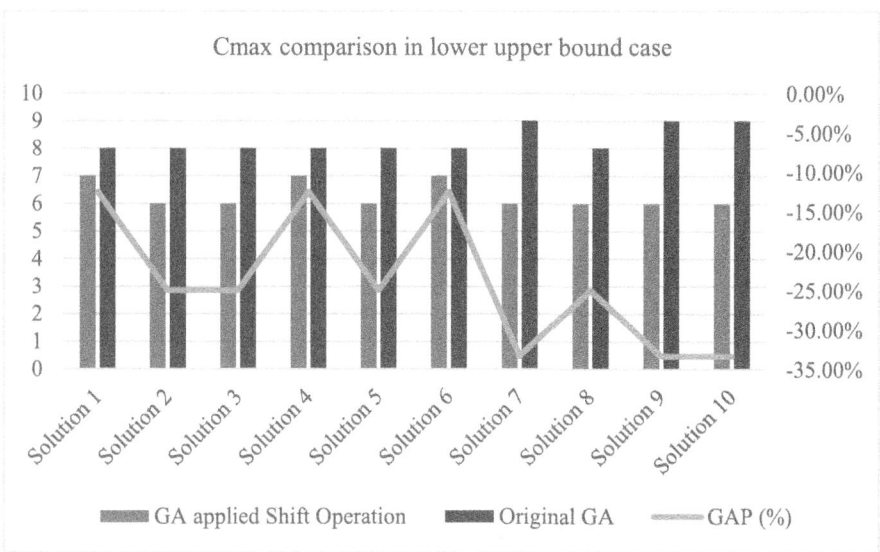

Fig. 11. Comparison Results of Cmax in lower upper bound case (High Iteration)

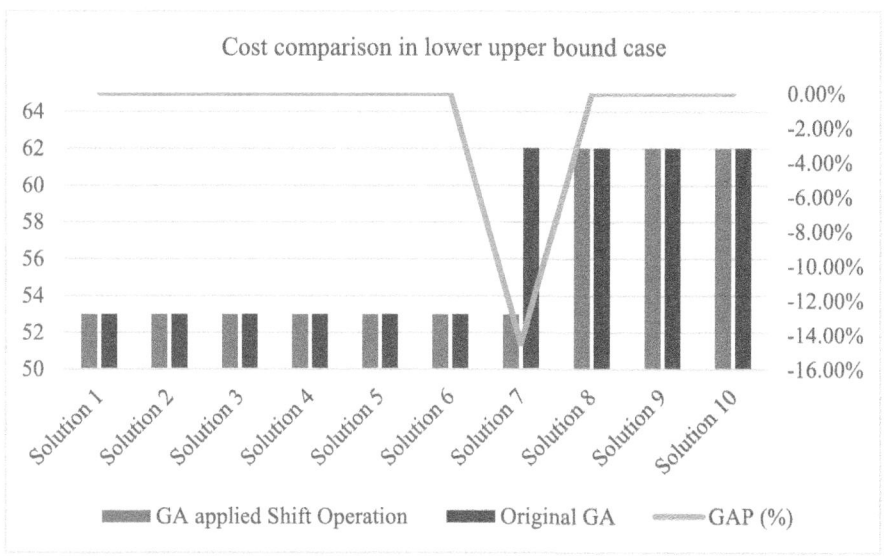

Fig. 12. Comparison Results of Cost in lower upper bound case (High Iteration)

In our effort, we try to apply the proposed shift operation flexible project scheduling to provide better project schedule in a sense of uncertain absorption. All the remaining tasks will be updated in a way that can predict new project schedule that minimize the overall cost, keep the makespan the same time with an expected probability. The result generated by SOFPS is better than generated by the original genetic algorithm. So, it is

Table 19. Delta observation

Running time (second)			
	Delta 1 = 2	Delta 2 = 4	GAP (%)
SOFPS	270.9166851	199.3333223	35.91%

quite promising for apply this to predict the result of project schedule under uncertainty environment.

5 Conclusions

In this paper, the Shift Operator Flexible Project Scheduling (SOFPS) algorithm is proposed to deal with the multi-mode resource constrained project scheduling (MMRCPSP) problem under uncertainty event. In the algorithm, a novel time-based robust shift operation is applied to deal with uncertainty events by utilizing historical data to calculate for each task. Chance constraint is established to ensure the result of makespan is not over a required Cmax with an expected threshold. When unexpected event does not happen yet, the first phase provides a minimal total makespan (Cmax) that the project schedule selects not only the appropriate mode but also determine the best resource type for each activity, considering renewable and non-renewable resources. When unexpected event happens, the second phase applies SOFPS to rearrange the activities to meet the required Cmax in an optimal cost objective. A very promising result is harvested when compares SOFPS with the original genetic algorithm. The result is better about 8% for the case that historical data has higher makespan upper bound than the current data and better about 20% for the case that historical data has lower makespan upper bound than the current data between the obtained project schedules and the original Genetic Algorithm

In future, some other new reactive approach and real-time update techniques could be developed to contribute to this operation to deal with uncertainty events.

Ackowledgement. This research is funded by Vietnam National University HoChiMinh City (VNU-HCM) under grant number B2023-28-06. This support is gratefully acknowledged.

References

1. Abido, M.A., Elazouni, A.: Modified multi-objective evolutionay programming algorithm for solving project scheduling problems. Expert Syst. Appl. **183**(115338), 1–10 (2021)
2. Balouka, N., Cohen, I.: A robust optimization approach for the multi-mode resource-constrained project scheduling problem. Eur. J. Oper. Res. **291**, 457–470 (2021)
3. Brcic, M., Katic, M., Hlupic, N.: Planning horizons based proactive rescheduling for stochastic resource- constrained project scheduling problems. Eur. J. Oper. Res. **273**, 58–66 (2019)
4. El-Abbasy, M.S., Elazouni, A., Zayed, T.: MOSCOPEA: multi-objective construction scheduling optimization using elitist non-dominated sorting generic algorithm. Autom. Construct. **71**, 153–170 (2016)

5. Luo, J., Vanhoucke, M., Coelho, J., Guo, W.: An efficient genetic programming approach to design priority rules for resource-constrained project scheduling problem. Expert Syst. Appl. **198**(116753), 1–20 (2022)
6. Ning, M., He, Z., Jia, T., Wang, N.: Metaheuristics for multi-mode cash flow balanced project scheduling with stochastic duration of activities. Autom. Constr. **81**, 224–233 (2017)
7. Ma, Z., Demeulemeester, E., He, Z., Wang, N.: A computational scheduling under two types of uncertain environments. Comput. Ind. Eng. **131**, 382–390 (2019)
8. Morin, P.A., Artigues, C., Hait, A.: Periodically aggregated resource-constrainted project scheduling problem. Eur. J. Ind. Eng. **11**(6), 792–817 (2017)
9. Morin, P.A., Artigues, C., Hait, A., Kis, T., Spieksma, F.C.R.: A project scheduling problem with periodically aggregated resource-constraints. Comput. Oper. Res. **141**(105688), 1–16 (2022)
10. Paraskevopoulos, D.C., Laporte, G., Repoussis, P.P.: Resource constrained routing and scheduling: review and research prospects. Eur. J. Oper. Res. **263**, 737–754 (2017)
11. Servranckx, T., Vanhoucke, M.: A tabu search procedure for the resource-constrained project scheduling problem with alternative subgraphs. Eur. J. Oper. Res. **273**, 841–860 (2019)
12. Servranckx, T., Vanhoucke, M.: Strategies for project scheduling with alternative subgraphs under uncertainty: similar and dissimilar sets of schedules. Eur. J. Oper. Res. **279**, 38–53 (2019)
13. Tian, J., Hao, X., Gen, M.: A hybird multi-objective EDA for robust resource constraint project scheduling with uncertainty. Comput. Ind. Eng. **130**, 317–326 (2019)
14. Tian, J., Hao, X., Murata, T.: Markov network based multi-objective EDA and its application for resource constrained project scheduling. IEEJ Trans. Electron. Inf. Syst. **136**(3), 290–298 (2016)
15. Wang, L., Zheng, X.L.: A knowledge-guided multi-objective fruit fly optimization algorithm for the multi- skill resource constrained project scheduling problem. Swarm Evol. Comput. **38**, 54–63 (2018)
16. Zahid, T., Agha, M.H., Schmidt, T.: Investigation of surrogate measures of robustness for project scheduling problems. Comput. Ind. Eng. **129**, 220–227 (2019)

Artificial Intelligence in Education: A Comprehensive Examination of Integration, Impact, and Future Implications

Kishori Kasat[1]([⊠]) [iD], Uday Sinha[1] [iD], Sakshi Juneja[1], Anurupa Ghatge[1] [iD], Nikhil Thorat[1] [iD], and Naim Shaikh[2] [iD]

[1] Symbiosis School for Liberal Arts (SSLA), Symbiosis International (Deemed University), Pune 400014, MH, India
kishori.kasat@ssla.edu.in

[2] Global Business School and Research Centre, Dr. D. Y. Patil Vidyapeeth (Deemed to be University), Pune, India

Abstract. The "Fourth Industrial Revolution" integrates digital, physical, and biological technologies and has a critical impact on industries and economies across the globe. In recent years, "Artificial Intelligence (AI)" technologies have become integral to the education sector too, promising to revolutionize the way students learn and teachers instruct. The integration of artificial intelligence (AI) in education is rapidly redefining the educational landscape, presenting a myriad of opportunities and challenges. The purpose of this research study is to examine the current state of AI integration in education, its profound impact on teaching and learning processes, and the future implications and challenges it holds for the education sector. A comprehensive research methodology using literature reviews and surveys with educators from the field is undertaken to provide deeper insights into the theme. The study finds a profound impact of integrating AI in education, providing educators with data-driven insights that inform their teaching strategies, leading to improved educational outcomes. AI-driven solutions offer the potential for personalized learning experiences, data-driven decision-making, and more efficient administrative processes in educational institutions. However, challenges such as privacy concerns, bias in AI algorithms, and issues related to data security need more vigilance. In conclusion, this research study contributes to the growing body of knowledge on AI in education, providing valuable insights for educators, policymakers, and researchers, highlighting the potential benefits and challenges associated with the integration of AI in education, and offering recommendations for its responsible and effective implementation.

Keywords: Artificial Intelligence (AI) · Challenges · Education · Ethical · Implications · Learning Experiences · Opportunities · Survey · Teaching

A. Mirzazadeh et al. (Eds.): ODSIE 2023, CCIS 2205, pp. 182–198, 2025.
https://doi.org/10.1007/978-3-031-81458-7_11

1   Introduction

1.1   Artificial Intelligence

"Artificial Intelligence" (AI) refers to the capabilities of modern computer systems that have evolved to perform several tasks, such as learning, problem-solving, reasoning, perception, language understanding, decision-making, and several others that typically require some human intelligence. While narrow (weak) AI systems can excel in performing specific functions (for example, voice recognition, playing online games, image recognition, etc.) conversely a general (strong) AI can perform complex and challenging tasks (for example, understanding, learning, and apply knowledge across diverse tasks which are similar to human intelligence). Currently, such AI based on their functionalities are evident in the form of machine learning (ML), natural language processing (NLP), expert systems, neural networks, deep learning, robotics, computer vision, and related systems. Artificial Intelligence (AI) therefore has potential and can have a significant impact in the education domain also introducing innovative approaches to teaching-learning, research, and administrative processes. Historical evidence suggests that AI is being utilized in the education domain in multiple ways, such as personalized learning in the form of adaptive learning and intelligent tutoring systems, automated assessments and grading, smart e-content, creation, gamification, online learning, interactive learning, virtual classrooms, language learning, professional development programs, data analysis, research, predictive analytics, and host of automated administrative tasks.

1.2   Integration of AI in Educational Systems

Integration of AI in educational systems can have several implications for educators, learners, and administrative governance systems. AI can enable personalized learning experiences by tailoring relevant educational content to individual student requirements which leads to personalized learning. Students have the opportunity to learn at their own pace, enjoy targeted support for difficult or complex components, and can also be challenged in areas where they excel. Automated assessment and grading systems powered by AI can save valuable time for educators allowing focus on rendering qualitative feedback, addressing individual learning requests, and higher engagement with students leading to efficient assessments and grading outcomes. Enhanced teaching practices can be achieved by teachers using AI tools to derive data-driven insights into student performance so that educators can identify critical areas for improvement, adjust their teaching pedagogies, and offer targeted training interventions to support student learning. AI can enhance the capabilities of virtual classrooms and online learning platforms through real-time language translation, interactive simulations, and AI-driven support contributing to a much more immersive and accessible learning experience. Streamlining administrative tasks using AI support is beneficial for educational institutions to achieve improved efficiency in enrollment, record-keeping, scheduling, and resource allocations, thus allowing institutions to focus on providing quality education. Predictive analytics can highlight students at risk of academic challenges. Early intervention strategies can be implemented to ensure additional support and resources, reducing any likelihood of dropout. AI promisingly contributes to making education systems more accessible and inclusive. Automatic transcription, translation services, and personalized accessibility features assist students with diverse requirements and can establish inclusive education. AI supports personalized professional development for teachers where they can access targeted training programs based on their needs, enhancing their teaching effectiveness in the classroom environment. Researchers can benefit from using AI to analyze large datasets for research and data analytics purposes. It can assist in identifying trends, patterns, and insights to inform educational policies and practices. Incorporating gamification elements and interactive content can increase student motivation and participation. AI contributes in a large way to create interactive and engaging learning experiences for all. A systematic review study [1] presents four key AI application areas for educational support services, administrative, and institutional activities, such as (a) prediction and profiling, (b) evaluation and assessment, (c) personalization and adaptive systems, and (d) intelligent systems for tutoring.

It is pertinent and crucial to approach the idea of integrating AI in educational systems with an ethical and thoughtful mindset. Conscious attention towards addressing serious challenges such as data privacy, biases in algorithms, and ensuring equitable access to several AI-driven educational resources will be indispensable to realizing the full potential of AI in shaping the future of education. AI will continue to impact education systems to refine and enhance personalized learning experiences, adapting content, pacing, and assessments to the individual user needs and learning styles of each student. AI tools will play a crucial role in creating and recommending smart educational content, leveraging advanced algorithms to customize materials based on student progress and preferences. AI-driven technologies, including virtual classrooms and online learning platforms, will positively contribute to making quality education more accessible to learners globally, overcoming geographical and socio-economic barriers. The continuous development of AI-driven learning assistants is becoming more sophisticated, providing students with real-time support, guidance, and feedback, both inside and outside of the classrooms. AI will assist teachers in various aspects of their regular work such as automating administrative tasks, providing data-driven insights into student performance, and allowing them to focus more on the human aspects of education. In recent times, AI has been integrated with augmented and virtual reality technologies to create immersive and interactive learning experiences, allowing students to engage with educational content in 3D spaces. AI will play a role in identifying and addressing gaps in students' skills, aligning education with the demands of the rapidly evolving labor market and technological ecosystems. There will be an increased focus on integrating discussions about ethical considerations and digital citizenship into the educational curriculum to ensure that students develop a responsible and informed approach to technology. AI-driven adaptive learning platforms will become integral to continuous learning and reskilling efforts, supporting individuals throughout their careers as they adapt to evolving job requirements. Educational institutions and policymakers will increasingly rely on AI-driven data analytics to make informed decisions about curriculum design, resource allocation, and policy formulation. The future may see greater collaboration between humans and AI systems, with AI serving as a complementary tool to human educators rather than a replacement, fostering a blended and collaborative learning environment. AI will assist in creating customized learning paths for students, allowing them to pursue their interests and strengths while still meeting educational standards. AI in educational settings could pose transformative impacts necessitating AI literacy, enhanced critical thinking skills, and proficiencies in prompt engineering [2]. In a recent systematic review of research to understand the role of AI in K-12 educational settings, the study [3] finds that students can experience an augmented learning environment through better curriculum design, evaluations, enhanced school operations, equity, and safety. Most teachers used AI tools including machine learning model demonstrations, behavior prediction, and academic performances.

Research Questions. From the above-discussed scenario about artificial intelligence in the education domain and our study objectives, this piece of investigation attempts to address the following research questions:

RQ1: What is the current state of AI integration in the case of an educational organization?
RQ2: How does the integration of AI in education impact teaching-learning processes?
RQ3: What are the key implications of such use of AI in Education?

Flow of the Study. Further, the flow of the study includes a literature review in Sect. 2, followed by the research methodology described in Sect. 3, data collected and analyzed in Sect. 4, managerial implications in Sect. 5, and conclusions drawn in Sect. 6.

2 Literature Review

2.1 Artificial Intelligence in Education

An in-depth review of past studies that investigate the impact of AI on teaching and learning, adoption trends, and emerging technologies in education was done. Significant developments can be seen in the integration of AI in education over the past twenty-five years. Evolutionary processes focus on classroom practices, collaborative tasks with educators and learners, and a few diversified domains, while revolutionary in the use of embedded technologies for supporting learners' everyday requirements, goals, practices, and communities [4]. Newer advancements in the field of robotics hold better promises for the usage of sensor gadgets to monitor human actions and surroundings. A study [5] also posits that future academic ecosystems will see educational 'cobots' that assist educators in smart classroom settings and the use of sensors that support learning. The application of Artificial Intelligence (AI) in higher education in current times across the globe is one of the emerging fields of educational technology. Rapid advancements in computing technology aid AI in educational settings and applications to enable teaching, learning, and decision-making. It can simulate human intelligence towards making better inferences, predictions, judgments, personalized guidance, support, and feedback to learners conversely helping educators or policy-makers in making effective informed decisions [6]. Future implications in the integration of AI for educational settings are associated with individualized learning, assessments, and precision educational requirements. As advanced data-driven models' approach and grow rapidly toward development, algorithm validation, and interpretations, the conclusions derived from educational analytics need to be applied with more caution. At the policy level, it must support lifelong learning, teacher learning, and privacy concerns for personal data [7]. In a recent study [8], it characterizes AI in education into three important paradigms, such as (a) AI-directed (learner-as-a-recipient), (b) AI-supported (learner-as-collaborator), and (c) AI-empowered (learner-as-a-leader). In another study [9] on how AI was being applied in the education sector, the content analysis revealed four key research trends employing swarm intelligence, the Internet of Things (IoT), deep learning, and neuroscience technologies. Most of the studies could be classified into (a) an application layer for reasoning, feedback, or adaptive learning, (b) a development layer for matching, classification, recommendation, or deep learning, and (c) an integration layer for role-playing, affection computing, gamification, or immersive learning. It can also be conveniently understood that the integration of Artificial Intelligence in Education is here to stay and several past research studies support its useful adoption and benefits from different techniques, such as medical informatics and education [10]; supporting collaborative learning and e-discussions [11]; chatbot applications [12]; emotionally aware AI smart classrooms [13]; academic performance predictions [14]; STEM education [15]; academic and administrative roles [16]; and several other educational domains.

2.2 Familiarity with AI Tools

Appreciating the importance of the education domain, educators and learners are invariably expected to be aware and familiar with different artificial intelligence tools and technologies, especially those that are relevant and applied in the education sector. These tools are needed to enhance teaching and learning experiences, provide customized educational content, and streamline most of the administrative tasks. The level of familiarity with various AI tools would include the efficiencies of (a) learning management systems (LMS), (b) adaptive learning systems, (c) chatbots and virtual assistants, (d) auto-grading applications, (e) natural language processing (NLP) applications, (f) extended realities (including augmented, virtual, and mixed), (g) academic data analytics, (h) e-content creation and recommendations, (i) assessments and evaluations, (j) intelligent tutoring, and (k) multiple continuous learning platforms.

According to a survey of teachers from various Universities in the Russian Federation, the study suggests that the degree of integrating various AI tools into the educational pedagogical processes depends on the level of awareness, readiness, and best practices of such usage by the teachers in their daily professional activities. However, these faculty also lack a systematic understanding of the organizational, teaching, and learning potential of AI tools. It is found that the practical application of AI tools in their pedagogies is limited to the use of specific AI technologies in teaching only specific subjects or disciplines [17].

2.3 Use of AI Educational Platforms

It has been ascertained that AI is being extensively adopted and used by educational systems in different forms transitioning from simple computer-assisted technologies to web-based and virtual (online) intelligent systems with embedded technology, use of humanoid robots, web-based chatbots, and several other technologies [18]. Emerging technologies such as the mobile internet, big data technologies, cloud computing, and related breakthroughs have transformed the educational sector providing advanced AI-enabled adaptive learning systems that are gaining more traction due to their ability to deliver quality learning content aligning with the individual learner needs [19]. Artificial intelligence-enabled educational platforms have found common use in (a) adaptive learning, (b) personalized learning paths, (c) data analytics providing performance insights, (d) language processing apps, (e) augmented and virtual reality experiences, (f) predictive analytics on student performance, (g) chatbots and virtual assistants, (h) continuous learning and professional development, (i) e-content creation, and (j) intelligent tutoring.

2.4 Perceived Benefits, Challenges, and Drawbacks

It is observed that while AI in education at its core is desired to support better student learning, experiences from several other AI domains indicate insufficiency in ethical intentions involved in it. Explicit issues relating to accountability, fairness, biases, transparency, autonomy, and inclusion are challenging concerns that need to be addressed to mitigate unintended consequences [20].

2.5 Role of Teacher (Educator), Background, and Experience

The role of a teacher in the effective and efficient integration of AI in education would depend upon the level of awareness, adoption, educational background, work experiences, and synergies between technology and human interactions. Teachers will need to harness the potential of artificial intelligence techniques for supporting collaborative learning and e-discussions with the learners [11].

2.6 Implications of AI in Education

One of the serious concerns and implications for the future of education systems is the growing shortage of global teachers. AI in education demands suitable designing of robotic personalities that can assume independent teacher roles and provide adequate economic benefits, social interactions, affect, and superior instructional deliveries [21]. The integration of AI in education can have several implications, both positive and challenging. Key positive implications would include (a) efficiencies and automation, (b) an adaptive learning environment, (c) personalized/customized learning facilities, (d) enhanced engagement among users, (e) data-driven decision-making and insights, (f) global access, (g) inclusive learning ecosystems, and (h) continuous learning. It could be challenging in terms of (a) student-teacher dynamics, (b) access, (c) educator preparedness, (d) biases in the AI algorithm, (e) security and ethical concerns, and (f) possibilities of job displacement for teachers. Continuous monitoring and assessments will be needed to identify and understand the various implications of integrating AI in education systems so that timely and corrective action can be initiated to attain positive outcomes.

A deeper review of past research literature on the topic of interest has provided several insights on key dimensions that affect and have serious implications for harnessing the potential of AI-integrated education systems. A conceptual framework for our study is developed using these constructs which is depicted below (see Fig. 1).

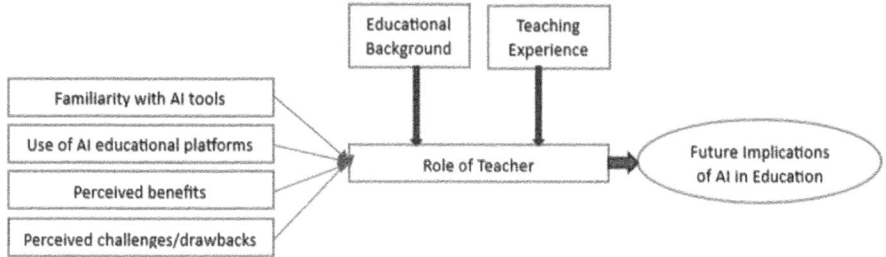

Fig. 1. Conceptual Framework for AI in Education. (Source: Authors).

3 Research Methodology

A suitable research methodology was adopted aligning with the conceptual framework for the study developed earlier. Delving deeper into the background of the study and the problem statement, three research questions were formulated. A systematic review of existing literature based on the topic of interest provided useful constructs to develop our conceptual framework. Based on the research questions and the constructs of the conceptual framework, a semi-structured questionnaire was developed to survey and elicit responses from educators, administrators, and other stakeholders using AI tools in the education domain. The questionnaire was designed using the Google Forms application which provides a seamless online dissemination, collection, and analysis of responses. The questionnaire was circulated to all UG/PG teachers who taught Management, Computer Sciences, Media and Communication, Humanities, and other subjects. Out of a total of 97 teacher respondents, only 51 completed responses (representing 53%) were selected and adequate for final analysis and synthesis. The collected responses were assessed to understand the level of AI adoption, perceptions about its impact and implication for users, and feedback on specific AI applications. Findings from the study are discussed in the results and discussion section, and conclusions are drawn from the entire study. The focus of this study is to understand the perception of AI in education by academicians. For this purpose, a primary survey was conducted using a structured questionnaire. Various questions based on potential benefits (Table 1), challenges, and drawbacks (Table 2) were collected using a Five-point Likert Scale, and a few open-ended questions were asked from various academicians from different disciplines and different universities teaching both at undergraduate and postgraduate levels.

4 Data Analysis

An online form was created using a tool for data collection. It was distributed to prospective respondents in the age group of 25–70 years. In the present study, a total of 51 respondents participated. Descriptive statistical analysis was used to discuss and results.

Table 1. Potential benefits measurement.

Not at all helpful	Slightly helpful	Moderately helpful	Very helpful	Extremely helpful
1	2	3	4	5

Source: Authors

Table 2. Drawbacks/Challenges measurement.

Not at all concerned	Slightly concerned	Somewhat concerned	Moderately concerned	Extremely concerned
1	2	3	4	5

Source: Authors

4.1 Descriptive Data

A descriptive statistical analysis technique was used to analyze and discuss the results. The following Table 3 provides a summary of the demographics of the respondents.

Table 3. Demographics of Respondents.

Dimension	Age (In years)	Count (Percentage) n = 51
Age	25–40	18 (35%)
	40–75	33 (65%)
Gender	Male	17 (33%)
	Female	34 (67%)
Organization type	Private	31 (61%)
	Public	5 (10%)
	Autonomous (Others)	15 (29%)
Teaching Level	Undergraduate	35 (69%)
	Postgraduate	16 (31%)
Disciplines	Management	9 (18%)
	Computer Science	5 (10%)
	Media and Communication	2 (4%)
	Humanities	26 (51%)
	Others	9 (18%)
Familiarity with AI	All	51 (100%)
Experienced and observed AI tools in the teaching-learning process	Yes	31 (61%)

(*continued*)

Table 3. (*continued*)

Dimension	Age (In years)	Count (Percentage) n = 51
	No	19 (37%)
AI can replace human teachers	Yes	2 (4%)
	No	40 (78%)
	Unsure	9 (18%)

Source: Survey responses from the online questionnaire (Collated by the Authors)

Table 3 suggests that all the respondents were familiar with Artificial Intelligence. Upon asking about their understanding of AI in education it was found that the understanding varied greatly with the concept, usage in academics, and for administrative purposes. Few respondents described AI as an algorithm-driven silicon-based interface (sometimes reductively called 'Tool') that is meant to facilitate, enhance, and fill gaps in uniquely human capacities as educators and learners. Some described AI as a tool that is empowering and enabling to a large degree when thought about in the context of education. It would empower academic pursuits in the context of learning and research and also enable educational management to become more efficient and optimize various processes involved. It has a has potential to transform content creation, delivery of knowledge, and learning. Few respondents described AI in the context of its benefits to students. With the rise of Artificial Intelligence in education, there are many different ways it is being used to help students learn. Chatbots are one example of AI educational apps that students might use soon. These are being increasingly implemented in classrooms where kids use iPads or laptops to chat with bots designed to help them understand specific topics such as math or reading comprehension. It helps students explore their creativity. However, few respondents showed apprehensions and fear toward AI as it can be considered as software not producing accurate information and can be potentially problematic. There is also a concern about the loss of a larger context of the information, which AI tools don't necessarily provide. However, despite the concerns expressed, 78% of the respondents believe that AI may not be able to replace the human teacher.

4.2 Potential Benefits

The potential benefits of adopting AI in education were explored and analyzed further as highlighted in Table 4.

Table 4. Potential benefits of AI in education.

Potential Benefits	Mean	Standard Deviation
Providing personalized learning experiences for students	3.2	1.1

(*continued*)

Table 4. (*continued*)

Potential Benefits	Mean	Standard Deviation
Time savings in planning of educational/training activities	3.5	1
Seamless content creation with good quality	3.2	1.2
Providing Real-time feedback to students	3.5	1.1
Enhanced administrative tasks (grading/assessments, announcements, scheduling, etc.)	3.7	1.1
Access to a wider range of learning resources	3.6	1
Automation of routine tasks, allowing more time for creative teaching	3.6	1
Improved accessibility for differently-abled students	3.6	1.1
Automatic student attendance monitoring	3.8	1.2

Source: Survey responses from the online questionnaire (Analyzed by the Authors)

The potential benefits of AI in the education dataset were further analyzed streamwise as highlighted in Table 5.

Table 5. Potential benefits of AI in education (Streamwise).

Potential Benefits	Mean Scores (Streamwise)				
	Management	Computer Science	Media and Communication	Humanities	Others
Providing Personalized learning experiences for students	3.3	4	2.6	3.2	3.2
Time savings in planning of educational/training activities	3.4	4.4	3.4	3.4	3.5
Seamless content creation with good quality	3.4	3.2	3.1	3.2	3.3
Providing Real-time feedback to students	3.7	4	3.3	3.5	3.6

(*continued*)

Table 5. (*continued*)

Potential Benefits	Mean Scores (Streamwise)				
	Management	Computer Science	Media and Communication	Humanities	Others
Enhanced administrative tasks (grading/assessments, announcements, scheduling, etc.)	3.2	4.2	3.8	3.7	3.8
Access to a wider range of learning resources	3.4	4	4.0	3.6	3.7
Automation of routine tasks, allowing more time for creative teaching	3.4	3.8	3.6	3.6	3.7
Improved accessibility for differently-abled students	3.5	3.4	3.6	3.6	3.6
Automatic student attendance monitoring	3.6	3.8	4.0	3.8	3.8

Source: Survey responses from the online questionnaire (Analyzed by the Authors)

Tables 4 and 5 indicate the potential benefits of AI in education as perceived by the respondents and measured on the five-point Likert scale. It is observed that the respondents collectively perceive AI as a very helpful tool with an average greater than 3.5 for most of the benefits. It showed that the respondents feel that AI can be helpful in time-saving in planning educational activities. It can enhance administrative tasks such as grading assessments, scheduling, automating routine tasks, and monitoring student attendance. It can also provide access to students with a wider range of learning resources.

Next, the potential benefits of AI in the education dataset were further analyzed teaching-level-wise as highlighted in Table 6 below.

The perception of AI for the academicians who teach at the level of undergraduate and postgraduate and also from public, private, and autonomous universities. The potential benefits perceived by the teachers on average do not differ much from the undergraduate and postgraduate levels.

Table 6. Potential benefits of AI in education (Teaching level-wise)

Potential Benefits	Mean Scores				
	Undergraduate	Postgraduate	Private	Public	Autonomous
Providing Personalized learning experiences for students	3.4	3.5	3.3	2.2	3.2
Time savings in the planning of educational/training activities	3.3	2.8	3.4	2.8	3.7
Seamless content creation with good quality	3.4	3.6	3.1	2.4	3.5
Providing Real-time feedback to students	3.6	3.8	3.5	3.2	3.8
Enhanced administrative tasks (grading/assessments, announcements, scheduling, etc.)	3.6	3.6	3.5	3.4	4.0
Access to a wider range of learning resources	3.6	3.6	3.5	2.8	4.0
Automation of routine tasks, allowing more time for creative teaching	3.4	3.8	3.5	3.0	3.7
Improved accessibility for differently-abled students	3.4	3.6	3.4	3.2	4.2
Automatic student attendance monitoring	3.8	3.6	3.6	3.0	4.3

Source: *Survey Responses from Questionnaire*

4.3   Challenges/Drawbacks

The potential challenges/drawbacks of using AI in education are analyzed in Tables 7 and 8 below. The responses indicate concerns that academicians have towards AI in education with an average greater than 3.5. The results showed that collectively respondents feel it is extremely concerning that AI can lead to loss of human interaction, privacy and data security issues, ethical concerns in decision-making by AI, and lack of human touch in the educator-learner interaction.

The above dataset exhibits if there is any difference in the average perception of the academician from across disciplines. It is observed that academician from a computer science background perceive AI to be more beneficial/very helpful and issues are less concerning for them whereas the perception does not vary widely across faculty from

Table 7. Potential challenges/drawbacks of AI in education.

Potential Challenges/drawbacks	Mean	Standard Deviation
Loss of human interaction in education	3.2	1.1
Privacy and data security concerns	3.5	1
Dependence on technology leads to skill gaps	3.2	1.2
Difficulty in adapting AI to different learning styles	3.5	1.1
Ethical (trust) concerns in decision-making by AI	3.7	1.1
Limited availability of quality AI tools	3.6	1
Lack of human touch in educator-learner academic interactions	3.6	1
Inadequacy of training about AI tools and usage	3.6	1.1
Poor mindset/attitude in embracing AI-assisted teaching-learning	3.8	1.2

Source: Survey responses from the online questionnaire (Analyzed by the Authors)

the other disciplines like management, humanities, and media find AI to be moderately helpful and are more concerning towards AI. The responses indicate that the role of the teacher will change with the integration of AI in education as teachers will be more considered facilitators, mentors, and advisors. Though AI will change the way of educating young minds the basic role of teacher cannot be replaced. Teachers might not be required for the creation or delivery of the content. Their role will be more prominent in guiding the students in research, innovation, and application of knowledge. Teachers will increasingly be called upon to introduce students to ways of determining the conditions of validity and reliability of information rather than simply delivering 'content'. They will also be expected to develop more pedagogically sound ways of delivering information: ways, perhaps, that can enhance retention, or simplify comprehension. We will also have to start focusing more sharply in classes on equipping students with higher-order mental operations such as creation, analysis, and classification and not merely identification. Questions of episteme are likely to become more important than raw information.

Next, the potential benefits of AI in the education dataset were further analyzed teaching-level-wise as highlighted in Table 9 below.

From the above analysis, however, the concerns are much reflected at the postgraduate level as far as privacy, data security, ethical issues, and lack of human touch are concerned. It is also observed that the public universities on average recognize lesser potential benefits and greater concerns as compared to the private and autonomous universities. This could be due to the lesser inclination and lack of technological advancement/awareness and reliability of traditional methods is more in public universities.

Table 8. Potential challenges/drawbacks of AI in education (Streamwise).

Potential Challenges/drawbacks	Mean Scores (Streamwise)				
	Management	Computer Science	Media and Communication	Humanities	Others
Loss of human interaction in education	3.2	3.2	3.8	3.7	3.6
Privacy and data security concerns	4.2	3.8	4.0	4.2	4.1
Dependence on technology leads to skill gaps	4.0	2.6	4.0	3.9	3.8
Difficulty in adapting AI to different learning styles	3.5	2.8	3.4	3.4	3.3
Ethical (trust) concerns in decision-making by AI	4.2	3.6	4.4	4.2	4.1
Limited availability of quality AI tools	3.7	3	3.4	3.8	3.7
Lack of human touch in educator-learner academic interactions	3.6	3.4	4.0	4.1	4.0
Inadequacy of training about AI tools and usage	3.6	3.4	4.0	3.7	3.6
Poor mindset/attitude in embracing AI-assisted teaching-learning	3.1	3.4	3.4	3.7	3.5

Source: Survey responses from the online questionnaire (Analyzed by the Authors)

5   Managerial Implications

5.1   Implications for Stakeholders in Education

The study has serious managerial implications for academicians, students, practitioners, administrators, and other stakeholders who engage AI in education. The strength of using AI in education offers several benefits, such as automation, efficiency, personalization, data-driven insights, global access, early intervention and support using data-driven insights, inclusivity, continuous learning, and opportunities for reskilling of users. It also inherits several challenges relating to ethical and privacy concerns, issues of biases, high implementation costs, need for relevant training on AI technology to all stakeholders, and attracts higher dependencies on the AI technology. It has numerous opportunities that

Table 9. Potential challenges/drawbacks of AI in education (Teaching levels).

Drawbacks/Challenges	Mean Scores				
	Undergraduate	Postgraduate	Private	Public	Autonomous
Loss of human interaction in education	3.6	3.6	3.7	3.4	3.8
Privacy and data security concerns	3.8	4.3	4.1	5.0	3.8
Dependence on technology leads to skill gaps	3.9	4	3.7	4.4	4.3
Difficulty in adapting AI to different learning styles	3.4	3.2	3.4	3.6	3.5
Ethical (trust) concerns in decision-making by AI	4.1	4.2	4.2	5.0	4.2
Limited availability of quality AI tools	3.6	3.9	3.7	3.8	3.9
Lack of human touch in educator-learner academic interactions	3.9	4.3	4.1	4.0	4.3
Inadequacy of training about AI tools and usage	3.7	3.5	3.5	3.4	4.3
Poor mindset/attitude in embracing AI-assisted teaching-learning	3.6	3.4	3.4	3.4	3.9

Source: Survey responses from the online questionnaire (Analyzed by the Authors)

provide users with innovative learning experiences, customized/personalized learning paths, global collaboration, and continuous advancements and helps align education with the growing future job market demands. AI in Education also poses serious threats in terms of ethical concerns, inequitable access, job displacement situations, regulation issues, and most prominently resistance to change.

5.2 Limitations of the Study

The study has limitations on the sample size and diversity of participants, thereby limiting the generalizability of the results. Further research with much larger and more diverse samples could help capture a wide range of perspectives and their experiences with AI in educational settings. The study focused on specific AI technologies or educational settings, influencing the results and conclusions drawn. Future studies that encompass different AI tools and platforms may exhibit varying impacts depending on factors such as subject area, grade levels, or institutional context. Finally, the study may need to address several privacy, ethical, and equity concerns associated with the usage of AI in Education comprehensively so that a holistic approach is implemented. Also, the limitations of the methodology, data collection strategies, and other analytical approaches such as mixed methods research design may enhance the validity and reliability of the study findings.

6 Conclusion

AI should be used for the creation and delivery of the content, however personalized attention is needed by a learner, which includes academic as well as emotional support to be extended through traditional methods of teaching. Teachers may shift their mundane routine tasks to AI apps, search content faster, and deliver some part of lectures through AI-assisted technology, their role in guiding the next generation of learners cannot be completely replaced. Students are likely to use AI to escape the assignment load by looking for a quick way out, and thus teachers will need to have policies about where AI is allowed and where it is not. e.g., formatting a draft, and creating visuals using AI with attribution of AI source may be allowed but substituting human assignments required to learn a subject deeper may not be allowed. Regulation of AI use without losing out on the important role played by the human interface in the original imagination of the teaching profession as facilitators of critical thought and skill training. Educators also believed that adequate training and knowledge of AI should be provided keeping in mind its advantages/challenges/drawbacks. Educators should know AI and AI detection tools understand the strengths and weaknesses of AI tools as and when they emerge. Also, have a strong ability to demonstrate these to the students so that they do not fall into the trap of thinking that technology is everything and human thinking means nothing.

There is a need to cultivate a certain attitude towards technological advances before they can develop knowledge or skills to leverage AI. New advances must not be instantly viewed as threats to an older or more established way of doing things before educators begin to learn skills. They also need to ensure that the resources that may be generated from AI are contextualized as well as balanced with more traditional sources of research. AI-driven solutions offer the potential for personalized learning experiences, data-driven decision-making, and more efficient administrative processes in educational institutions. However, challenges such as privacy concerns, bias in AI algorithms, and issues related to data security need more vigilance.

References

1. Zawacki-Richter, O., Marín, V.I., Bond, M., Gouverneur, F.: Systematic review of research on artificial intelligence applications in higher education–where are the educators? Int. J. Educ. Technol. High. Educ. **16**(1), 1–27 (2019)
2. Walter, Y.: Embracing the future of Artificial Intelligence in the classroom: the relevance of AI literacy, prompt engineering, and critical thinking in modern education. Int. J. Educ. Technol. High. Educ. **21**(1), 15 (2024)
3. Martin, F., Zhuang, M., Schaefer, D.: Systematic review of research on artificial intelligence in K-12 education (2017–2022). Comput. Educ. Artif. Intell. **6**, 100195 (2023). https://doi.org/10.1016/j.caeai.2023.100195
4. Roll, I., Wylie, R.: Evolution and revolution in artificial intelligence in education. Int. J. Artif. Intell. Educ. **26**, 582–599 (2016)

5. Timms, M.J.: Letting artificial intelligence in education out of the box: educational cobots and smart classrooms. Int. J. Artif. Intell. Educ. **26**, 701–712 (2016)
6. Hwang, G.J., Xie, H., Wah, B.W., Gašević, D.: Vision, challenges, roles and research issues of artificial intelligence in education. Comput. Educ. Artif. Intell. **1**, 100001 (2020). https://doi.org/10.1016/j.caeai.2020.100001
7. Luan, H., et al.: Challenges and future directions of big data and artificial intelligence in education. Front. Psychol. **11**, 1–11, 580820 (2020). https://doi.org/10.3389/fpsyg.2020.580820
8. Ouyang, F., Jiao, P.: Artificial intelligence in education: the three paradigms. Comput. Educ. Artif. Intell. **2**, 100020 (2021). https://doi.org/10.1016/j.caeai.2021.100020
9. Zhai, X., et al.: A review of artificial intelligence (AI) in education from 2010 to 2020. Complexity **2021**, 1–18 (2021)
10. Lillehaug, S.I., Lajoie, S.P.: AI in medical education—another grand challenge for medical informatics. Artif. Intell. Med. **12**(3), 197–225 (1998)
11. McLaren, B.M., Scheuer, O., Mikšátko, J.: Supporting collaborative learning and e-discussions using artificial intelligence techniques. Int. J. Artif. Intell. Educ. **20**(1), 1–46 (2010)
12. Okonkwo, C.W., Ade-Ibijola, A.: Chatbots applications in education: a systematic review. Comput. Educ. Artif. Intell. **2**, 100033 (2021). https://doi.org/10.1016/j.caeai.2021.100033
13. Kim, Y., Soyata, T., Behnagh, R.F.: Towards emotionally aware AI smart classroom: current issues and directions for engineering and education. IEEE Access **6**, 5308–5331 (2018)
14. Rodríguez-Hernández, C.F., Musso, M., Kyndt, E., Cascallar, E.: Artificial neural networks in academic performance prediction: systematic implementation and predictor evaluation. Comput. Educ. Artif. Intell. **2**, 100018 (2021). https://doi.org/10.1016/j.caeai.2021.100018
15. Xu, W., Ouyang, F.: The application of AI technologies in STEM education: a systematic review from 2011 to 2021. Int. J. STEM Educ. **9**(1), 1–20 (2022)
16. Ahmad, S.F., Alam, M.M., Rahmat, M.K., Mubarik, M.S., Hyder, S.I.: Academic and administrative role of artificial intelligence in education. Sustainability **14**(3), 1101 (2022). https://doi.org/10.3390/su14031101
17. Sysoyev, P.V.: Artificial intelligence in education: awareness, readiness and practice of using artificial intelligence technologies in professional activities by university faculty. Vysshee Obrazovanie v Rossli **32**(10), 9–33 (2023)
18. Chen, L., Chen, P., Lin, Z.: Artificial intelligence in education: a review. IEEE Access **8**, 75264–75278 (2020)
19. Kabudi, T., Pappas, I., Olsen, D.H.: AI-enabled adaptive learning systems: a systematic mapping of the literature. Comput. Educ. Artif. Intell. **2**, 100017 (2021). https://doi.org/10.1016/j.caeai.2021.100017
20. Holmes, W., et al.: Ethics of AI in education: towards a community-wide framework. Int. J. Artif. Intell. Educ. 1–23 (2021)
21. Edwards, B.I., Cheok, A.D.: Why not robot teachers: artificial intelligence for addressing teacher shortage. Appl. Artif. Intell. **32**(4), 345–360 (2018)

LEO Satellites: Enhancing Connectivity and Data Collection Across Industries

Nimet Selen Yesilkaya[1] (ID) and Irem Bayraktar[2](✉) (ID)

[1] Mechanical Engineering Department, Mersin University, 33100 Mersin, Turkey
`yesilkaya@mersin.edu.tr`
[2] Aerospace Engineering Department, Tarsus University, 33400 Mersin, Turkey
`irem_bayraktar@tarsus.edu.tr`

Abstract. Low Earth Orbit (LEO) satellites have become essential in various industries, transforming connectivity, data collection, communication, and navigation. Orbiting closer to Earth offers advantages such as lower latency and cost-effective deployment, making LEO satellites crucial for modern industrial infrastructure. With a wide range of applications, they play a vital role in monitoring atmospheric layers and near-space environments, from communication and observation to exploration. Their consistent delivery of high-precision data over extended periods is a key strength. This study highlights the increasing importance of satellite technology, driven by advantages like lower transmission power requirements and improved coverage, especially in polar regions. Understanding these benefits is essential for maximizing satellite technology's potential. To harness LEO satellites effectively, a thorough understanding of orbital parameters and precise control is necessary. This research thoroughly examines orbit analysis and Earth coverage, providing insights into coverage areas and capabilities. It also explores data exchange dynamics with ground stations and their practical implications. In summary, this study offers a detailed exploration of LEO satellite technology, shedding light on its diverse contributions to various sectors. The proposed strategic ground station network not only greatly extends communication durations with LEO satellites but also enables more frequent updates to satellite commands.

Keywords: LEO Satellite · Space Industry · Earth Coverage · Data Transmission

1 Introduction

Satellite systems have evolved into indispensable technologies in modern life, with diverse applications encompassing telecommunications [1–4], earth and space observations [5], global resource monitoring [6], military observation [7], precision terrestrial navigation [8], microgravity science [9], and more [10–12]. To fulfill their primary missions and ensure optimal functioning, satellites are managed by a ground control station. This station transmits commands and software updates to the satellite via Earth and receives attitude and telemetry data from it [13]. Networks of multiple interconnected

© The Author(s), under exclusive license to Springer Nature Switzerland AG 2025
A. Mirzazadeh et al. (Eds.): ODSIE 2023, CCIS 2205, pp. 199–213, 2025.
https://doi.org/10.1007/978-3-031-81458-7_12

fixed control stations facilitate bidirectional data transfer. Control stations collect information during the time the satellite is within line of sight, spanning from signal reception to loss, based on transit estimation parameters. Various factors, including antenna gain, network transmission speed, and geographical obstacles, contribute to the complexity of this process [14]. The launch of Low Earth Orbit (LEO) satellites into their designated orbits is a meticulously orchestrated process that involves advanced technology and precise execution. Typically situated at altitudes ranging from 160 to 2,000 km above the Earth's surface, LEO satellites are carried into space atop rockets. The satellites are encapsulated within the rocket's payload fairing, providing protection during the initial stages of the launch [15]. As the rocket ascends through the Earth's atmosphere, the fairing is jettisoned to expose the satellite.

In contrast to geostationary satellites, LEO satellites traverse the sky rapidly, offering a limited window of visibility to observers on Earth [16]. As these satellites orbit, they establish communication with the control station during specific uplink-downlink opportunities. The timing of these opportunities hinges on the orbital geometry, control station location, and mission characteristics [17]. However, during these crucial periods, known or unexpected restrictions and challenges may arise, impacting the seamless uploading or downloading of requested data. To accurately track satellites in space and anticipate sensor coverage [18], real-time position data and future trajectory predictions are essential. This predictive capability relies on knowledge of both the satellite's current position and its orbital elements [19]. Utilizing these orbital elements allows the creation of a plausible historical trajectory for the satellite. The next critical step involves strategically positioned and well-prepared control stations ready to engage with the approaching satellite. Failure to establish this connection could render the satellite nothing more than orbital debris, as commands cannot be transmitted, and telemetry data remains unattainable. On the other hand, Low Earth Orbit (LEO) satellites offer a brief window of visibility to control stations during their transit [20]. Consequently, the establishment of multiple control stations in diverse locations becomes imperative, facilitating network connectivity for data sharing [21, 22].

The Air Force Satellite Control Network represents a global configuration of space/ground connectivity resources, featuring both common user and dedicated control nodes. Its mission encompasses the support, reception, processing, and dissemination of spacecraft telemetry, tracking, and control requirements [23]. Comprising two operational control nodes and seventeen telemetry, monitoring, and control antennas strategically positioned in nine geographical locations worldwide, this network ensures comprehensive coverage and functionality. The European Space Agency's tracking station network constitutes a global system of ground stations, establishing crucial links between orbiting satellites and the European Space Operations Center [24]. At its core, the ESTRACK network comprises thirteen terminals strategically positioned at nine stations across six countries. Each ESTRACK station accommodates one or more terminals, housing antennas and their associated signal processing equipment [25]. These stations exhibit versatility by supporting multiple missions, and the ESTRACK network allows for the shared utilization of resources with other agencies and satellite operators. In a distinct capacity, NASA's Tracking and Data Relay Satellite System provides support for NASA satellites [26]. Comprising a constellation of nine geosynchronous satellites and associated ground systems, it functions as a robust transfer system between platforms and ground facilities. Though not classified strictly as a ground support network, it plays a crucial role in NASA's satellite operations. It's essential to note that these network systems, whether belonging to the European Space Agency or NASA, are substantial government or space agency-based projects requiring robust protection due to their significant costs. In contrast, educational ground station networks present disparate security and reliability requirements compared to their commercial or governmental counterparts. The simplicity of connecting these educational ground stations via the internet offers a distinctive approach. The internet protocol serves as a global interface for these ground stations, facilitating seamless connectivity and simplifying the inclusion of additional participants [27].

The Global Training Network for Satellite Operations, a project spearheaded by the European Space Agency and overseen by the Training Office [28], is encapsulated in the GENSO initiative. This endeavor was conceived with the aim of establishing a global network interconnecting diverse ground stations across the world [29]. The primary objective is to implement a versatile multi-software system that endows participants with ground station capabilities through the network. The focal point of GENSO's efforts revolves around the architectural design of ground station networks, with a specific emphasis on incorporating features such as authentication, user management, and resource allocation.

By leveraging internet-based traffic tunneling, GENSO endeavors to extend the communication time window with Low Earth Orbit (LEO) satellites. This extension aims to surpass the conventional 30-min per day limit, theoretically enabling continuous communication for up to 24 h per day. This expanded timeframe not only pertains to telemetry downloads but also encompasses the transmission of remote commands. The primary goals of the GENSO network encompass providing satellite operators with remote access to real-time mission data, enabling limited remote control of participating ground stations, and establishing a global standard for training ground station software. Through these objectives, GENSO seeks to enhance operational efficiency and foster a standardized approach to satellite ground station training on a global scale.

This study conducted a detailed trajectory analysis of LEO applications to evaluate their communication times with control stations. Findings revealed a significant limitation in LEO satellites—due to their low orbit, they experienced constrained communication time windows with control stations. To address this, the proposal suggests establishing a network of interconnected control stations. This network not only significantly extends communication times with LEO satellites but also enables more frequent updates of satellite commands, enhancing the overall dynamism and flexibility of missions. However, it emphasizes the substantial increase in costs associated with deploying multiple control stations, highlighting the need for a balanced and cost-effective solution to ensure the sustainable deployment and operation of such networks.

2 Overview and Motivation

The fundamental goal of the majority of satellites is to either observe specific points on Earth's surface or establish communication with them. However, owing to their restricted line of sight, satellites are only capable of covering a specific portion of the Earth at any given moment. In stark contrast to Low Earth Orbit (LEO) satellites, Geostationary Orbit (GEO) satellites possess the unique capability of simultaneously covering approximately one-third of the Earth's surface. Despite this distinct advantage, GEO satellites encounter a notable drawback characterized by extended periods of low coverage.

To mitigate this challenge, GEO satellites play a pivotal role in the establishment of inter-satellite links. In scenarios where the distance between two satellites spans around 45,000 km—necessitating an optical link from LEO to GEO and an optical and/or radio frequency (RF) link from GEO to the ground station—LEO satellites acquire the capability to transmit telemetry data even when not directly in the line of sight with the control station. This strategic implementation not only amplifies the coverage of LEO satellites but also emancipates them from reliance on traditional ground stations. Consequently, the satellite-to-satellite communication time escalates to approximately 45%, enabling direct communication between two satellites for nearly half of their orbital period. This innovative approach not only enhances the overall efficiency of satellite communication but also exemplifies a shift towards a more autonomous and adaptable satellite network. The establishment of inter-satellite links opens up new possibilities for optimizing data transfer, reducing dependency on ground infrastructure, and bolstering the resilience and effectiveness of satellite missions in various applications, including telecommunications, Earth observation, and global resource monitoring.

In contrast to Geostationary Earth Orbit (GEO) satellites, a growing number of contemporary satellite systems favor Low Earth Orbit (LEO) or Medium Earth Orbit (MEO) constellations characterized by circular orbits. While these systems often adopt the tried-and-true Walker's Delta6 constellations, designed to ensure continuous global coverage, they face challenges when attempting to construct regional coverage systems. Complications emerge in scheduling when multiple satellites operate in varying orbits, and numerous control stations are strategically located in diverse geographical locations.

During the satellite visibility period—when a satellite is within the line of sight of a control station for a limited time window—if a second satellite enters the control station's reception zone, the control station must make critical decisions. It must determine whether to persist with the existing satellite or establish a new connection. In these instances, the overlapping visibility period of the satellite with lower priority needs to be canceled. This scheduling challenge becomes paramount for optimizing communication and resource utilization in satellite constellations characterized by multiple orbits and dispersed control stations.

This intricate scheduling challenge is encapsulated by the Satellite Range Scheduling problem, which involves creating a schedule for operating multiple satellites over a defined time period. Efficiently coordinating the operation of multiple orbiting satellites heavily relies on the quality of coordination among ground stations. Integrating ground stations into a well-coordinated network not only extends contact times with satellites but also augments the overall reliability of satellite operations. Key to this coordination are requirements such as implementing an automatic scheduling program or procedure. Such tools enable swift and flexible scheduling while establishing priorities for both control stations and users, laying the foundation for optimized and effective satellite constellations.

The CubeSat Programming System is meticulously designed to cater exclusively to Low Earth Orbit (LEO) satellites, ensuring the occurrence of only single intersections between two satellites at any given control station. Employing advanced optimization techniques, this system employs a combination of a Branch-and-Bound constraint search and a Hill Climbing algorithm to derive efficient solutions. While maintaining equal priorities across all ground stations, user-assigned priorities dictate the hierarchy of request priorities. In scenarios where conflicting visibility times arise between two satellites at a control station, the tracking time allocated to the first satellite accounts for only a fraction of its total visibility time. However, this sophisticated system encounters certain limitations. It does not account for tracking satellites with visibility times shorter than a predetermined period. Moreover, when a satellite becomes visible at multiple control stations simultaneously, it is monitored by the control station endowed with the highest priority. Additionally, a mandated minimum time interval between the conclusion of one satellite tracking session and the initiation of another is rigorously enforced.

The overarching objective of the CubeSat Programming System is to conduct a comprehensive coverage analysis for a satellite system that aligns with the principles of technological cooperation, joint development, and joint use. The core mission revolves around meeting the requirements for high-resolution image acquisition, specifically tailored for global military intelligence. This mission underscores the importance of transcending geographical restrictions and exploring potential civil applications such as monitoring forest lands, detecting illegal construction activities, swiftly assessing damage in the aftermath of natural disasters, identifying cultivation areas, and generating detailed geographical map data.

In essence, the investigation conducted by this system is laser-focused on identifying scenarios where a LEO satellite, armed with cutting-edge technical features, can execute the most effective mission within the predefined geographical region of interest. This approach not only underscores the system's commitment to robust performance but also showcases its versatility in addressing diverse applications, from military intelligence to civil and environmental monitoring.

3 Numerical Test and Results

The North American Aerospace Defense Command (NORAD), an integral binational agency jointly operated by the United States and Canada under the U.S. Air Force Space Command, assumes a pivotal role in real-time space object tracking. At the core of NORAD's mission is the provision of early warning and aeronautical control to safeguard North America. In the meticulous management of airspace, NORAD orchestrates a sophisticated network that encompasses satellites, ground-based radar systems, weather radar installations, and a fleet of fighter aircraft. This intricate network is strategically designed to neutralize potential threats emanating from both air and spacecraft, thereby enhancing the overall security of the North American region.

NORAD's genesis traces back to its original establishment with the primary objective of safeguarding North America against the looming threats posed by long-range bombers and tactical ballistic missiles. Over the years, as technology has advanced and the spectrum of potential threats has evolved, NORAD has consistently adapted and

expanded its capabilities to stay at the forefront of defense and security. In contemporary times, the NORAD system remains indispensable, serving as a vigilant guardian of
regional security by continuously monitoring and responding to activities in space.

The collaborative nature of NORAD, involving both the United States and Canada,
reflects a shared commitment to the defense of North American airspace. The organization's multifaceted approach, incorporating a fusion of satellite technologies, ground-
based radar systems, weather monitoring tools, and airborne assets, underscores the
comprehensiveness of its space surveillance capabilities. As space becomes an increasingly critical domain, NORAD's ongoing mission plays a crucial role in maintaining
situational awareness, responding swiftly to emerging threats, and upholding the security
of the airspace over North America.

In the context of object tracking, the NORAD system operates by generating two-line
item sets for each object in orbit. The intricacies of these sets provide a comprehensive
depiction of the satellite's essential characteristics and orbital parameters. The first line,
meticulously detailed in Table 1, is a string of 69 characters meticulously crafted to
convey crucial descriptors specific to each satellite. Simultaneously, the second line, as
depicted in Table 2, is another string of 69 characters that encapsulates fundamental
Kepler elements essential for precise orbit calculations.

These NORAD-generated data sets serve as the cornerstone for the mathematical
underpinnings of object tracking simulations. Leveraging sophisticated Matlab codes
within this study, these NORAD data sets are instrumental as initial calculation values.
This utilization significantly contributes to the attainment of a more realistic and accurate
simulation of the satellites' orbital movements and trajectories.

The Matlab codes employed in this study employ a chain reaction approach to calculate the movement of each successive satellite. This approach builds upon the information
derived from the preceding satellite, creating a seamless and continuous simulation that
accurately reflects the dynamic nature of satellites in orbit. The utilization of NORAD
data sets as the foundation for these simulations ensures that the mathematical models
align closely with real-world observations, providing a robust framework for the study
of satellite dynamics and behavior in the Earth's orbital environment.

To compute the orbital period (T) in seconds, the study utilizes the average motion
outlined in Table 2 (dataset 2/8). This meticulous approach ensures accurate and dynamic
orbital simulations, highlighting NORAD's integral role in maintaining space situational
awareness and enhancing regional security measures.

$$T = \frac{24 \times 3600}{\text{mean motion}} \tag{1}$$

Using the computed orbital period, the semi-major axis (a) is determined through
the following formula.

$$a^3 = \left(\frac{T}{2\pi}\right)^2 \mu \tag{2}$$

Table 1. First line of NORAD element set.

#	Column	Description
1/1	1	Line Number
1/2	3–7	Satellite Number
1/3	8	Classification
1/4	10–11	International Designator
1/5	12–14	International Designator
1/6	15–17	International Designator
1/7	19–20	Epoch Year
1/8	21–32	Epoch
1/9	34–43	1st Derivative of Mean Motion
1/10	45–52	2nd Derivative of Mean Motion
1/11	54–61	BSTAR Drag Term
1/12	63	Ephemeris Type
1/13	65–68	Element Number
1/14	69	Letters, Blanks, Periods, etc

Table 2. Second Line of NORAD element set.

#	Column	Description
2/1	1	Line Number
2/2	3–7	Satellite Number
2/3	9–16	Inclination
2/4	18–25	Right Ascension of Ascending Node
2/5	27–33	Eccentricity
2/6	35–42	Argument of Perigee
2/7	44–51	Mean Anomaly
2/8	53–63	Mean Motion
2/9	64–68	Revolution Number at Epoch
2/10	69	Checksum

The standard gravitational parameter for Earth is $\mu = 398600\,km^3/sn^2$. Utilizing the mean anomaly value provided in data set 2/7 from Table 2, the eccentric anomaly (E) and real anomaly (θ) are derived.

$$Mean\ Anomaly = E - e \cdot sinE \qquad (3)$$

$$\tan\frac{\theta}{2} = \sqrt{\frac{1+e}{1-e}}\tan\frac{E}{2} \tag{4}$$

In this context, eccentricity (e) is extracted from data set 2/5. Further, employing the slope values (i) from data set 2/3, the right ascension (Ω) of the ascending node from data set 2/4, and the perigee argument (ω) from data set 2/6, the position vector ($\overrightarrow{R}$) and its magnitude, R (the radius measured from the Earth's center to the corresponding location) are calculated as outlined below.

$$\overrightarrow{R} = \begin{bmatrix} R\cos\theta & R\sin\theta & 0 \end{bmatrix}^T \tag{5}$$

$$R = \frac{a(1-e^2)}{1+e\cos\theta} \tag{6}$$

The position vector is initially expressed in the perifocal frame and necessitates conversion to the geocentric (x, y, z) frame. To achieve this, the Q transformation matrix, as provided by Eq. (7), is applied.

$$Q = \begin{bmatrix} \cos\Omega\cos\omega - \sin\Omega\sin\omega\cos i & -\cos\Omega\sin\omega - \sin\Omega\cos\omega\cos i & \sin\Omega\sin i \\ \sin\Omega\cos\omega + \cos\Omega\sin\omega\cos i & -\sin\Omega\sin\omega + \cos\Omega\cos\omega\cos i & -\cos\Omega\sin i \\ \sin\omega\sin i & \cos\omega\sin i & \cos i \end{bmatrix} \tag{7}$$

Commencing the simulation cycle with a previously computed first real anomaly value, the satellite is set into motion within the geocentric frame. The true anomaly and time are directly proportional, requiring time changes for each alteration in the true anomaly. To ascertain the spacecraft's position over time, the eccentric anomaly must be recalculated utilizing the new true anomaly, as specified in the following formulation.

$$E = \tan^{-1}\left(\frac{\sin\theta\sqrt{1-e^2}}{e+\cos\theta}\right) \tag{8}$$

Considering Earth's rotation, the position vector undergoes a vector multiplication by the Earth's angular velocity matrix, R_d.

$$R_d(t) = \begin{bmatrix} \cos(\omega_d t) & \sin(\omega_d t) & 0 \\ -\sin(\omega_d t) & \cos(\omega_d t) & 0 \\ 0 & 0 & 1 \end{bmatrix} \tag{9}$$

where, $\omega_d = 7{,}292 \times 10^{-5}$ rad/sec. Consequently, the satellite's position along its orbit can be determined using these expressions, accounting for a rotating Earth. Conversely, subsatellite points, known as nadirs, are essential on the Earth's surface, as depicted in Fig. 1. Obtaining this point involves subtracting the satellite's altitude from the Earth's center radius (R). Subsequently, a new ground track vector ($\overrightarrow{G}$) is obtained, with a magnitude equal to R_d. Subsatellite point coordinates, represented as latitude and longitude, are then calculated as follows.

$$E_{nadir} = \sin^{-1}\left(\frac{G_z}{R_d}\right) \tag{10}$$

$$B_{nadir} = \cos^{-1}\left(\frac{G_x}{R_d \cos E_{nadir}}\right) \qquad (11)$$

Once subsatellite coordinates are determined, assessing whether the control station falls within the satellite coverage area becomes straightforward. The known coordinates of the subsatellite point and the control station facilitate the calculation of the distance between the two points using the large distance equation. Given that the distance between ground latitudes is approximately $1,11 \times 10^5$ meters, dividing the range coverage by this value yields the maximum area that the satellite can cover as a function of the degree of latitude.

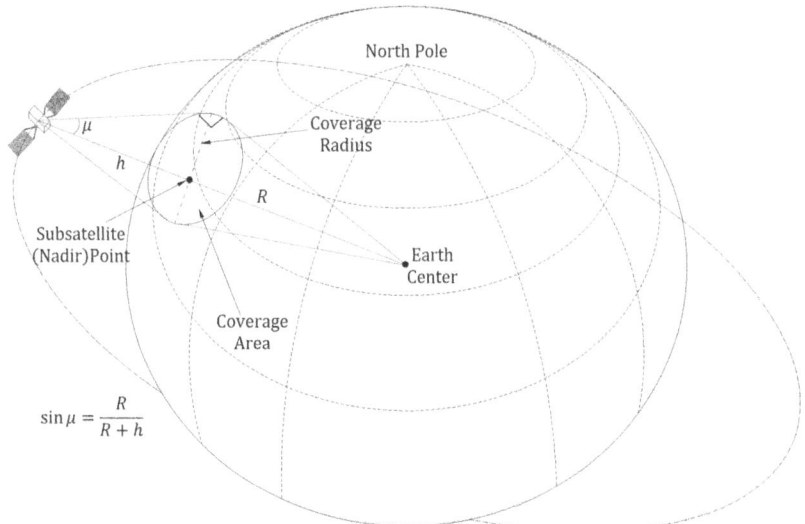

Fig. 1. Coverage parameters of the satellite.

Until now, the assumption has been that the satellite's orbit is known. The subsequent step involves altering its orbit above the Earth's surface and reevaluating the sub-satellite locations to estimate the revised coverage area. Illustrated in Fig. 2 is the hypothetical scenario with three control stations located at different positions on the Earth's surface, each at an altitude of 550 km, presenting an exemplary ground track for a polar-orbiting satellite. The control stations are situated in Mersin (36°48' N 34°38' E), Sydney (33°52' S 151°13' E), and Lima (12°03' S 77°02' W). For communication initiation, the satellite must either pass over or be in close proximity to a control station. It is assumed, as previously mentioned, that communication starts only when the altitude (ε) is higher than 5°, as this coverage area is predominantly linked to the satellite's altitude.

Referring to Fig. 2, Figs. 3, 4, and 5 depict the azimuth and elevation angles of the satellite as observed from the ground stations. In Fig. 3, the ground station in Mersin is positioned at the center of the circles, also representing the 90° elevation angle. Examining the satellite's coverage area reveals that, after completing 21 orbits around the Earth, the satellite passed west of Mersin twice, resulting in only these two line-of-sight passes

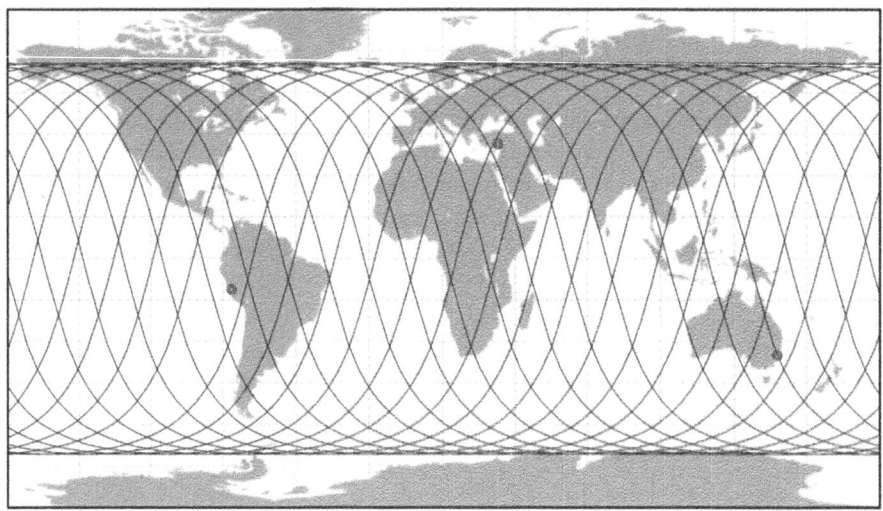

Fig. 2. Locations of control stations and satellite orbital tracks (21 orbits).

over Mersin throughout its orbit. The first, relatively lengthier transition spans 13 min and 21 s, while the second transition takes 8 min and 33 s. Calculating these connection interval times when the altitude is higher than 0, and then recalculating for altitudes 5° higher, reveals that communication can be established for 8 min and 47 s during the first pass and 3 min and 41 s during the second pass. Consequently, while the control station antennas must maintain visibility of the satellite, communication cannot be sustained for approximately 274 s during the first pass and 292 s during the second pass.

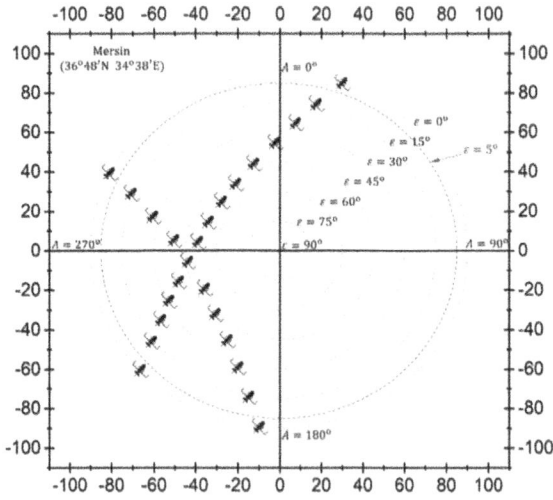

Fig. 3. Transitions around Mersin control station.

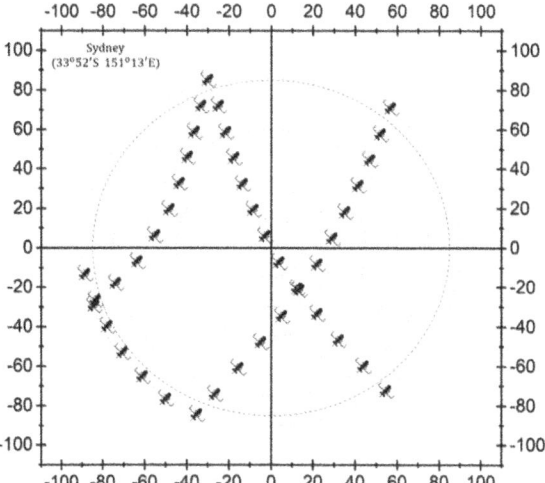

Fig. 4. Transitions around Sydney control station.

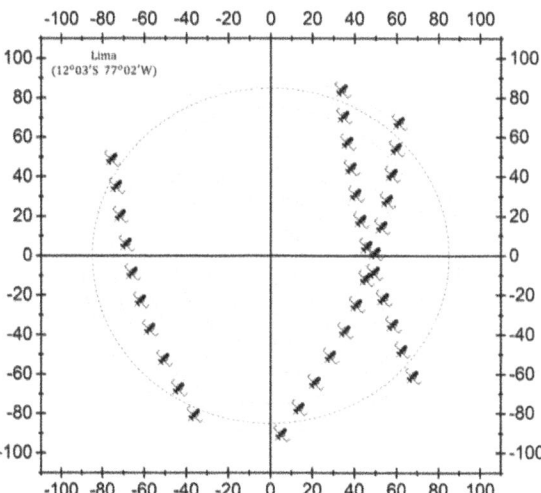

Fig. 5. Transitions around Lima control station.

Similar to Fig. 3, Fig. 4 illustrates the orbital tracks over Sydney. Following the satellite's completion of 21 orbits around the Earth, it is evident that it orbited the control station in Sydney a total of 4 times. However, upon closer inspection, one of these passes falls below the threshold value of 5°, rendering communication with the control station unattainable during that specific pass. In essence, there are effectively 3 communicable transitions. It is noteworthy that one of these passes occurred directly above the control station, resulting in the longest observed transition lasting 16 min and 5 s.

Figure 5 displays the satellite's three passes over the city of Lima, with communication times during these passes spanning 10 min and 21 s, 10 min and 56 s, and 12 min and 44 s, respectively. Consequently, the acquisition time is evidently closely tied to the satellite's altitude. Throughout the acquisition period, the satellite is assumed to download telemetry data to the control station and subsequently delete the data from its storage. The download speed is assumed to be constant, utilizing an X-Band transmitter with a speed of 20 Mbps for all three ground stations. Additionally, it is assumed that the network data exchange rate with the control station is 100 Mbps. These speeds can be adjusted for more accurate simulations. Notably, the file sizes of captured images are contingent on various factors such as product type, spectral options, pixel size, coverage area, and bit depth. For instance, the throughput of a multispectral image captured at 10 megapixels, covering an area of 20 × 20 km, is approximately 420.32 MB per image.

The comparison between the cost savings from inter-satellite communications and the benefits of multiple control stations involves evaluating various factors to determine which option offers the most advantageous balance of cost-effectiveness and performance. Inter-satellite communication typically involves leveraging existing satellite infrastructure to relay data between satellites, potentially reducing the need for additional ground infrastructure. This approach can lead to significant savings in terms of initial setup costs and ongoing operational expenses associated with constructing and maintaining multiple ground stations. On the other hand, multiple control stations offer benefits such as enhanced communication reliability and flexibility. With multiple ground stations distributed across different locations, there is increased redundancy, reducing the risk of communication failures due to factors like adverse weather conditions or technical issues at a single site. Additionally, having multiple control stations allows for greater flexibility in directing satellite operations and responding to dynamic mission requirements. Ultimately, the decision between inter-satellite communication and multiple control stations depends on the specific requirements and priorities of the satellite mission. A comprehensive cost-benefit analysis considering factors such as initial setup costs, ongoing operational expenses, communication reliability, flexibility, and mission objectives is essential to determine the most suitable communication approach for a given scenario.

Satellites operating in different orbits can indeed deliver the required communications performance, but their effectiveness depends on various factors including their orbital characteristics, mission objectives, and the specific communication technologies employed. Satellites in different orbits serve different purposes and have unique advantages and challenges. LEO satellites offer lower latency and better coverage for certain applications like broadband internet, while GEO satellites provide continuous coverage over a fixed area but with higher latency due to their higher altitude. Signal interference can occur in satellite communication systems due to various factors such as atmospheric conditions, electromagnetic interference, or intentional jamming. However, advanced signal processing techniques and frequency allocation regulations help mitigate these issues to a certain extent.

4 Conclusion

The trajectory analysis of Low Earth Orbit (LEO) applications was meticulously undertaken to assess their communication times with control stations. The results of this numerical study unveiled a notable limitation inherent in these satellites—due to their low orbit, they experienced constrained communication time windows with control stations. In response to this challenge, a straightforward yet effective approach is proposed: the establishment of a network comprising numerous interconnected control stations. This strategic network not only significantly prolongs communication times with LEO satellites but also affords the opportunity for more frequent updates of satellite commands. Such agility in command updates enhances the overall dynamism and flexibility of the satellite's mission. However, it is imperative to acknowledge that the deployment of multiple control stations entails a substantial increase in associated costs. This cost consideration underscores the need for a balanced and cost-effective solution to ensure the sustainable deployment and operation of such networks.

In pursuit of a pragmatic alternative, inter-satellite communication emerges as a viable and economically feasible option. The proposition involves leveraging multiple satellite communication networks, whether characterized as homogeneous (involving communication solely between LEO satellites) or heterogeneous (encompassing communication between a LEO satellite and counterparts in Medium Earth Orbit (MEO) or Geostationary Earth Orbit (GEO)). This alternative approach strikes a delicate balance between ensuring efficient communication and addressing cost considerations, offering a nuanced solution to extend communication times with control stations. By exploring this avenue, the satellite industry can optimize communication efficiency without disproportionately escalating operational costs, thereby fostering a sustainable and technologically viable approach for future satellite deployments and missions.

In terms of future work, further investigation could be conducted to explore the optimization of the proposed network topology for the strategic ground station network. This could involve studying different configurations and placement strategies to maximize coverage and minimize costs. Additionally, research could be directed towards developing advanced communication protocols and technologies to enhance the efficiency and reliability of inter-satellite communication networks. Moreover, studies focusing on the integration of emerging technologies such as artificial intelligence and machine learning algorithms for autonomous satellite communication management could be explored to streamline operations and reduce human intervention. Finally, assessing the environmental impacts and sustainability considerations of deploying and operating such satellite communication networks could be an essential aspect of future research to ensure the long-term viability and eco-friendliness of these systems.

References

1. Piltyay, S.I., Sushko, O.Y., Bulashenko, A.V., Demchenko, I.V.: Compact Ku-Band iris polarizers for satellite telecommunication systems. Telecomm. Radio Eng. **79**(19), 1673–1690 (2020). https://doi.org/10.1615/TelecomRadEng.v79i19.10
2. Landoni, M.: Innovation policy in progress, institutional intermediation in public procurement of innovation: satellite telecommunications in Italy. R&D Manag. **47**(4), 583–594 (2017). https://doi.org/10.1111/radm.12246
3. Hoyhtya, M., et al.: Database-assisted spectrum sharing in satellite communications: a survey. IEEE Access **5**, 25322–25341 (2017). https://doi.org/10.1109/ACCESS.2017.2771300
4. Pultarova, T.: Telecommunications – space tycoons go head to head over mega satellite network. Eng. Tech. **10**(2), 20–30 (2015). https://doi.org/10.1049/et.2015.0221
5. Bakirci, M., Bayraktar, I.: Assessment of YOLO11 for ship detection in SAR imagery under open ocean and coastal challenges. In: 21st International Conference on Electrical Engineering, Computing Science and Automatic Control (CCE), October 23–25, Mexico City, Mexico, pp. 1–6 (2024). https://doi.org/10.1109/CCE62852.2024.10770926
6. Coxen, C.L., Frey, J.K., Carleton, S.A., Collins, D.P.: Species distribution models for a migratory bird based on citizen science and satellite tracking data. Global Ener. Conserv. **11**, 298–311 (2017). https://doi.org/10.1016/j.gecco.2017.08.001
7. Kodheli, O., et al.: Satellite communications in the new space era: a survey and future challenges. IEEE Commun. Surv. Tutor. **23**(1), 70–109 (2021). https://doi.org/10.1109/COMST.2020.3028247
8. Bakirci, M., Bayraktar, I.: Boosting aircraft monitoring and security through ground surveillance optimization with YOLOv9. In: 12th International Symposium on Digital Forensics and Security (ISDFS), April 29–30, San Antonio, TX, USA, pp. 1–6 (2024). https://doi.org/10.1109/ISDFS60797.2024.10527349
9. Montenbruck, O., et al.: GNSS satellite geometry and attitude models. Advan. Space Res. **56**(6), 1015–1029 (2015). https://doi.org/10.1016/j.asr.2015.06.019
10. Donaldson, D., Storeygard, A.: The view from above: applications of satellite data in economics. J. Econo. Persp. **30**(4), 171–198 (2016). https://doi.org/10.1257/jep.30.4.171
11. Bakirci, M., Bayraktar, I.: Transforming aircraft detection through LEO satellite imagery and YOLOv9 for improved aviation safety. In: 26th International Conference on Digital Signal Processing and its Applications (DSPA), March 27–29, Moscow, Russian Federation, pp. 1–6 (2024). https://doi.org/10.1109/DSPA60853.2024.10510106
12. Sanctis, M.D., Cianca, E., Araniti, G., Bisio, I., Prasad, R.: Satellite communications supporting internet of remote things. IEEE Intern. Thin. J. **3**(1), 113–123 (2016). https://doi.org/10.1109/JIOT.2015.2487046
13. Yin, J., et al.: Satellite-based entanglement distribution over 1200 kilometers. Science **356**(6343), 1140–1144 (2017). https://doi.org/10.1126/science.aan3211
14. Giambene, G., Kota, S., Pillai, P.: Satellite-5G integration: a network perspective. IEEE Network **32**(5), 25–31 (2018). https://doi.org/10.1109/MNET.2018.1800037
15. Bakirci, M.: Design and aerodynamic analysis of a rocket nose cone with specific fineness ratio. In: IEEE 6th International Conference on Actual Problems of Unmanned Aerial Vehicles Development (APUAVD), October 19–21, Kyiv, Ukraine, pp. 80–85 (2021). https://doi.org/10.1109/APUAVD53804.2021.9615407
16. Abburu, S., Golla, S.B.: Satellite image classification methods and techniques: a review. Int. J. Comput. Appl. **119**(8), 20–25 (2015). https://doi.org/10.5120/21088-3779
17. Ali, I., Cawkwell, F., Dwyer, E., Barrett, B., Green, S.: Satellite remote sensing of grasslands: from observation to management. J. Plant Ecol. **9**(6), 649–671 (2016). https://doi.org/10.1093/jpe/rtw005

18. McDowell, J.C.: The low earth orbit satellite population and impacts of the SpaceX Starlink constellation. Astrophy. J. Lett. **892**(2), 1–11 (2020). https://doi.org/10.3847/2041-8213/ab8016

19. Qu, Z., Zhang, G., Cao, H., Xie, J.: LEO satellite constellation for internet of things. IEEE Access **5**, 18391–18401 (2017). https://doi.org/10.1109/ACCESS.2017.2735988

20. Su, Y., Liu, Y., Zhou, Y., Yuan, J., Cao, H., Shi, J.: Broadband LEO satellite communications: architectures and key technologies. IEEE Wirel. Commu. **26**(2), 55–61 (2019). https://doi.org/10.1109/MWC.2019.1800299

21. You, L., Li, K.X., Wang, J., Gao, X., Xia, X.G., Ottersten, B.: Massive MIMO transmission for LEO satellite communications. IEEE J. Select. Are. Commu. **38**(8), 1851–1865 (2020). https://doi.org/10.1109/JSAC.2020.3000803

22. Bakirci, M., Cetin, M.: Improving position-time trajectory accuracy in vehicle stop-and-go scenarios by using a mobile robot as a testbed. J. Control Eng. Appl. Inf. **25**(3), 35–44 (2023). https://doi.org/10.61416/ceai.v25i3.8365

23. Xia, S., Jiang, Q., Zou, C., Li, G.: Beam coverage comparison of LEO satellite systems based on user diversification. IEEE Access **7**, 181656–181667 (2019). https://doi.org/10.1109/ACCESS.2019.2959824

24. Liu, S., et al.: LEO satellite constellations for 5G and beyond: how will they reshape vertical domains? IEEE Commu. Magaz. **59**(7), 30–36 (2021). https://doi.org/10.1109/MCOM.001.2001081

25. Ge, H., Li, B., Jia, S., Nie, L., Wu, T., Yang, Z.: LEO enhanced global navigation satellite system (LeGNSS): progress, opportunities, and challenges. Geo-Spat. Inform. Sci. **25**(1), 1–13 (2022). https://doi.org/10.1080/10095020.2021.1978277

26. Deng, R., Di, B., Zhang, H., Kuang, L., Song, L.: Ultra-dense LEO satellite constellations: how many LEO satellites do we need? IEEE Trans. Wirel. Commu. **20**(8), 4843–4857 (2021). https://doi.org/10.1109/TWC.2021.3062658

27. Bakirci, M.: A novel swarm unmanned aerial vehicle system: incorporating autonomous flight, real-time object detection, and coordinated intelligence for enhanced performance. Traitement du Signal **40**(5), 2063–2078 (2023). https://doi.org/10.18280/ts.400524

28. Khalife, J., Neinavaie, M., Kassas, Z.M.: The first carrier phase tracking and positioning results with Starlink LEO satellite signals. IEEE Trans. Aerosp. Electr. Syst. **58**(2), 1487–1491 (2022). https://doi.org/10.1109/TAES.2021.3113880

29. Wu, Z., Jin, F., Luo, J., Fu, Y., Shan, J., Hu, G.: A graph-based satellite handover framework for LEO satellite communication networks. IEEE Commu. Lett. **20**(8), 1547–1550 (2016). https://doi.org/10.1109/LCOMM.2016.2569099

Exploring Robotic Arm Dynamics in Mobile Platforms for Space Industrial Applications

Cahit Taslicali⬤ and Abdullah Demiray$^{(\boxtimes)}$ ⬤

Department of Aeronautical Engineering, Istanbul Technical University, 34469 Istanbul, Turkey
{ctaslicali,demiraya22}@itu.edu.tr

Abstract. Robotic arms are essential tools in the space industry, enabling a wide range of tasks from handling payloads and conducting repairs to sample collection and construction. Their importance in the space industry continues to grow as technology advances and space activities expand. In this research endeavor, an in-depth exploration of the intricacies involved in the operation of a robotic arm within a dynamic, non-fixed reference frame is undertaken. The research commences with the development of a sophisticated mathematical model meticulously crafted to encompass the dynamic characteristics of the platform to which the robotic arm is intricately linked. This model serves as the bedrock for systematic analysis of the inputs that govern the robotic arm's movements and actions. Through rigorous numerical analysis of this model, a comprehensive survey of various operational methods is undertaken, subjecting each to a meticulous examination of their distinct advantages and disadvantages. Moreover, the inquiry extends to an in-depth investigation into the reaction forces exerted upon the platform during the execution of each operational mode. As the investigation unfolds, data and analysis are drawn upon to identify the conditions under which a robotic arm, operating on a mobile base, can be consistently controlled with optimal precision and reliability. Ultimately, the findings underscore the importance of tailoring control modes to align with the specific application of the robotic arm. This adaptable approach offers the potential for significant benefits, enhancing the performance, efficiency, and overall utility of robotic arm systems across a wide spectrum of industrial applications.

Keywords: Robotics · Dynamic Model · Control · Space Industry

1 Introduction

Unmanned systems have emerged as indispensable tools in space research, offering unprecedented capabilities in exploration, data acquisition, and manipulation [1, 2]. Space rovers, equipped with advanced sensors and mobility mechanisms, play a crucial role in planetary exploration, allowing for the study of celestial bodies in detail and the collection of valuable scientific data [3]. These autonomous rovers can navigate diverse terrains, analyze surface compositions, and transmit real-time information back to ground control stations [4, 5]. Robotic manipulators, on the other hand, are instrumental in performing intricate tasks in the challenging environment of outer space.

A. Mirzazadeh et al. (Eds.): ODSIE 2023, CCIS 2205, pp. 214–227, 2025.
https://doi.org/10.1007/978-3-031-81458-7_13

These manipulators, often integrated into spacecraft or stationed on planetary surfaces, enable precise sample collection, maintenance of equipment, and intricate repairs [6, 7]. Together, these unmanned systems revolutionize space research, enhancing our understanding of the cosmos and paving the way for future exploration missions. Their ability to operate autonomously in harsh and distant environments contributes significantly to the advancement of scientific knowledge and the quest for new discoveries beyond Earth.

Transporting unmanned systems into space is a complex yet essential aspect of space exploration. Rockets serve as the primary vehicles for launching these sophisticated machines beyond Earth's atmosphere. The intricacies of this process involve careful engineering to ensure the safety and functionality of the unmanned systems during the violent acceleration and harsh conditions of launch [8]. Payload integration and securing these systems within the rocket's payload fairings demand precision and meticulous planning. Once in space, these unmanned systems are deployed to execute their designated tasks, whether it be planetary exploration, satellite deployment, or maintenance activities. The efficiency and reliability of the launch process directly impact the success of space missions, emphasizing the critical role of rockets in enabling the deployment and operation of unmanned systems in the vast expanse of outer space.

Embarking on space missions with human crews involves substantial risks and considerable expenses. Given the current landscape, there exists a consensus within the scientific community advocating for the substitution of humans with space robots in various extraterrestrial activities. Robots, being highly adaptable machines, benefit from rapid advancements in information technology, allowing for swift modifications and improvements. Their allure as substitutes for humans in space endeavors is further amplified by their versatility and efficiency. Tasks ideally suited for robotic execution encompass spacecraft rendezvous and docking in orbit, in-space spacecraft assembly, orbital servicing, maintenance support, scientific experiments, and planetary exploration missions [9]. The trajectory of space exploration is steering towards the utilization of intelligent robots and robotic systems equipped with local decision-making capabilities, reducing dependency on ground station commands [10, 11]. Notably, space-based robotic manipulators grapple with distinct dynamics compared to their ground-based counterparts, operating on mobile and seemingly weightless platforms amidst microgravity environments [12]. Unlike their stationary ground-based counterparts, these manipulators adapt to the dynamic conditions of orbit, where the absence of Earth's gravity introduces a unique set of challenges [13].

The direct reliance on human capabilities within space stations is fraught with substantial risks and limitations, prompting the imperative deployment of robotic arms equipped to address these challenges and unlock a multitude of applications. Operating within the confines of space stations, these robotic structures not only confer notable convenience but also play a vital role in advancing space research and exploration. The significance of robotic arms is particularly pronounced in meeting the diverse requirements of space stations, especially in intricate tasks like maintenance and repair. In light of the inherent risks confronted by humans in the hostile space environment, the integration of robotic arms becomes indispensable for safeguarding the sustainability of space exploration and the well-being of astronauts. These structures establish a safer working environment within the space station, significantly alleviating the burdens placed on astronauts amid perilous and demanding conditions. Beyond simplifying pivotal operations such as maintenance and repair, robotic arms bolster operational efficiency, facilitating accelerated outcomes. The incorporation of cameras into these robotic structures further enhances their utility, enabling meticulous inspections of the space station's exterior and facilitating prompt fault detection. To encapsulate, robotic arms play a pivotal role in streamlining various operations encompassing research, security, material transfer, and maintenance within space studies. Their presence empowers astronauts to optimize their limited time effectively, allowing them to predominantly focus on research, exploration, and fulfilling their mission duties. Consequently, robotic arms assume a critical position in ensuring the seamless functioning of a space station, driven by their paramount importance in enhancing safety, operational efficiency, and overall mission success.

In the expansive field of orbital activities, endeavors unfold both externally and internally to the spacecraft, encompassing tasks such as the relocation of satellites between orbits or the meticulous docking with spacecraft designated for maintenance operations [14]. Within this orbital operational framework, a robotic manipulator may be strategically docked or attached to a spacecraft, executing intricate tasks. Operating as the base for the robotic manipulator, the spacecraft dynamically adjusts its position in harmony with the manipulator's movements during task execution. However, this synchronized interplay may perturb the spacecraft's attitude and translational motion, prompting the implementation of precise attitude/orbit control measures [15].

In contrast to their autonomous counterparts, telerobots involve human operators stationed at a central control hub, orchestrating commands for the robot. The contemporary shift towards autonomous robots accentuates the imperative need for stringent safety protocols to safeguard the success of orbital operations. Ensuring mission success entails a multifaceted approach, including the mitigation of collision risks in robotic space endeavors, meticulous regulation of manually applied forces, and the prevention of hardware and software failures through comprehensive risk analysis, dynamic scrutiny, and robotic control mechanisms designed for both detection and prevention [16, 17]. In essence, a pivotal determinant for the triumph of every space-related missions in low gravity environments lies in the comprehensive examination of inertial platforms' responses to the nuanced movements of robotic entities.

In control systems, ensuring the stability of the robotic platform is paramount during the active phases of onboard robots. Whether configured as a spacecraft with an integrated robotic manipulator or serving as the foundational robotic base for a spacecraft equipped with a manipulator, the platform demands a sophisticated control subsystem. This subsystem is indispensable for mitigating excitations arising from reaction torques and forces generated by the intricate activities of the robot ensemble [18]. Embracing both attitude and/or orbit control, this comprehensive subsystem acts as a linchpin for upholding spacecraft stability amid manipulator operations. At the core of this stability preservation are active control mechanisms that rely on a seamless collaboration between sensors and actuators. Sensors play a pivotal role in discerning the platform's precise positions and orientations, with the resultant data meticulously assessed by the software responsible for spacecraft's behavior. Ingeniously designed controllers then take the reins, orchestrating the actuators with precision to intervene and sustain platform stability [19]. The innovative proposition of this study introduces a groundbreaking concept—a multi-link robotic arm strategically programmed to locomote synchronously yet in opposing directions. The overarching objective is to eradicate reaction torques on the robot platform, thereby significantly curtailing the control efforts needed to maintain spacecraft stability during manipulator activities.

In the field of space robotics, a wealth of literature provides in-depth analyses of modeling and dynamics, with a particular focus on configuration and attitude control for exploration stations featuring extended and multi-tasked robotic manipulators designed to manipulate small masses [20, 21]. The methodologies employed in these studies leverage various attitude control techniques in conjunction with Lagrangian mechanics, unraveling the intricate dynamics of the system. Playing a pivotal role in guiding the rotation-vibration motion of the space station, the linear quadratic regulator (LQR) technique stands out. The control architecture, showcased in these analyses, demonstrates a keen ability to account for the elastic flexibility inherent in manipulator links, exerting a significant influence on the attitude movement of the entire system. Another noteworthy study [22] explores meeting and docking scenarios involving non-cooperative targets, employing dual quaternions in tandem with position, orientation, and motion estimation to unravel relative attitude motion dynamics. Comprehensive overviews of dynamics and control in satellite-mounted robots are expounded upon in selected studies [23].

These investigations address intricate challenges such as the dynamics of robots mounted on unstable bases, the nuanced response of satellites to robot maneuvers, and the consequential issues of weak control and acceleration during the coordination of considerable masses, offering valuable insights for microgravity experiments [24]. Further contributions to the field include techniques presented in [25], which delve into the reorientation of free-flying multibody structures through the movement of appendages. Moreover, [25] discusses and experimentally validates the calculation of joint torques for manipulator endpoint control, under the assumption of knowledge regarding vehicle thrust forces. Introducing an innovative concept, [26] puts forth the idea of a simulated roboarms designed to ease the complete mechanics of these robotic platforms. This simulated counterpart, affixed at the system's center of mass, mirrors the real manipulator, facilitating a streamlined derivation of motion. The exploration of autonomous debris capture using robots is the focal point of [27], which employs cutting-edge techniques in advanced machine vision, smart sensing, and operability tailored specifically for space robots. This diverse array of literature collectively contributes to advancing the understanding and capabilities of space robotic systems.

2   Mathematical Model and Manipulator Control

The representative CAD drawing of the integrated robotic arm (manipulator) within the scrutinized space station is visually presented in Fig. 1. The comprehensive methodology employed in this study encompasses both mathematical modeling and numerical simulations, focusing on the dynamic analysis of a robotic manipulator seamlessly integrated into the space station. The utilization of the Differential Kinematic Redundancy (DKR) technique is instrumental in effecting spacecraft attitude control. The simulations are designed with a focus on active state control, delving into the analysis of control efforts under two distinct scenarios. The first scenario involves the robotic manipulator in a non-operational state, while the second scenario explores the control dynamics during manipulator operation. A third scenario is specifically tailored to a unique pose of the robotic platform, where the joints are strategically commanded in different headings, effectively nullifying corresponding attitudes of the system. Thus, this intriguing finding underscores the significance of meticulous equipment placement design, showcasing that thoughtful arm placements can effectively cancel out responses on the entire system. This implies that the placement of the manipulators has minimal impact on the fixed position, with the exception of rotational movements induced by the arms. Such insights highlight the importance of precision in equipment design for optimal space station functionality and stability.

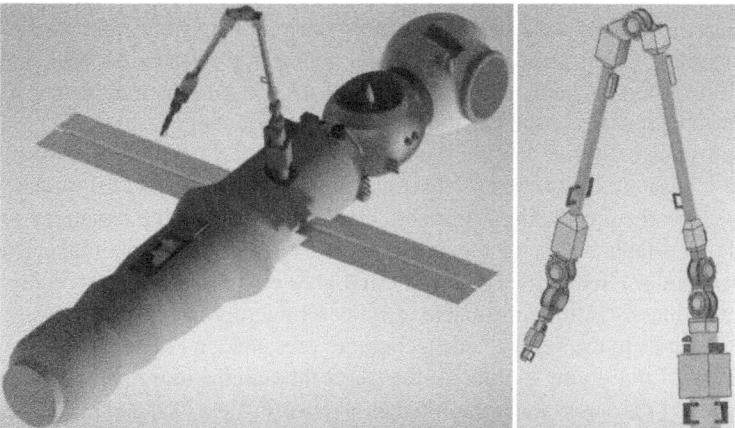

Fig. 1. Representative CAD drawing of the integrated robotic arm (manipulator) within the space station.

Figure 2 presents the physical representation of a robotic manipulator seamlessly integrated into the space station. In this context, the axes x, y, and z correspond to the bank, pitch, and yaw axes, respectively, while $x_{1,2}$, $y_{1,2}$ and $z_{1,2}$ denote the arm axes of the robotic manipulator. It is crucial to emphasize that the consideration of external perturbations was limited to the gravity-gradient torque, originating from the first-order approximation of gravitational potential energy. The mathematical foundation for this analysis involves the Lagrangian mechanics for non-stationary reference frame, yielding modified Euler representation that intricately capture the coupling between the rotational motion of the spacecraft and the motion of the manipulator arms. Leveraging Lagrange's approach for global reference frame, the equations governing the motion of the robotic manipulator are derived. The expression mentioned above for non-stationary reference frame is as follows:

$$\frac{d}{dt}\frac{\partial E_k}{\partial \omega} + \hat{\omega}\frac{\partial E_k}{\partial \omega} = T_e + T_c \tag{1}$$

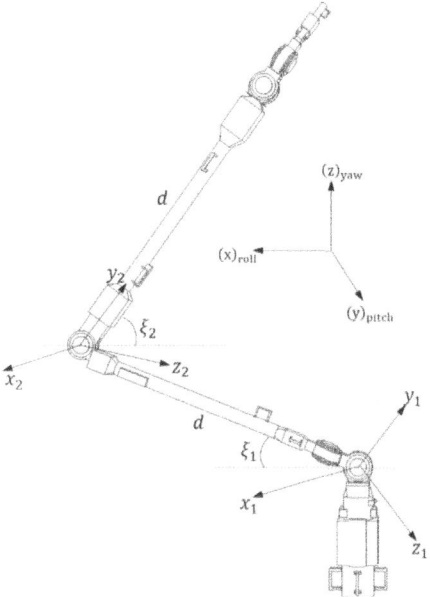

Fig. 2. Model parameters.

This formulation serves as a critical framework for elucidating the intricate dynamics governing the robotic manipulator's motion within the space station environment. In this context, the partial derivative of kinetic energy with respect to angular velocity, denoted as $\partial E_k/\partial \omega$, yields the angular momentum vector. Here, E_k represents kinetic energy, $\widehat{\omega}$ stands for a skewed velocity term involving the elements of the global velocity coefficient, and T_e and T_c denote the vectors representing external and control torques, respectively. The Lagrangian expression for generalized coordinates is articulated by the following equation:

$$\frac{d}{dt}\frac{\partial \mathcal{L}}{\partial \dot{\beta}_m} - \frac{\partial \mathcal{L}}{\partial \beta_m} = F_{\beta_m} \qquad (2)$$

The above equation encapsulates the intricate interplay between kinetic energy, angular momentum, external torques, and control torques within the dynamic framework, providing a comprehensive understanding of the system's behavior and facilitating further analysis of the robotic manipulator's motion in the space station setting. In this context, the Lagrange function, denoted as $\mathcal{L}$, incorporates the degrees of freedom associated with each connection of the robotic manipulator, represented by β_m, where m takes values of 1, 2, and 3. The generalized forces, denoted as F_{β_m}, are integral components in this formulation. The Lagrange function encompasses both kinetic V_k and potential energy V_p. For the specific focus of this study, where potential energy is absent, the Lagrange function simplifies to kinetic energy. As the parts of the arm undergo movement, the total inertia matrix, represented by Y_k, undergoes dynamic changes, manifesting in the following manner:

$$Y_k = Y_i + q_i r_s q_j \tag{3}$$

This dynamic transformation in the inertia matrix captures the evolving characteristics of the robotic manipulator system as its links execute motion, providing essential insights into the system's behavior and response during operational scenarios. In this expression, J_s represents the inertia matrix corresponding to the rotational dynamics of the entire system. The transformation matrix, denoted as Q, establishes the relationship between the system frame—comprising the bank, pitch, and yaw axes—and the frame of the connections in the robotic arm. Additionally, I_a stands for the inertia matrix specific to the robotic arms within the system. The kinematic equations governing the linear dynamics of the system, particularly when small attitude angles are considered, are articulated as follows:

$$\begin{bmatrix} \omega_x \\ \omega_y \\ \omega_z \end{bmatrix} = \begin{bmatrix} 1 & 0 & -s\beta \\ 0 & c\alpha & s\alpha c\beta \\ 0 & -s\alpha & c\alpha c\beta \end{bmatrix} \begin{bmatrix} \dot{\alpha} \\ \dot{\beta} \\ \dot{\gamma} \end{bmatrix} - R \begin{bmatrix} c\beta s\gamma \\ s\alpha s\beta s\gamma + c\alpha c\gamma \\ c\alpha s\beta s\gamma - s\alpha c\gamma \end{bmatrix} \tag{4}$$

These set of equations delineate the intricate relationships between the rotational dynamics of the entire system, the orientation of the robotic arm connections, and the associated inertia matrices, forming a foundational basis for understanding the system's behavior during dynamic operations. In this context, R represents the orbital ratio, a constant value applicable to circular orbits, while α, β, ve γ denote the bank, pitch, and yaw angles. Under these conditions, the differential kinematic equations governing the motion of the orbiting spacecraft are expressed as follows:

$$\begin{bmatrix} \dot{\alpha} \\ \dot{\beta} \\ \dot{\gamma} \end{bmatrix} = c^{-1}\beta \begin{bmatrix} c\beta & s\alpha s\beta & c\alpha s\beta \\ 0 & c\alpha c\beta & -s\alpha c\beta \\ 0 & s\alpha & c\alpha \end{bmatrix} \begin{bmatrix} \omega_x \\ \omega_y \\ \omega_z \end{bmatrix} + R c^{-1}\beta \begin{bmatrix} s\beta \\ c\gamma c\beta \\ s\gamma s\beta \end{bmatrix} \tag{5}$$

3 Results

The derived mathematical model was implemented in a computer simulation, considering two distinct scenarios. The first scenario involves the robotic manipulator in a non-operational state, while the second scenario simulates the robotic arm opening from 0 to 180 degrees. In this configuration, the movement of the robotic arm does not passively counteract reactions, resulting in the base responding to arm retractions. The objective of these simulations is to investigate the attitude-control parameters required to maintain the space station's attitude within predefined limits. By exploring these scenarios, insights into the effectiveness of the control strategy during different manipulator states and movements are gained, providing valuable information for optimizing spacecraft stability during robotic operations.

In Fig. 3 (left), the action control for the station is depicted with the robotic manipulator turned off, accompanied by the corresponding control effort shown in Fig. 3 (right).

The achieved control within 10 s is evident. Notably, it's observed that controlling the bank angle is markedly easier than managing the pitch and yaw angles, aligning with expectations given the robotic manipulator's configuration. While the bank angle control is achieved swiftly, it is apparent that this control demands significantly greater effort compared to the other attitude angles. Figure 4 presents the pose control and corresponding efforts taking into account the operation of the platform, i.e., while the manipulator is in motion. Results indicate that the corresponding effort is almost same as the magnitude larger in comparison with the scenario where the entire platform remain stationary. This increase in control effort during manipulator arm movement aligns with expectations. It's noteworthy that the largest effort required to manage the pose is 0.104 Nm when the arms are static and 0.583 Nm when the arms are in motion. These findings shed light on the intricate interplay between robotic manipulator actions and the ensuing control efforts needed to maintain space station stability.

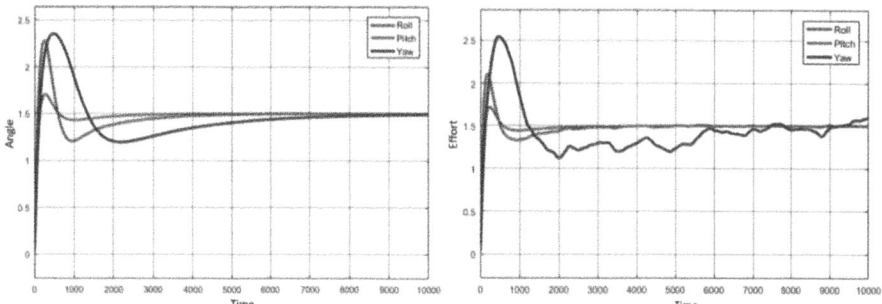

Fig. 3. Attitude control (left) and control effort (right) for the inactive manipulator.

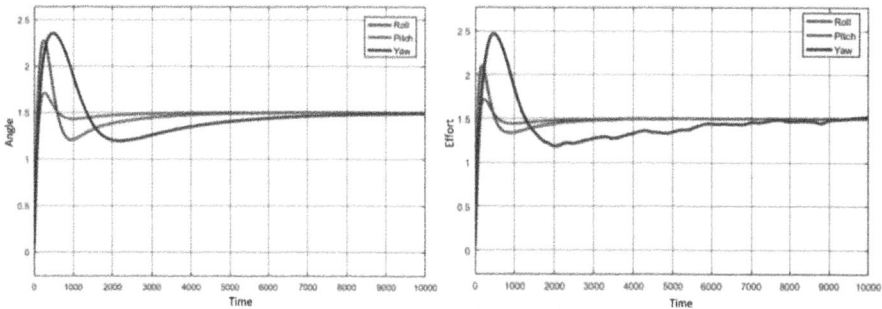

Fig. 4. Attitude control (left) and control effort (right) for the active manipulator.

A notable observation is that the total duration necessary to align the platform's pose with the predicted numbers is longer when the entire platform is in motion. To facilitate a clearer comparison, Fig. 5 presents curves for bank, pitch, and yaw corresponding to both scenarios. The procedure leading to these results commences with setting the bank, roll, pitch, and yaw to 0 positions. The first robot ring operates in positive direction (counterclockwise), however the other one operates synchronously the opposite direction. This orchestrated movement ensures the equilibrium of the entire system when there is no peripheral distortions on the system, following a passive approach. Remarkably, this passive approach obviates the need for control actions to counteract the impact of corresponding force on the base of the platform. The simultaneous and opposite movement of the components effectively eliminates this reaction, affirming the stability of the platform. This nuanced comparison sheds light on the temporal intricacies associated with achieving desired spacecraft attitudes under the influence of robotic manipulator operations.

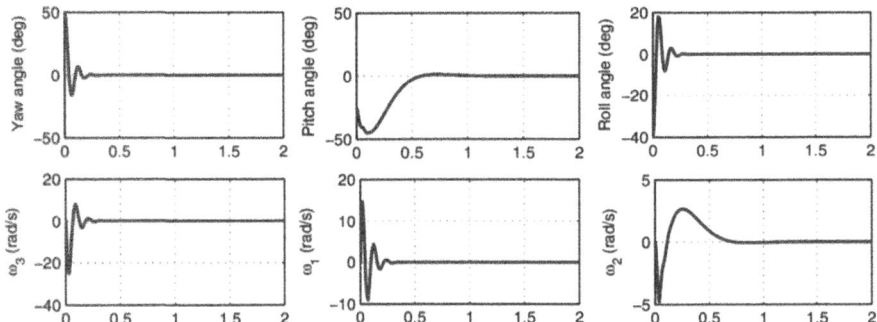

Fig. 5. Comparison of changes in roll, pitch and yaw angles and their corresponding control efforts when the manipulator is inactive and active.

4 Discussion

The presented study explored the implementation and analysis of a derived mathematical model within a computer simulation framework, focusing on two distinct scenarios to assess the effectiveness of a control strategy in maintaining the space station's attitude during robotic operations. As mentioned previously, the first scenario involves the robotic manipulator in a non-operational state, while the second simulates the robotic arm opening fully up to π radians.

The simulation results reveal crucial insights into the dynamic interplay between the robotic manipulator's actions and the corresponding control efforts required to sustain spacecraft stability. As illustrated, the attitude control of the station with the manipulator turned off, accompanied by the control effort. Notably, the control of the bank angle is observed to be significantly more straightforward than managing pitch and yaw angles. This observation aligns with expectations given the configuration of the robotic manipulator. The swift achievement of bank angle control contrasts with the increased effort required, emphasizing the complexity of the control task.

Further investigation delved into the impact of the movement of the entire platform on required pose control and the associated control efforts. The results highlight a notable increase in control effort, approximately one order of magnitude higher, when the manipulator is in motion compared to a stationary state. This escalation in control effort during manipulator arm movement is a critical finding, shedding light on the intricate dynamics of spacecraft stability during robotic operations. The maximum control effort is about 0.1 Nm when the arms are static and exceeds 0.5 Nm when in motion, emphasizing the substantial challenge posed by dynamic manipulator actions. The remarkable fivefold difference in energy consumption underscores the imperative to tailor energy use according to the specific scenario in both cases. This substantial variation not only highlights the need for precise energy management but also serves as a pivotal factor that should be meticulously incorporated into the comprehensive mission details. Within the broader context of the space mission, understanding and accounting for this considerable distinction in energy requirements are pivotal for optimizing operational efficiency, ensuring mission success, and making informed decisions regarding resource allocation and sustainability.

An intriguing aspect worthy of examination is the prolonged duration necessary to align the spacecraft's attitude with predicted values during the operational phase of the robotic manipulator. A more in-depth investigation delves into this phenomenon, offering a comprehensive comparison of the bank, pitch, and yaw curves for both scenarios presented on the same graph. The temporal intricacies inherent in this analysis bring to light the subtle nuances associated with the spacecraft's attitude control.

It becomes evident that the passive approach, involving the orchestrated movement of the manipulator arms from zero to 180 degrees, introduces a notable time delay in achieving the desired spacecraft attitudes. This orchestrated motion, meticulously designed to maintain platform stability without necessitating active control actions, contributes to the extended alignment process. The deliberate synchronization of manipulator links in a counterclockwise and clockwise movement pattern aims to ensure platform stability in the absence of external disturbances, following a passive operational paradigm.

This nuanced observation sheds light on the temporal intricacies involved in spacecraft attitude adjustments under the influence of dynamic manipulator operations. The deliberate, synchronized movement introduces a temporal lag, emphasizing the need for a meticulous understanding of the time-dependent aspects associated with achieving and maintaining the desired spacecraft orientations. This finding contributes to a deeper understanding of the temporal dynamics in robotic operations, providing valuable insights for refining control strategies and optimizing the overall efficiency of spacecraft maneuvers during robotic activities.

Moreover, the reaction forces generated by the movement of the robotic arm on a mobile base distribute dynamic loads throughout the system. These loads vary depending on factors such as arm movement speed, direction, and payload manipulation. They directly impact the mechanical stresses experienced by the robotic arm and the mobile base. High reaction forces can induce significant mechanical stresses on critical components such as joints, actuators, and supporting structures. Analyzing these stresses is essential for identifying potential points of failure and designing components with adequate strength and durability. The reaction forces also influence the motion dynamics of the robotic arm system on a mobile base. These forces affect the system's acceleration, deceleration, and overall stability during operation. Thus, their impact is essential for developing control strategies that mitigate undesirable effects such as vibrations, oscillations, and instability, ensuring precise and smooth operation of the robotic arm. On the other hand, efficient utilization of energy and resources is closely tied to the reaction forces exerted on the platform. High reaction forces may require increased energy consumption to maintain stability and control, affecting overall operational efficiency. Optimizing operational modes to minimize reaction forces can enhance energy efficiency and prolong system autonomy, especially in resource-constrained environments such as space missions. Moreover, determination of the influence of reaction forces is paramount for ensuring the safety of personnel and equipment involved in robotic arm operations on mobile bases. Excessive reaction forces can pose risks of structural damage, component failure, and even personnel injury. Implementing safety protocols and designing fail-safe mechanisms are critical for mitigating potential hazards associated with high reaction forces.

The research presented in this study makes a substantial contribution by providing crucial insights into the optimization of spacecraft stability during robotic operations. The findings not only highlight the significance of this optimization but also underscore the intricate challenges inherent in controlling spacecraft attitude when influenced by dynamic manipulator actions. The study places a strong emphasis on the pivotal need to consider both static and dynamic states in spacecraft control strategies. This dual consideration is fundamental to comprehensively addressing the complexities associated with robotic operations in space. Understanding how the control dynamics differ between static and dynamic scenarios becomes paramount in developing effective strategies for maintaining spacecraft stability under varying conditions.

Furthermore, the nuanced understanding derived from this research serves as a solid foundation for refining control algorithms. The complexities introduced by dynamic manipulator actions necessitate sophisticated algorithms capable of adapting to the changing dynamics of the system. This study reveals key information that can inform the development and enhancement of control algorithms, ensuring their efficacy in diverse operational scenarios. Ultimately, the overarching goal is to enhance the overall performance of robotic systems in the realm of space exploration. By addressing the challenges identified in this study, researchers and engineers can work towards creating more robust and adaptive systems that contribute to the success and reliability of spacecraft operations. This research lays the groundwork for advancing the field of space robotics, fostering innovation, and paving the way for future missions with heightened precision and efficiency.

5 Conclusion

This investigation centers on the dynamic analysis of robotic manipulator operations in orbital environments, specifically addressing scenarios where the manipulator is affixed to an unstable base. Within the framework of attitude dynamics simulations, a comprehensive mathematical model has been formulated to represent a spacecraft with characteristics akin to a robotic manipulator. The primary objective is to scrutinize the effect of robotic manipulator arm movements on pose determination efforts and the temporal requirements for resetting minimal attitude angles to zero. The analysis encompasses a comparative assessment of situations involving both stationary and moving robotic arms. Simulations unequivocally demonstrate an increase in control effort when the robotic manipulator is in motion, corroborating the complex interplay between arm movements and attitude control dynamics. Furthermore, it is established that, under the same conditions, the time needed to nullify small attitude angles is prolonged when the entire platform is in motion. The simulations also specifically explore scenarios where arm movements are simultaneously and orchestrated in reverse directions, strategically designed to counteract corresponding forces on the entire system. Notably, the simulation outcomes affirm that, in this synchronized configuration, the attitude remains unaffected through the movement of the platform, underscoring the efficacy of this approach in maintaining platform stability.

Future studies could explore deeper into the dynamic analysis of robotic manipulator operations in various orbital environments, considering factors such as different gravitational fields, space debris interaction, and varying atmospheric conditions. Because, understanding how these factors influence robotic arm systems can enhance their reliability and performance in space missions. Additionally, research efforts could focus on optimizing control strategies for robotic manipulator arms in dynamic environments. This includes developing advanced control algorithms that can adapt in real-time to changes in the environment and arm movements to minimize control effort and maximize efficiency. Further investigation is warranted into the relationship between robotic manipulator arm movements, pose determination efforts, and attitude control dynamics. Understanding how arm movements affect pose determination accuracy and the temporal requirements for attitude angle resetting is crucial for improving overall system performance. Moreover, future research could also explore the integration of robotic arm systems with autonomous navigation and control systems. This integration would enable the robotic arms to autonomously adapt their movements based on environmental cues and mission objectives, enhancing their capabilities for tasks such as space exploration and infrastructure maintenance. Conducting experimental validation and testing in simulated and real-world environments is essential to validate the findings of theoretical studies. This includes testing robotic arm systems in space analog environments and on orbiting platforms to assess their performance and reliability in actual operating conditions.

References

1. Bakirci, M., Cetin, M.: Improving position-time trajectory accuracy in vehicle stop-and-go scenarios by using a mobile robot as a testbed. J. Control Eng. Appl. Inform. **25**(3), 35–44 (2023). https://doi.org/10.61416/ceai.v25i3.8365
2. Caruso, M., Bregant, L., Gallina, P., Seriani, S.: Design and multi-body dynamic analysis of the Archimede space exploration rover. Acta Astronaut. **194**, 229–241 (2022)
3. Sallabarger, C.: Canadian space robotic activities. Acta Astronaut. **41**(4), 239–246 (1997). https://doi.org/10.1016/S0094-5765(98)00082-4
4. Bakirci, M.: Data-driven system identification of a modified differential drive mobile robot through on-plane motion tests. Electrica **23**(3), 619–633 (2023). https://doi.org/10.5152/electrica.2023.22164

5. Bakirci, M., Toptas, B.: Kinematics and autoregressive model analysis of a differential drive mobile robot. IEEE 4th International Congress on Human-Computer Interaction, Optimization and Robotic Applications, June 9–11, Ankara, Turkey, pp. 1–6 (2022). https://doi.org/10.1109/HORA55278.2022.9800071

6. Sawada, H., Ui, K., Mori, M., Yamamoto, H., Hayashi, R., Mayunaga, S.: Micro-gravity experiment of a space robotic arm using parabolic flight. Adv. Robot. 18(3), 247–267 (2014). https://doi.org/10.1163/156855304322972431

7. Xue, Z., Liu, J., Wu, C., Tong, Y.: Review of in-space assembly technologies. Chin. J. Aeronaut. 34(11), 21–47 (2021)

8. Bakirci, M.: Design and aerodynamic analysis of a rocket nose cone with specific fineness ratio. In: IEEE 6th International Conference on Actual Problems of Unmanned Aerial Vehicles Development, October 19–21, Kyiv, Ukraine, pp. 80–85 (2021). https://doi.org/10.1109/APUAVD53804.2021.9615407

9. Xu, W., Liang, B., Xu, Y.: Survey of modeling, planning, and ground verification of space robotic systems. Acta Astronaut. 68(11), 1629–1649 (2021). https://doi.org/10.1016/j.actaastro.2010.12.004

10. Boumans, R., Heemskerk, C.: The European robotic arm for the international space station. Robot. Auton. Syst. 23(1), 17–27 (1998). https://doi.org/10.1016/S0921-8890(97)00054-7

11. Bakirci, M.: A novel swarm unmanned aerial vehicle system: incorporating autonomous flight, real-time object detection, and coordinated intelligence for enhanced performance. Traitement du Signal 40(5), 2063–2078 (2023). https://doi.org/10.18280/ts.400524

12. Wen, Z., Wang, Y., Luo, J., Kuijper, A., Di, N., Jin, M.: Robust, fast and accurate vision-based localization of a cooperative target used for space robotic arm. Acta Astronaut. 136, 101–114 (2017). https://doi.org/10.1016/j.actaastro.2017.03.008

13. Gu, Y.L., Xu, Y.: A normal form augmentation approach to adaptive control of space robot systems. Dyn. Control. 5, 275–294 (2015). https://doi.org/10.1007/BF01968678

14. Abad, A.F., Ma, O., Pham, K., Ulrich, S.: A review of space robotics technologies for on-orbit servicing. Prog. Aerosp. Sci. 68, 1–26 (2014). https://doi.org/10.1016/j.paerosci.2014.03.002

15. Liu, S., Wu, L., Lu, Z.: Impact dynamics and control of a flexible dual-arm space robot capturing an object. Appl. Math. Comput. 185(2), 1149–1159 (2017). https://doi.org/10.1016/j.amc.2006.07.035

16. Wu, Y.H., Yu, Z.C., Li, C.Y., He, M.J., Hua, B., Chen, Z.M.: Reinforcement learning in dual-arm trajectory planning for a free-floating space robot. Aerosp. Sci. Technol. 98, 1–14 (2020). https://doi.org/10.1016/j.ast.2019.105657

17. Shi, L., Kayastha, S., Katupitiya, J.: Robust coordinated control of a dual-arm space robot. Acta Astronaut. 138, 475–489 (2017). https://doi.org/10.1016/j.actaastro.2017.06.009

18. Yoshida, K.: Achievements in space robotics. IEEE Robot. Autom. Mag. 16(4), 20–28 (2019). https://doi.org/10.1109/MRA.2009.934818

19. Oda, M., Inaba, N., Fukushima, Y.: Space robot technology experiments on NASDA's ETS-VII satellite. Adv. Robot. 13(3), 335–336 (1998). https://doi.org/10.1163/156855399X00784

20. Liu, Y., Chen, J., Liu, J., Jing, X.: Nonlinear mechanics of flexible cables in space robotic arms subject to complex physical environment. Nonlinear Dyn. 94, 649–667 (2018). https://doi.org/10.1007/s11071-018-4383-y

21. Shi, L., Jayakody, H., Katupitiya, J., Jin, X.: Coordinated control of a dual-arm space robot: novel models and simulations for robotic control methods. IEEE Robot. Autom. Mag. 25(4), 86–95 (2018). https://doi.org/10.1109/MRA.2018.2864717

22. Wang, M., Luo, J., Yuan, J., Walter, U.: Coordinated trajectory planning of dual-arm space robot using constrained particle swarm optimization. Acta Astronaut. 146, 259–272 (2018). https://doi.org/10.1016/j.actaastro.2018.03.012

23. Takahashi, R., Ise, H., Konno, A., Uchiyama, M., Sato, D.: Hybrid simulation of a dual-arm space robot colliding with a floating object. In: IEEE International Conference on Robotics and Automation, Pasadena CA, pp. 1–6 (2018). https://doi.org/10.1109/ROBOT.2008.454 3367

24. Yan, L., Xu, W., Hu, Z., Liang, B.: Multi-objective configuration optimization for coordinated capture of dual-arm space robot. Acta Astronaut. **167**, 189–200 (2020). https://doi.org/10.1016/j.actaastro.2019.11.002

25. Imaida, T., Yokokohji, Y., Doi, T., Oda, M., Yoshikawa, T.: Ground-space bilateral teleoperation of ETS-VII robot arm by direct bilateral coupling under 7-s time delay condition. IEEE Trans. Robot. Autom. **20**(3), 499–511 (2014). https://doi.org/10.1109/TRA.2004.825271

26. Chatterjee, A., Chatterjee, R., Matsuno, F., Endo, T.: Augmented stable fuzzy control for flexible robotic arm using LMI approach and neuro-fuzzy state space modeling. IEEE Trans. Industr. Electron. **55**(3), 1256–1270 (2018). https://doi.org/10.1109/TIE.2007.896439

27. Chen, G., Wang, Y., Wang, Y., Liang, J., Zhang, L., Pan, G.: Detumbling strategy based on friction control of dual-arm space robot for capturing tumbling target. Chin. J. Aeronaut. **33**(3), 1093–1106 (2020). https://doi.org/10.1016/j.cja.2019.04.019

Digital Transformation of Supply Chain and Logistics Systems

Blockchain Based Privacy Preservation and Misbehavior Analysis in Financial Supply Chain

M. R. Sumalatha, Aditya Kumar, Nethra Janardhanan$^{(\boxtimes)}$, and S. Abhinash

Department of Information Technology, Madras Institute of Technology, Anna University, Chennai, India

Abstract. The financial supply chain is vital for economic growth, enabling business expansion and innovation. However, it's challenged by issues like lack of standardization, fraud risks, and limited transparency. The Privacy Preserving Blockchain Misbehavior Detection (PPBCMD) scheme aims to tackle these through blockchain technology for secure, transparent data sharing and storage. Cryptographic techniques like Zero Knowledge Proofs (ZKP) proves the authenticity of data without disclosing additional information, and Partial Homomorphic Encryption (PHE) to ensure confidentiality and perform computations on encrypted data provide a fine-grained approach to data security, ensuring that sensitive financial data remains secure and confidential. By employing privacy-trained ML models that utilize differential privacy using laplace noise on encrypted data, the proposed system ensures the confidentiality of sensitive fields while effectively detecting any instances of misbehavior within the financial supply chain characterized by unusually large transactions and those happening at atypical times. The system was validated using the PaySim dataset, with the Random Forest algorithm achieving a 97.59% F1 score after noise while preserving the sensitive financial data fields.

Keywords: Blockchain · Privacy Preservation · Zero Knowledge Proofs · Partial Homomorphic Encryption · Differential Privacy

1 Introduction

The financial supply chain, a critical component of the global economy, plays a crucial role in ensuring the smooth flow of capital and funds, facilitating business transactions between suppliers and customers. It serves as the backbone for maintaining healthy cash flow, managing risk, and ensuring timely payments to suppliers and vendors. The efficient functioning of the financial supply chain is vital for sustaining transparency, accountability, and trust among different parties involved. However, the current system faces numerous obstacles such as transparency issues, data breaches, fraudulent activities, and compliance challenges. Ensuring the trustworthiness of management and facilitating effective supervision of storage facilities have become increasingly arduous. This has resulted in financial institutions displaying a tendency to reduce lending and

© The Author(s), under exclusive license to Springer Nature Switzerland AG 2025
A. Mirzazadeh et al. (Eds.): ODSIE 2023, CCIS 2205, pp. 231–248, 2025.
https://doi.org/10.1007/978-3-031-81458-7_14

expedite loan recovery. The situation is compounded by hindrances in the movement of substantial quantities of goods, leaving traders constrained by limited capital and facing challenges in obtaining loans. As a result, the progress of business endeavors is impeded. However, the emergence of blockchain technology provides a promising solution for addressing challenges related to transparency and oversight, enabling consensus among participants. By leveraging blockchain's decentralized and tamper-resistant properties, the financial supply chain can potentially overcome obstacles and pave the way for more secure, efficient, and trustworthy operations. In parallel, traditional methods of computation and cryptographic systems exhibit limitations when it comes to securing financial data. Furthermore, cryptographic systems such Blockchain-based proxy re-encryption with equality test (BPREET) [1], Cheon-Kim-Kim-Song (CKKS) [2], Attribute-based encryption (ABE) [3], and Iseki's FHE [4] have concerns regarding security and high computational costs. The performance and scalability of Fully Homomorphic Encryption (FHE) and Multi-Party Computation (MPC) in general require improvement to handle large-scale financial data computations. In addition, the advent of a public blockchain-based invoice finance platform [5] marks progress in data confidentiality and the use of blockchain for a reputation system but suffers from its dependence on PGP encryption, which falls short in supporting privacy-preserving computations or independent verification. Moving forward, a hybrid data storage approach [6] was proposed to merge a centralized database with blockchain, aiming to mitigate high data load and boost privacy in blockchain traceability systems, yet it faces challenges with the central database's vulnerability to tampering and the limitations of SHA256 not supporting homomorphic properties. A comparative study of partial homomorphic encryption methods [7] emphasized the computational efficiency and security advantages of Paillier and ElGamal cryptosystems over unpadded RSA due to their semantic security.

Hence, the development of novel cryptographic primitives like Zero-Knowledge Proofs (ZKP) and Partial Homomorphic Encryption (PHE) is needed to enhance security guarantees while minimizing computational overhead. Against this backdrop, it becomes evident that establishing standardization and interoperability protocols for cryptographic systems is crucial to enable their integration and deployment across diverse financial supply chain networks. Furthermore, the emergence of blockchain technology offers a promising solution to address transparency and oversight challenges, as it enables consensus. Through harnessing the decentralized nature and tamper-resistant characteristics of blockchain, the financial supply chain holds the potential to surmount barriers and establish a pathway towards enhanced security, efficiency, and reliability in its operations. Building upon this potential, we propose the implementation of a Privacy Preserving Blockchain Misbehavior Detection (PPBCMD) system. By integrating privacy preservation measures with robust detection mechanisms, this system will further strengthen the integrity and trustworthiness of the financial supply chain, ensuring the identification and mitigation of fraudulent activities while safeguarding sensitive information. The contributions of this paper are as follows:

1) We propose the implementation of a permissioned financial supply chain on the Ethereum blockchain, utilizing smart contracts. This approach fosters trust and transparency between suppliers and vendors, facilitating efficient and secure transactions.

2) Secondly, to ensure the confidentiality and integrity of transaction data, advanced cryptographic techniques such as Zero Knowledge Proofs and Partial Homomorphic Encryption were used. By utilising these technologies, we can achieve data encryption while preserving authenticity, confidentiality, and privacy.

3) Lastly, we propose the development of a robust misbehavior detection model. This model would be trained on encrypted data using differential privacy based on laplace noise and machine learning algorithms. By employing these techniques, we can effectively identify and flag fraudulent transactions, enhancing the overall security and reliability of the financial supply chain.

2 PPBCMD Architecture and System Design

A novel Privacy Preservation BlockChain Misbehaviour Detection (PPBCMD) system is framed in this work. Figure 1, depicts the architecture diagram of our proposed solution. Each components of the diagram will be discussed further.

2.1 React Frontend

The first step in the design flow is the frontend component that involves React and Redux. The frontend is responsible for taking form submissions and establishing a WebSocket connection with Ganache, which is a personal blockchain for Ethereum development. The frontend is built using React and Redux, which are popular JavaScript libraries for building user interfaces and managing application state. The frontend allows users to submit financial transactions which are of 2 types - Purchase type which is a form to purchase commodities from selected vendors and Loan type which is a form to request capital from finance providers. The form submission details will require buyer name, vendor he selects, product name, product quantity, hour of day. The amount is generated after selecting the product specifics. On submitting the form, the above data is then encrypted using the Paillier homomorphic encryption and zero-knowledge proofs using zk-SNARKs, Circom, and Groth library of NodeJS. The encrypted data is then sent to the smart contract on the blockchain and the funds are successfully transferred from the buyer to the vendor. To interact with the blockchain, the frontend establishes a WebSocket connection with Ganache. This allows the frontend to send and receive data from the blockchain in real-time. The WebSocket connection is established using the Web3.js library, which is a JavaScript library for interacting with Ethereum. The smart contracts are coded using Solidity programming language and migrated to Ganache test network. Its functions are to handle purchase form, triggering of purchase, shipping and receiving events, bill generation, loan form with handling of loan request and loan approved events, whitelist addition, detecting if the transaction is abnormal or normal and more. To interact with the smart contracts on the blockchain, the frontend uses the Metamask wallet. Metamask is a browser extension that allows users to interact with Ethereum-based applications using their Ethereum wallet. The frontend sends transactions to the smart contract using Metamask, which signs the transaction and sends it to the blockchain.

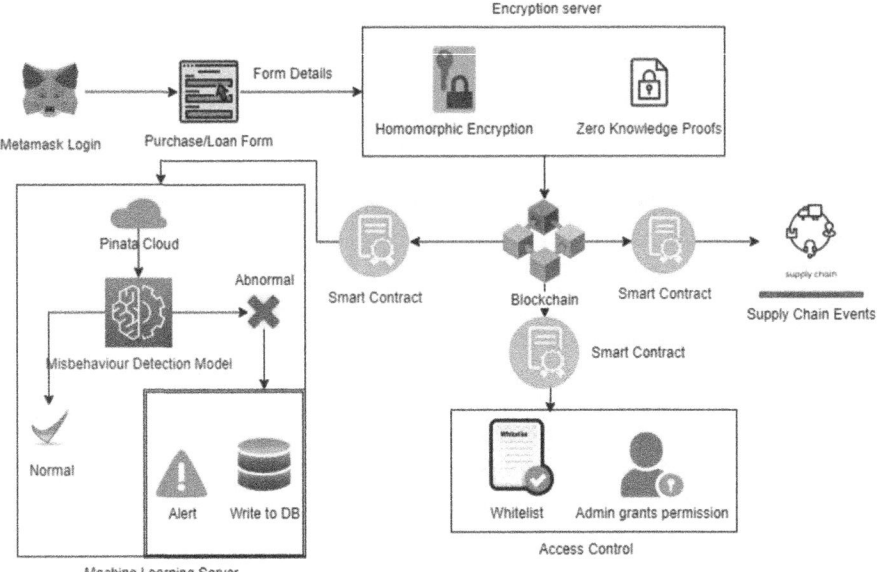

Fig. 1. PPBCMD Architecture.

2.2 Encryption Server

The second step in the PPBCMD system design is data encryption. The financial trans-actions between buyers and vendors are encrypted using two types of encryption tech-niques - Paillier homomorphic encryption for the amount and hour of the day, and zero-knowledge proofs using zk-SNARKs, Circom, and Groth library of NodeJS for encrypting the remaining fields. Paillier homomorphic encryption is a type of encryp-tion that allows for computations to be performed on encrypted data without decrypting it. This is useful for financial transactions, as it allows for calculations to be performed on the encrypted data without revealing the sensitive information. Zero-knowledge proofs using zk-SNARKs, Circom, and Groth library of NodeJS are used to encrypt the remain-ing fields. This ensures that the sensitive data is protected and cannot be accessed by unauthorized parties.

The working of the encryption algorithms in this scenario is as follows. Paillier homomorphic encryption is used to encrypt the amount and hour of the day in the financial transactions. The encryption process involves generating a public key and a private key. The public key is used to encrypt the data, while the private key is used to decrypt the data. The encrypted data is then sent to the smart contract on the blockchain. Zero-knowledge proofs using zk-SNARKs, Circom, and Groth library of NodeJS are used to encrypt the remaining fields in the financial transactions. The encryption process involves generating a proof that the data is valid, without revealing the data itself. This is done using complex mathematical algorithms that generate a proof that the data is valid, without revealing any information about the data itself. The proof is then generated and verified using the groth16 library. Once the verification of proof is completed in the NodeJS backend, the generated proof is submitted to the blockchain to depict the zk-SNARK encrypted fields.

2.3 Smart Contracts for Access Control and Supply Chain

The next step in the PPBCMD design is the smart contract. The encrypted data is loaded into a smart contract on the blockchain. The smart contract contains the rules and logic for processing the financial transactions. When the transaction with the account is received into the smart contract, first it is checked if the account is added to the whitelist. If it is added, then the transaction is allowed to proceed else an alert box is displayed stating to request admin for approval. There is an admin master account that takes care of approving the whitelist request in the form of another smart contract whose contents are stored on the blockchain.

In the case of submitting a purchase/loan form, the contract checks whether the required input fields are submitted with a valid buyer and seller account to transfer the funds along with encrypted details. The smart contract containing the fraud detection module is called and checked if the pattern is abnormal. If the pattern is classified as abnormal by the machine learning model, an alert box similar to a flag is generated and shown in the webpage which can be checked later by the regulatory authorities. If regulatory authorities conclude that some abnormal transaction has taken place, they can take appropriate action against the miscreants. In the case of normal classification, the purchase event is emitted and triggers the shipping smart contract and the received smart contract in the form of events. All the events are stored in the blockchain and also retrieved from the chain and viewed in the webpage to show overview of the account activity in the form of events/transactions history.

2.4 Misbehavior Detection Model

The fourth step in the system design is the misbehavior detection model as shown in Fig. 2. The machine learning model is designed to identify unusual patterns or outliers in the data that may indicate fraudulent activity. The misbehavior detection model is trained on differential privacy applied data using Laplace mechanism with various algorithms such as XGBoost, Random forest, Decision tree and Extra tree to ensure that the privacy of the sensitive data is protected and detected efficiently on the encrypted data. The fundamental use of the Laplace noise mechanism is to protect individual privacy in the PaySim data by adding carefully calibrated noise from a Laplace distribution, thus providing differential privacy protection that conceals the presence or absence of individual records in a dataset. The machine learning model is loaded into the smart contract and is used to detect fraud in the financial transactions. The smart contract processes the encrypted data using the machine learning model retrieved from Pinata IPFS and outputs whether the transaction is fraudulent or not. This ensures that fraudulent activity is detected and prevented, protecting the financial supply chain from losses in multiple scenarios.

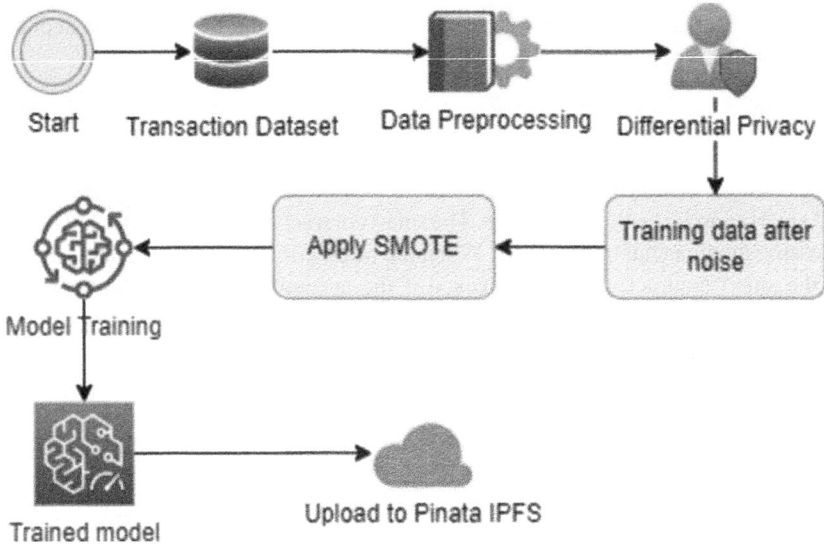

Fig. 2. Misbehaviour Detection Model Training.

3   Algorithm and Data Flow

3.1   Purchase Form Submission to Blockchain

The user submits the purchase form with required fields product, product quantity, amount, vendor address, hour of day, buyer name and vendor name. Before storing to on-chain, 2 kinds of encryptions are performed offering multiple levels of data security. Zero Knowledge Proofs ensures authenticity and allows for easy validation of data by other nodes without them needing to have knowledge of what data is present.

Once encryption is completed, the smart contract is called to check if the user has access control privileges. If user is not possessing access control, the contract exits and a prompt is generated to apply for the whitelist. Else, the data can be tested by the misbehavior detection function in smart contract and output is obtained [Algorithm 4]. If the output states that data is abnormal and not matching with the pattern, an alert is prompted stating that abnormal behavior is detected and the data is added to SQLite DB in a table for flagged records. The orderID is generated using the keccak256 hash of the fields and state variables are stored. The funds are transferred between parties and the event in the next stage of the supply chain is emitted which is ShippingEvent() followed by ReceivedEvent() in the exact order.

Algorithm 1 Algorithm for Purchase Form Flow

Require: Buyer Account Bu_{acc}, Backend Server *server*

   Inputs: *fields* ← Buyer account Bu_{acc}, Vendor account V e_{acc}, Buyer name Bu_n, Vendor name V e_n, Amount *amt*, Product Name Pr, Quantity *qty*, Hour of day *hr*

   amt,hr ← POST **/encryptHE** on *server* - performing homomorphic encryption on | fields *amt, hr*

   proof ← POST **/encryptZKP** on *server* - performing zk-snark encryption on remaining fields

   Smart contract Purchase() method called:

   procedure PURCHASE(buyerAccount,VendorAccount,amt,hr,proof)

      State variables:

      Whitelist contract $W_{contract}$

      List of accounts added to whitelist: W_{arr}

      List of Purchase orders: *orders*

      List of ZKP Proofs: *zkp*

      if Bu_{acc} **not in** W_{arr} **then**

         output("User not in whitelist")

         exit

      else

         emit Event MLPredict() of ML contract

         valid ← predictOutput()

         if *valid* is false **then**

            Alert("abnormal detected")

            addtoFlaggedDB(fields)

         end if

         orderId ← keccak256(abi.encodePacked(buyerAccount,vendorAccount,amt, hr));

         order ← Order(orderId, buyerAccount,vendorAccount,amt, hr)

         zkp.push(proof)

         orders.push(order)

         payable(vendorAccount).transfer(msg.value);

         emit Event ShippingEvent()

      end if

   end procedure

   end

3.2 Algorithms for Encryption

Homomorphic Encryption Using Paillier. Homomorphic encryption ensures confidentiality and easy misbehavior analysis as it allows computations to be performed on the encrypted data without the need to decrypt it first. Thus, in this case, it is perfect to use in the case of machine learning to test on encrypted data. First, homomorphic encryption is performed to encrypt the 2 fields upon which the model will test i.e. amount and hour of day [Procedure 1]. Paillier generates public and private keys and encrypts the 2 fields. But since conversion should be done to BigInt and stored, it exceeds the gas limit. Thus, it should be converted to a suitable format like base64 encoding. The encrypted fields are pushed to a byte array and converted to string using Buffer.from() method. In this way, substantial gas fees are saved when storing on the chain.

```
procedure EncryptHE(fields)
    pubKey, privateKey ← paillier.generateRandomKeys(2048)
    encryptedAmount ← pubKey.encrypt(BigInt(amt)).toString(16)
    encryptedHr ← pubKey.encrypt(BigInt(hr)).toString(16)
    byteArrayAmt ← [], byteArrayhr ← []
    for i from 0 to encryptedAmt.length do
        byteArrayAmt.push(parseInt(encryptedAmt.substr(i, i + 2), 16))
    end for
    for i from 0 to encryptedHr.length do
        byteArrayHr.push(parseInt(encryptedHr.substr(i, i + 2), 16))
    end for
    encryptedBase64amt ← Buffer.from(byteArray).toString('base64')
    encryptedBase64hr ← Buffer.from(byteArrayhr).toString('base64')
    return encryptedBase64amt, encryptedBase64hr
end procedure
```

Encryption using Zero Knowledge Proofs. Next, the rest of the fields are stored using ZK-Snarks [Procedure 3]. The prerequisite for ZKP encryption using snarkJS library is that an arithmetic circuit containing the private signals, output commitment signals, the constraints of the circuit should be defined. **[Procedure 2].**

Here private signals are the data to be encrypted, and commitment is the public output visible to all which is used to satisfy constraints. Once the circuit is defined, it is compiled to obtain a circuit.json file. This file is used to obtain Proving and Verification keys with which a zero knowledge proof can be generated and verified using the groth16 library. If the verification returns true, the proof has been successfully generated and the flow proceeds further in the function.

3.3 Misbehaviour Detection Algorithm Flow

After training and saving the machine learning model as a pickle file, it is then uploaded to the Pinata cloud storage platform from the Colab notebook using the 'pinatapy' library. The CID returned from the upload is stored in the ML smart contract as a state variable. Once the purchase function is called, the ML contract is called in the flow for misbehavior detection.

A Flask websocket server is set up in the backend with controller function taking inputs as fields and the CID of uploaded pickle model. The model is retrieved from pinata and unpickled in the server. It then makes a prediction against the data passed in the system to get a value of 0 normal or 1 abnormal. The prediction output is passed to the ML contract again from the Flask Server. From the contract, the output is passed back to the purchase contract.

procedure Circuit Creation

    Private signals: → *buyerName, vendorName, productName, productQuantity*

    Commitment Signals: → *buyerCommitment, vendorCommitment, product-Commitment, quantityCommitment*;

    Hash function: component *hash = SHA256()*

Compute commitments using the hash function:

hash.in[0] == buyerName
hash.out[0] == buyerCommitment;
hash.in[1] == vendorName;
hash.out[1] == vendorCommitment;
hash.in[2] == productName;
hash.out[2] == productCommitment;
hash.in[3] == productQuantity;
hash.out[3] == quantityCommitment;

Define the constraints of the circuit:

    enforce *validBuyer* = (hash.in[0] == *buyerName* and *buyerCommitment* == hash.out[0]);

    enforce *validVendor* = (hash.in[1] == *vendorName* and *vendorCommitment* == hash.out[1]);

    enforce *validProduct* = (hash.in[2] == *productName* and *productCommitment* == hash.out[2]);

    enforce *validQuantity* = (hash.in[3] == *productQuantity* and *quantityCommitment* == hash.out[3]);

Define the input signal:

component InputSignal(private a, private b, private c, private d)

Define main component and connect input signal to output signal:

    component main = InputSignal(buyerName, vendorName, productName, productQuantity)

    main.buyerName → buyerCommitment;
    main.vendorName → vendorCommitment;
    main.productName → productCommitment;
    main.productQuantity → quantityCommitment;

    circuit ← snarkjs.Circuit(circuitPath)
    provingKey, verificationKey ← snarkjs.keygen(circuit)

end procedure

---

procedure   EncryptZKP(*buyerName,*   *vendorName,*   *productName, productQuantity*)

Require: Proving key *provingKey*, Verification key *verKey*

    witness ← *buyerName, vendorName, productName, productQuantity*

    proof ← groth16.genProof(provingKey,witness)

    booleanV erification ← groth16.verifyProof(verKey, proof)

    if *booleanV erification* **is** false **then**

        exit

    else

        return proof

    end if

end procedure

Algorithm 2 Misbehaviour Detection Algorithm Flow

Require: Purchase() function emitted ML() event, Pickle file uploaded to Pinata Cloud IPFS
    Inputs: *fields* ← buyerAccount Bu_{acc}, vendorAccount Ve_{acc}, encryptedAmount *amt*, encryptedHourOfDay *hr*, Proof *proof*
    procedure tEST(fields)
Require: CID of model *cid*, Flask server *fl*
        emit Event testData(fields,cid)
        in Flask server:
        pickledModel ← GET https://gateway.pinata.cloud/ipfs/(cid)/
        with open(pickledModel, 'rb') as file: *model* ← pickle.load(file)
        prediction ← *model*.predict(fields)
        calls function predictOutput(prediction) of ML Smart Contract
        procedure PREDictOutPut(prediction)
            if *prediction* **is** 0 **then**
                output "normal"
            else
                output "abnormal"
            end if
            return prediction
        end procedure
    end procedure

---

4 Implementation

In the proposed method, a whitelist was used to allow only approved entities to access a specific system or service. This approach is a proactive security measure that aims to reduce the risk of security breaches by limiting the attack surface. Whitelisting blocks untrustworthy or unauthorized entities from accessing a system or service. With a whitelist in place, the potential for malicious activity is significantly reduced as only approved entities are granted access or privileges. Figure 3 and Fig. 4 shows User request whitelist access from admin to make purchases and a master admin account able to add the user to the whitelist using smart contracts. After requesting access to the whitelist, the user is authorized to make transactions. If user wasn't added to the whitelist, an alert would generate asking him to apply to the whitelist. The supply chain of events can be viewed in the events dashboard for a particular user when he/she is logged in. In the below figure, we have simulated some blockchain flow without encryption to show the traceability of our supply chain. The events emitted in our chain are Order Event, Shipped, Received. In Fig. 5, we have showed an example of a product ordered and shipped. It is clear it is the same product by viewing the order ID hash, buyer, vendor and other fields. In this way, the integrity aspect of the supply chain is satisfied. Now, the user can make transactions such as a purchase request or a loan request. We have multiple vendors selling multiple commodities at different price points to simulate a real-world supply chain. The data goes to the encryption server written in NodeJS performing Paillier homomorphic encryption and encryption using ZK-snarks to encrypt the data before moving it on the chain. This is done as performing huge encryption on chain would consume too much gas. In the figure, we have made a purchase request of 1 glove from Apollo Hospital and we can see the new form of the data stored in Ganache after being submitted to the chain, checked if it is part of the whitelist, and if it is, run through the misbehavior detection model for abnormal transaction flagging. Some scenarios, our misbehavior model can flag anomalous transactions which it classifies by anomalous according to the patterns it has learnt in the training phase by searching for outliers in the existing learned patterns.

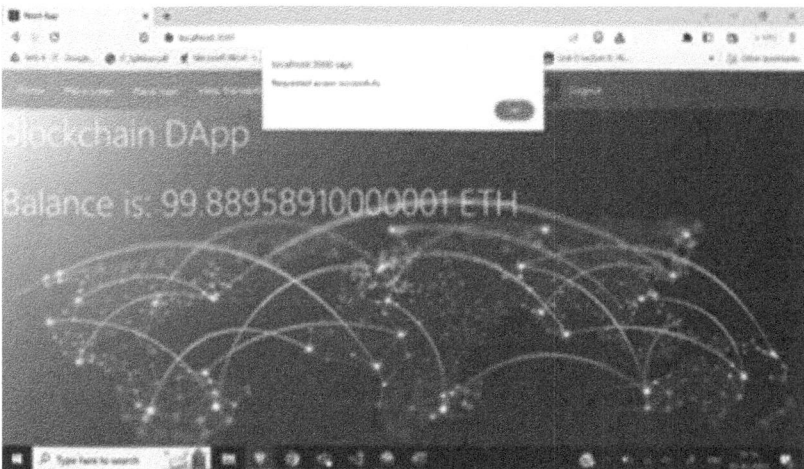

Fig. 3. Requesting access to whitelist.

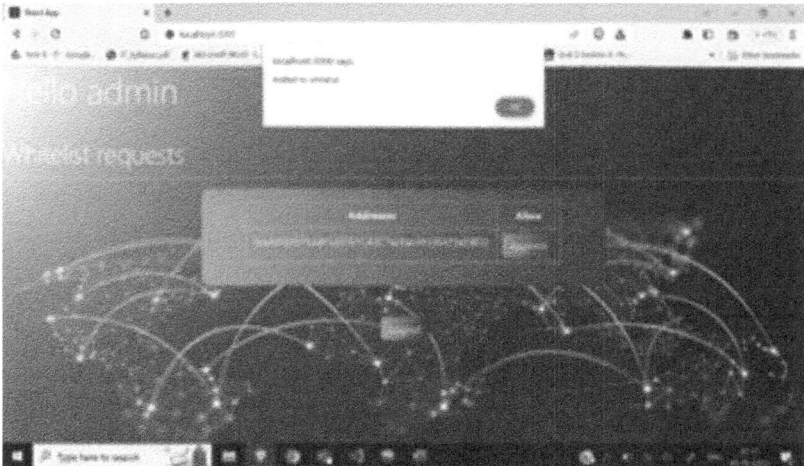

Fig. 4. Adding requested account to whitelist by admin.

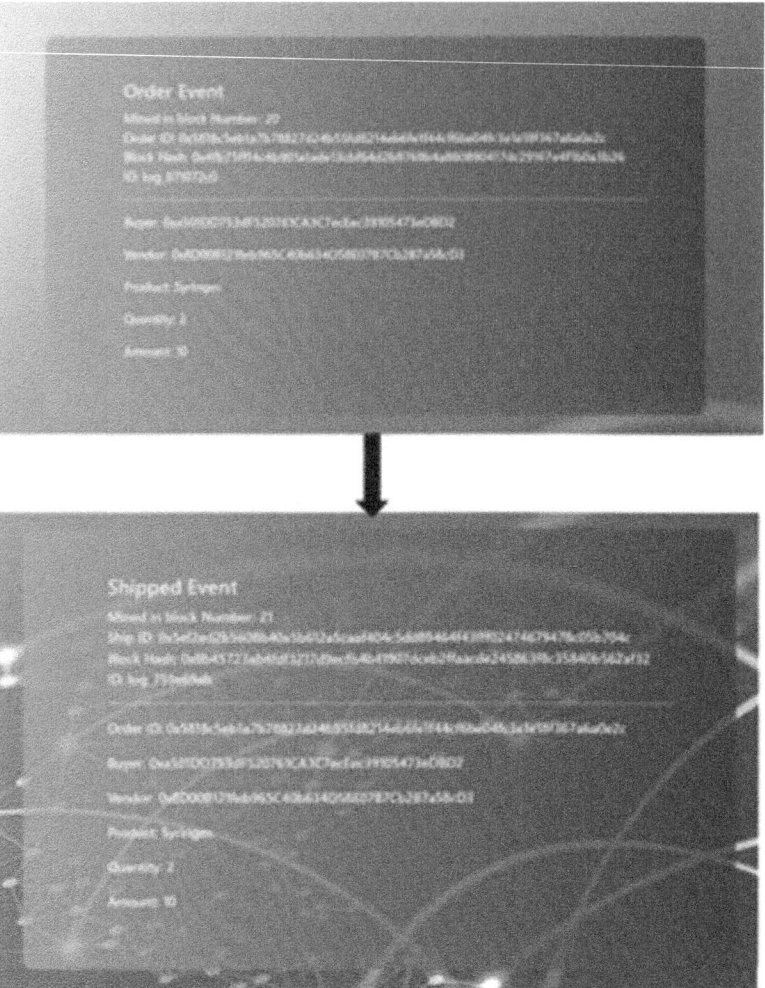

Fig. 5. Supply chain events for user in events dashboard.

5   Results and Discussion

The PPBCMD system outperformed several state of the art blockchain based systems offering functionalities like data encryption, confidentiality and privacy, access control using machine learning model to flag anomalous transactions by identifying potential fraud characterized by transactions significantly higher than usual amounts in the dataset and activities occurring at irregular hours, and alerting the admin authority. The contracts used were for Purchase, Whitelist, Loan Request, Loan Approved, Ship and Received pertaining to the Purchase and Loan functionalities of the system as well as predictOutput contract for the machine learning prediction.

The analysis extracted were of cost of each contract in terms of gas used converted to ETH and rupees, quantitative analysis of encryption used covering time taken to encrypt in terms of input size and input key pairs displaying the system's secure properties as well as qualitative analysis comparing the PP-BCMD system with other existing state of the art systems and reviewing its features. Metrics such as accuracy and F1 score of the 4 algorithms were used to evaluate the machine learning model and also depicted in the form of tables and bar graphs showing the comparison of the model before adding differential privacy and after its addition. Also, several other existing system's machine learning metrics implementing similar use cases are displayed.

5.1 Cost Analysis

In 2023, assuming a gas price of 40 gwei, 1 gwei roughly translates to 0.000000001 ether. The cost in terms of ether and rupees are calculated for each of the smart contract functions. The costs are retrieved from Ganache blockchain after the transaction is executed calling the particular contract function (Table 1).

Table 1. Gas costs of the smart contract functions.

Function	Gas used	Cost (Ether)	Cost (INR)
Contract Creation	1784532	0.0356	8917.46
Purchase	540394	0.0135	3377.46
Ship	230099	0.0069	1725.74
Received	296470	0.0118	2964.7
loanRequest	241731	0.0084	2114.54
loanApprove	320942	0.0160	4011.77
requestAddToWhiteList	442052	0.0162	4063.24
adminApproveWhitelist	359008	0.0125	3141.57
predictOutput	198035	0.0079	1980.35

Thus, we observe that the functions in total consume about 0.142 ETH in gas fees which is reasonably low compared to other state of the art systems. Thus, the quantitative cost analysis is performed.

5.2 Quantitative Analysis of Paillier Encryption Module

Analysis is performed to find the time taken for encrypting the amount, hour of day fields using paillier homomorphic encryption by starting a timer when the homomorphic encryption function starts and timer stopped when the control returns back to the function after encryption. The public and private key pairs are generated using input bits. The more the input bits, the stronger the computation and therefore, longer time for encryption and decryption. Since encryption time is not short when input size bits are 2048, we can conclude that the data is encrypted strongly (Table 2).

Table 2. Time taken to encrypt and decrypt Paillier homomorphic encryption (insecs).

No. of key pairs (bits)	Time to encrypt	Time to decrypt
128	0.0069	0.0047
256	0.0345	0.0186
512	0.1824	0.0652
1024	1.1909	0.3827
2048	8.4163	2.2177
4096	17.9891	5.7053
8192	43.3921	14.0339

5.3 Quantitative Analysis of ZKP Encryption Module

The encryption time for zero knowledge proof takes longer than homomorphic encryption according to analysis conducted below suggesting an efficient encryption algorithm. The time for proof generation and proof verification increases with increase in the input data size (in bytes). The data is obtained by submitting forms and starting the timer when the proof generation starts and stops to obtain time for proof generation and similar for calculating the proof verification. The test cases given were of varying size to the buyerName field. Initially, a small buyerName of 10 bytes was submitted to the form, then 50 bytes, 100 bytes, 150 bytes and 200 bytes were submitted and time calculated. Thus, the encryption is analyzed thoroughly and metrics presented.

5.4 Qualitative Analysis of PPBCMD System

In the references reviewed, most works did not implement all the 5 criteria of Data Encryption, Confidentiality, Privacy, Integrity and Access Control. They relied on a single source of storage for their data as well which leads to a failure in fault-tolerance (Table 3).

Table 3. Time taken to encrypt Proof (secs) Vs. Input Data Size (bytes).

Input Data Size (bytes)	Time to generate	Time to verify
10	3.6091	2.2706
50	7.2086	5.9330
100	13.9008	10.2870
150	19.0502	14.6776
200	25.3771	18.5400

In our proposed work, we are satisfying all the 4 criteria and using multiple modes of storage for different types of data. The access control criteria is satisfied by whitelisting by admin, data encryption ensured using 2 kinds of encryptions, confidentiality ensured by homomorphic encryption for the amount and hour of day fields, IPFS to upload media and access with its CID, and privacy ensured by use of zero knowledge proofs which allow validation of data through the blockchain by other nodes without underlying information of the transferred data (Table 4).

Table 4. Qualitative Analysis of PPBCMD Model.

Reference Work	Blockchain	Consensus	Encryption	Confidentiality	Integrity	Privacy	Storage
[8]	Ethereum (public)	PoW	x	✓	x	x	Central Database
[9]	Ethereum (private)	PoA	✓	✓	✓	✓	Central Database
[10]	Ethereum (private)	Not stated	✓	x	x	✓	IPFS
[11]	Ethereum (private)	Not stated	x	x	x	x	Central Database
[12]	Bitcoin	PoW	x	✓	✓	✓	Blockchain
Proposed Work	Ethereum (public)	PoW	✓	✓	✓	✓	Blockchain with IPFS

5.5 Analysis of Misbehaviour Detection Model

Accuracy Scores Before and After Laplace Noise The results were evaluated on the PaySim dataset which is a publicly available dataset that simulates mobile money transactions. The dataset consists of over 5 million mobile money transactions. The features of the transaction dataset consisting of 11 columns which is used for detecting fraudulent transactions include Step, type, amount, nameOrig, oldbalanceOrg, newbalanceOrig, nameDest, oldbalanceDest, newbalanceDest, isFraud, isFlaggedFraud. From Table 5 and Table 6, we can infer that Random Forest has higher training and testing accuracies standing at 98.89% & 98.13% before noise respectively, and 97.25% & 96.28% after noise respectively.

Table 5. Accuracy Scores Before Laplace Noise (in %).

XGBoost	Random Forest	Decision Tree	Extra Tree
Training98.04	98.89	98.15	97.54
Testing97.78	98.13	97.44	96.70

F1 Scores Before and After Laplace Noise. We list the F1 scores of different algorithms for the proposed method before and after adding laplace noise, from this we can

Table 6. Accuracy Scores After Laplace Noise (in %).

XGBoost	Random Forest	Decision Tree	Extra Tree
Training96.71	97.25	95.19	94.66
Testing94.77	96.28	94.98	93.82

infer that the Random Forest algorithm has the highest F1 score of 99.48% before noise and the highest F1 score of 97.59% after noise (Figs. 6 and 7).

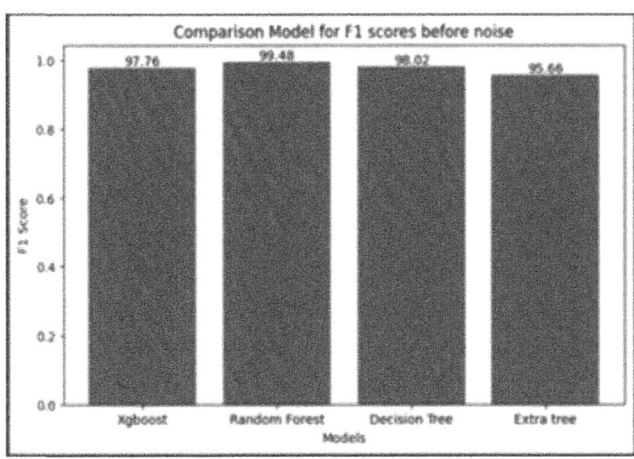

Fig. 6. F1 scores before Laplace noise.

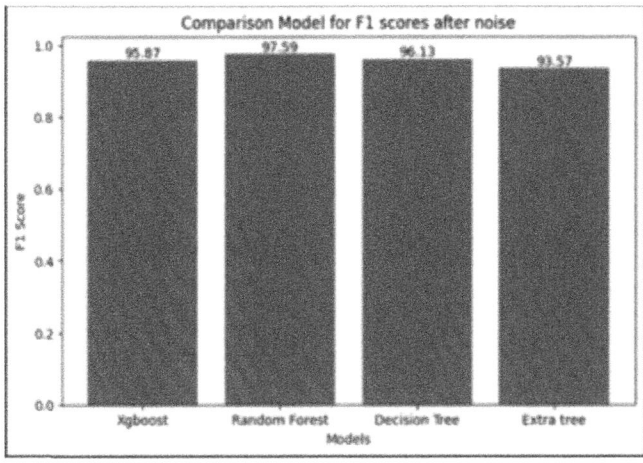

Fig. 7. F1 scores after Laplace noise.

6 Conclusion and Future Work

6.1 Conclusion

In conclusion, the system design for a financial supply chain system that incorporates blockchain, supply chain, encryption, and machine learning modules is a complex and innovative solution that addresses the challenges of secure and efficient financial transactions. The use of blockchain technology and smart contracts ensures that transactions are tamper-proof and transparent, while the use of encryption and machine learning models ensures that sensitive data is protected and fraudulent activity is detected and prevented. Overall, the system design for a financial supply chain system that incorporates blockchain, supply chain, encryption, and machine learning modules is a promising solution for secure and efficient financial transactions.

6.2 Future Work

The future work for this system design involves several areas of improvement. One area of improvement is the scalability of the system. As the number of transactions increases, the system may become slower and less efficient. To address this, the system can be optimized for scalability by using sharding or other techniques to distribute the workload across multiple nodes. Another area of improvement is the integration of more advanced machine learning models. While the current system uses several machine learning algorithms, there may be more advanced models that can improve the accuracy of fraud detection. For example, deep learning models such as neural networks may be able to identify more complex patterns in the data. In addition to the existing anomaly detection model, the model can incorporate rule based models, clustering models to detect additional abnormal scenarios to detect money laundering, hawala and other illegal activities to be monitored.

References

1. Chen, B., He, D., Kumar, N., Wang, H., Choo, K.-K.R.: A blockchain-based proxy re-encryption with equality test for vehicular communication systems. IEEE Trans. Netw. Sci. Eng. **8**(3), 2048–2059 (2021)
2. Miao, Y., Liu, Z., Li, H., Choo, K.-K.R., Deng, R.H.: Privacy-preserving byzantine-robust federated learning via blockchain systems. IEEE Trans. Inf. Forensics Secur. **17**, 2848–2861 (2022)
3. Wang, M., Guo, Y., Zhang, C., Wang, C., Huang, H., Jia, X.: Medshare: a privacy-preserving medical data sharing system by using blockchain. IEEE Trans. Serv. Comput. **16**(1), 438–451 (2023)
4. Nakadai, N., Iseki, T., Hayashi, M.: Improving the security strength of Iseki's fully homomorphic encryption. In: Proceedings of the 35th International Technical Conference on Circuits/Systems, Computers and Communications (ITC-CSCC), pp. 299–304 (2020)
5. Guerar, M., Merlo, A., Migliardi, M., Palmieri, F., Verderame, L.: A fraud-resilient blockchain-based solution for invoice financing. IEEE Trans. Eng. Manag. **67**(4), 1086–1098 (2020)

6. Yang, X., Li, M., Yu, H., Wang, M., Xu, D., Sun, C.: A trusted blockchain-based traceability system for fruit and vegetable agricultural products. IEEE Access **9**, 36282–36293 (2021)
7. Mohammed, S.J., Taha, D.B.: Performance evaluation of RSA, ElGamal, and Paillier partial homomorphic encryption algorithms. In: Proceedings of the 2022 International Conference on Computer Science and Software Engineering (CSASE), pp. 89–94 (2022)
8. Azaria, A., Ekblaw, A., Vieira, T., Lippman, A.: Medrec: using blockchain for medical data access and permission management. In: 2016 2nd International Conference on Open and Big Data (OBD), pp. 25–30 (2016)
9. Dwivedi, A.D., Srivastava, G., Dhar, S., Singh, R.: A decentralized privacy-preserving healthcare blockchain for IoT. Sensors **19**(2), 326 (2019)
10. Miyachi, K., Mackey, T.K.: Hocbs: a privacy-preserving blockchain framework for healthcare data leveraging an on-chain and off-chain system design. Inf. Process. Manag. **58**(3), 102535 (2021)
11. Maslove, D.M., Klein, J., Brohman, K., Martin, P.: Using blockchain technology to manage clinical trials data: a proof-of-concept study. JMIR Med. Inform. **6**(4), e11949 (2018)
12. Ouaddah, A., Elkalam, A.A., Ouahman, A.: Fairaccess: a new blockchain-based access control framework for the internet of things. Secur. Commun. Netw. **9** (2017)
13. Jumonji, S., Sakai, K., Sun, M., Ku, W.: Privacy-preserving collaborative filtering using fully homomorphic encryption. In: Proceedings of the IEEE 38th International Conference on Data Engineering (ICDE), Los Alamitos, CA, USA, pp. 1551–1552 (2022)
14. Zhao, L., Wang, Q., Zou, Q., Zhang, Y., Chen, Y.: Privacy-preserving collaborative deep learning with unreliable participants. IEEE Trans. Inf. Forensics Secur. **15**, 1486–1500 (2020)
15. Sun, X., Zhang, P., Liu, J.K., Yu, J., Xie, W.: Private machine learning classification based on fully homomorphic encryption. IEEE Trans. Emerg. Top. Comput. **8**(2), 352–364 (2020)
16. Du, M., Chen, Q., Xiao, J., Yang, H., Ma, X.: Supply chain finance innovation using blockchain. IEEE Trans. Eng. Manag. **67**(4), 1045–1058 (2020)
17. Shahid, A., Almogren, A., Javaid, N., Al-Zahrani, F.A., Zuair, M., Alam, M.: Blockchain-based agri-food supply chain: a complete solution. IEEE Access **8**, 69230–69243 (2020)
18. Li, X., Yan, H., Cheng, Z., Sun, W., Li, H.: Protecting regression models with personalized local differential privacy. IEEE Trans. Depend. Secure Comput. **20**(2), 960–974 (2023)
19. Chen, Z., Ni, T., Zhong, H., Zhang, S., Cui, J.: Differentially private double spectrum auction with approximate social welfare maximization. IEEE Trans. Inf. Forensics Secur. **14**(11), 2805–2818 (2019)
20. Ha, T., Dano, T.K.: Investigating local differential privacy and generative adversarial network in collecting data. In: 2020 International Conference on Advanced Computing and Applications (ACOMP), pp. 140–145 (2020)
21. Bonkoungou, S., Roy, N.R., Ako, N.H.A.-E., Batra, U.: Credit card fraud detection using ml: a survey. In: 2023 International Conference on Intelligent and Innovative Technologies in Computing, Electrical and Electronics (IITCEE), pp. 732–738 (2023)
22. Mathew, J.C., Nithya, B., Vishwanatha, C.R., Shetty, P., Priya, H., Kavya, G.: An analysis on fraud detection in credit card transactions using machine learning techniques. In: 2022 Second International Conference on Artificial Intelligence and Smart Energy (ICAIS), pp. 265–272 (2022)
23. Aditi, A., Dubey, A., Mathur, A., Garg, P.: Credit card fraud detection using advanced machine learning techniques. In: 2022 Fifth International Conference on Computational Intelligence and Communication Technologies (CCICT), pp. 56–60 (2022)

Analysis of Blockchain Barriers and Enablers for Improving the Transparency of Agriculture Supply Chain

Houda Dahbi$^{(\boxtimes)}$, Abla Chaouni Benabdellah, and Amine Belhadi

BearLab, Rabat Business School, International University of Rabat, Rocade Rabat-Salé, Technopolis, Rabat, Morocco
`houda.dahbi@uir.ac.ma`

Abstract. Blockchain technology (BT) in agriculture offers transformative solutions for challenges like supply chain traceability, food safety, data security, financial inclusion, land rights, and market access for small farmers. By providing transparent, decentralized, and secure systems, it is revolutionizing the agricultural industry. This paper investigates the analysis of blockchain barriers and enablers for improving transparency in the agriculture supply chain (ASC). Subsequently, it employs the Technique for Order of Preference by Similarity to Ideal Solution (TOPSIS) to rank these enablers in order of significance for effective BT implementation in ASC. Additionally, the paper explores a sensitivity analysis to ascertain the robustness of the TOPSIS method results. This research guides decision-makers, agricultural stakeholders, and technologists in making strategic choices for improving transparency and sustainability in the African agricultural sector.

Keywords: Blockchain technology · Sustainability · Enablers · Decision making · Sensitivity analysis · Barriers

1 Introduction

The agriculture supply chain faces challenges such as climate change, resource scarcity, operational inefficiency, regulatory compliance, product quality, logistics, and consumer demands. Therefore, confronting these challenges requires using efficient methods [1]. Blockchain technology is increasingly becoming a game-changer in the agricultural supply chain, offering a new level of transparency and efficiency. As a decentralized and secure digital ledger, blockchain provides an immutable record of transactions, from farm inputs to the end consumer [2]. Blockchain technology is revolutionizing agriculture by ensuring traceability, reducing fraud, improving food safety, and streamlining processes. It enhances trust and efficiency through smart contracts and real-time data on produce conditions and logistics. Despite challenges like complexity, costs, and security, enablers like digital infrastructure development, training, and institutional support can help effectively adopt blockchain in agriculture [3]. With the agricultural sector rapidly evolving and the need for greater transparency and efficiency, the adoption of blockchain technology is becoming crucial. However, this integration raises a fundamental question: What are the major barriers of the adoption of blockchain in agriculture? What are the aim enablers to overcome these barriers?

This issue requires in-depth exploration of aspects such as technological complexity, costs, cultural and organizational resistance, as well as data security issues. At the same time, it is crucial to identify the strategies and supports needed for blockchain integration. These include digital infrastructure development, educational initiatives, policy support, and collaborations, even in the face of budgetary constraints. The successful adoption of blockchain in the agricultural supply chain requires a thorough understanding of these key elements. Previous studies have highlighted the importance of new technologies, such as BT, in strengthening the resilience of the agricultural supply chain [4, 5]. Some have discussed BT's role in building resilience strategies to ripple effects within the supply chain [6, 7]. Others have demonstrated the significant link between BT adoption and supply chain performance [4, 8]. The literature on blockchain adoption in agriculture lacks comprehensive analysis of barriers and enablers, despite increasing interest in the technology. In particular, there is a gap in the systematic evaluation and prioritization of these enablers, especially in the context of budgetary constraints. Additionally, numerous studies have examined the application of blockchain in agriculture. However, has been limited researchers on the analysis and categorization of BT enablers for ASC using TOPSIS methods. This approach is essential to guide policymakers and stakeholders in the agricultural sector towards the strategic adoption of blockchain.

The objective of this study is to clarify the challenges and opportunities associated with integrating blockchain technology in the agricultural sector. It aims to identify the main barriers, such as technological complexity, high costs, resistance to change and data security concerns that hinder its adoption. At the same time, the study concentrates on identifying key enablers that can facilitate this adoption. These include the development of digital infrastructure, educational initiatives, institutional support, and collaborations. The study applies analytical methods such as TOPSIS with the goal of developing strategies for successful blockchain integration. These strategies focus on enhancing transparency and efficiency in the agricultural supply chain, while also considering budgetary constraints. In doing so, the study aspires to provide strategic recommendations for policy makers and key stakeholders in the agricultural sector.

This paper is structured as follows: in Sect. 2, the related works are provided. The research methodology is presented in Sect. 3, while Sect. 4 is dedicated to identify enablers and barriers to BT adoption. In Sect. 5, the results and discussion of the TOPSIS method, and sensitivity analysis are provided to examine the robustness of the method applied. The implications are discussed in Sect. 6, while the conclusion is presented in the final section.

2   Related Works

Previous studies have shown that BT can mitigate disruptions in supply chains by developing more robust business models [9]. Furthermore, BT has been shown to be effective in improving recovery strategies in the event of disruptions, thereby contributing to greater supply chain resilience [10]. Recently, new studies have conducted reviews to highlight the general benefits that blockchain (BT) can offer to supply chain performance. Specifically, they focus on how BT can enhance the resilience of these supply chains [1]. Despite increasing research interest and reliance on new technologies for improving supply chain resilience, current literature shows a gap. It lacks in-depth exploration of the correlation between dynamic blockchain capabilities and agricultural supply chain resilience [3].

Therefore, studies have been undertaken to create a comprehensive approach. This approach aims to facilitate the adoption of blockchain technology in the agricultural supply chain [1]. Among these approaches, one notable method is the Multi-Criteria Decision-Making (MCDM). This technique has been employed by several researchers in scientific literature. For example, the author of [11] used AHP and Dematel for studying the main barriers to implementing BT in sustainable circular digital supply chains. The adoption of Blockchain Technology (BT) to enhance Sustainable Supply Chain Performance (SSCP) in agriculture was investigated through the application of fuzzy VIKOR (VIšekriterijumsko KOmpromisno Rangiranje), as discussed by the author in [3]. Moreover, the author [12] used SWARA (Step-wise Weight Assessment Ratio Analysis) and CoCoSo (Combined Compromise Solution) for studying Identification and evaluation of BT implementation solutions to achieve sustainability in the SC.

Recent research has improved our understanding of a variety of SC issues and decision-making processes [8]. Furthermore, a range of research studies on various topics has been conducted. These studies aim to explore the barriers to blockchain technology (BT) adoption in the agricultural supply chain [8]. Accordingly, the analysis of the literature on blockchain adoption in agriculture lacks comprehensive analysis of barriers and enablers, despite increasing interest in the technology. In particular, there is a lack in the systematic evaluation and prioritization of these enablers, especially in the context of budgetary constraints. Additionally, numerous studies have examined the application of blockchain in agriculture [3]. However, has been limited researchers on the analysis and categorization of BT enablers for ASC using TOPSIS methods.

3   Research Methodology

Recent research studies have expanded our knowledge of various supply chain challenges. They have also provided insights into the processes involved in decision-making [11]. Additionally, a multitude of studies have been conducted to examine the impact of integrating blockchain technology in agriculture supply chains. The aim is to enhance their resilience, sustainability, and digitization [3]. Figure 1 displays the comprehensive research methodology used in this study. It details the step-by-step process for assessing and examining the adoption of blockchain technology in ASC.

Phase 1: Related works. During this step, a literature analysis was explore and discuss both historical and contemporary research focusing on the barriers to adopting blockchain technology in supply chains. Keywords used included "blockchain barriers in a sustainable supply chain" and "blockchain barriers in agriculture supply chain". The research process evolved iteratively, starting with brainstorming sessions led by the researchers. As new keywords emerged in the literature, they were incorporated into the

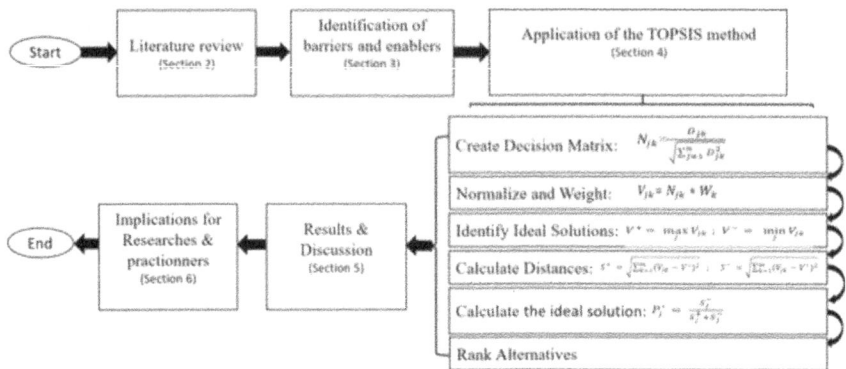

Fig. 1. Research methodology.

research process following the snowball methodology. The keywords and terms used in the search were also combined using Boolean logic (AND/OR) to expand the scope of the search.

Phase 2: Identification of barriers and enablers. The initial step involves the collection of data. It aims to classify the enablers and determine their importance in the adoption of blockchain (BT) in the agricultural sector. The model's success and the relevance of its outcomes hinge on the insights of experts from various backgrounds. Thus, four professionals from different sectors agreed to participate in this research. Additionally, three academics specializing in BT and supply chain management joined the study through direct visits and email exchanges. These seven members, each with over eight years of experience, formed a comprehensive team.

Phase 3: Application of the TOPSIS. Is a multi-step process for multi-criteria decision making. It begins with normalizing the decision matrix to make different measures comparable. Next, a positive ideal solution (the best case for each criterion) and a negative ideal solution (the worst case) are constructed. The distance of each alternative to these ideal solutions is calculated, often using Euclidean distance. Finally, a similarity score for each alternative is determined by comparing these distances.

Phase 4: Results and implications. A sensitivity analysis was carried out to test the robustness of the results. Accordingly, for implication for researchers and practitioners section consisted of offering some managerial and practical implications.

4 Enablers and Barriers Identification

Following the outcomes of the literature review, a thorough compilation of barriers for the adoption of blockchain technology in agricultural supply chains was formulated. Consequently, a set of enablers, which are instrumental in overcoming these barriers, has been discovered and identified by specialists and experts. Four distinct categories for enablers have been determine, as depicted in Fig. 2.

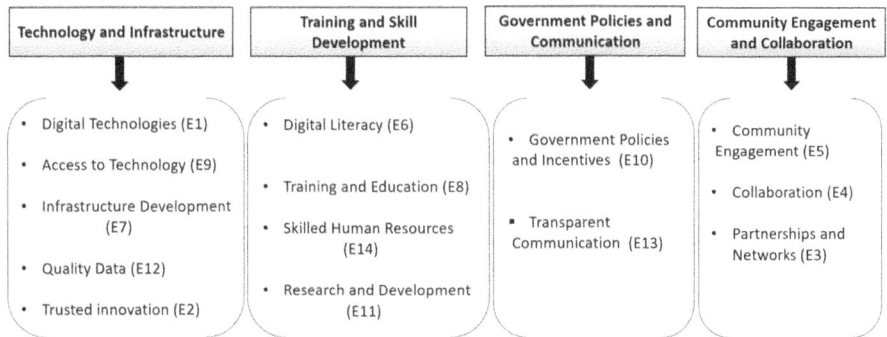

Fig. 2. Enablers to BT adoption for ASC.

- **Digital technologies (E1):** Enhance resilience and recovery in farming communities, Support the assurance of food security [13].
- **Trusted innovation (E2):** Fosters credibility, efficiency, and sustainability in agriculture [6].
- **Partnerships and networks (E3):** Foster social engagement, communication and networking [6].
- **Collaboration (E4):** Encourage participatory methods and designate mentors to sustain cooperation [3].
- **Community Engagement (E5):** Builds support and trust through local interaction, enhancing social sustainability in agriculture [1].
- **Digital Literacy (E6):** Enhance farmers' digital skills, improve data management in farming [13].
- **Infrastructure Development (E7):** Enhance internet connectivity, upgrade transportation systems [14].
- **Training and Education (E8):** Offer technology training to farmers, facilitate successful technology adoption [15].
- **Access to Technology (E9):** Provide farmers with modern technologies, include IoT devices, drones, and precision tools [4].
- **Government Policies and Incentives (E10):** Implement supportive government policies, offer subsidies and incentives [11].
- **Research and Development (E11):** Foster ongoing research and development, innovate solutions and practices [1].
- **Quality Data (E12):** Ensure accuracy of data, maintain high data quality [15]
- **Transparent Communication (E13):** Promote open communication, ensure transparency [6].
- **Skilled Human Resources (E14):** Employ qualified personnel, competently manage BT technologies, ensure effective implementation [15].

More clearly, the list of barriers to Blockchain Technology (BT) implementation in agriculture supply chain is categorized into four main groups (Fig. 3):

- **Resource Constraints and Security and Data:** Lack of infrastructure, insufficient financial resources, and data security concerns can seriously hinder the integration of blockchain into agricultural supply chains [8].

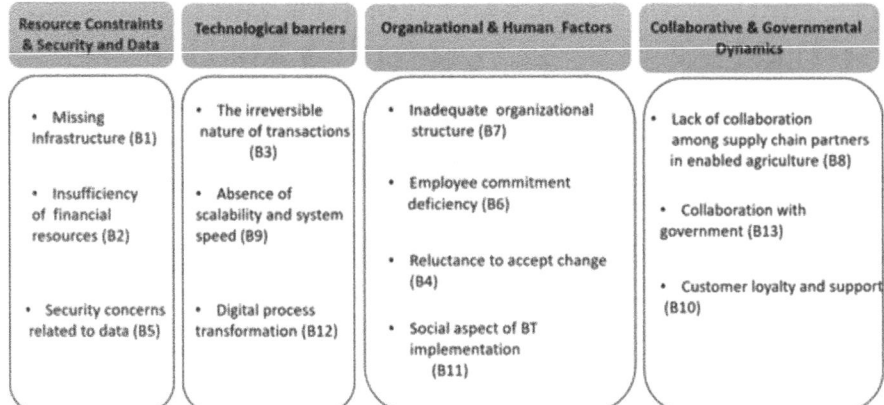

Fig. 3. Barriers to BT adoption for ASC.

- **Technological barriers:** lack of scalability and speed of systems and transformation of digital processes are technical barriers that can slow down the adoption of blockchain in agriculture [3].
- **Organizational and Human Factors:** Resistance to change, insufficient employee engagement and unsuitable organizational structures constitute major barriers to the integration of blockchain in agriculture [13].
- **Collaborative and Governmental Dynamics:** The collaboration between the application chain partners and the collaboration comes with collaborative and collaborative faction partners that influence the blockchain adoption in the agricultural sector [14].

5   Results and Discussion

The use of the TOPSIS approach for adopting blockchain technology in the African industry reveals key findings. According to Table 1, skilled human resources (E14), collaboration (E4), and research and development (E11) emerge as the most critical and influential enablers. In contrast, access to technology (E9), community engagement (E5), and infrastructure development (E7) rank as the least important enablers for blockchain adoption in the African context.

Notably, digital technologies (E1) are crucial for educating farmers, managing climate-related risks, enhancing product certification, and facilitating data-driven decision-making. Trusted innovation (E2) is identified as vital for ensuring the transformation's credibility, efficiency, and sustainability, laying a strong foundation for the integration of blockchain, which can lead to modernization, increased transparency, and growth within the agricultural sector.

Collaborative efforts (E4) are recognized for bolstering blockchain credibility by fostering a diverse stakeholder ecosystem. Community engagement (E5) is critical for gaining local support and aligning blockchain solutions with the specific needs of the farming community. Digital literacy (E6) is essential to encourage active participation in the blockchain movement.

Table 1. TOPSIS final results.

Enabler	Score	Rank
E1	0.07587	6
E2	0.07551	8
E3	0.08797	4
E4	0.84909	2
E5	0.0498	13
E6	0.08275	5
E7	0.06248	12
E8	0.06426	10
E9	0.04525	14
E10	0.07568	7
E11	0.09329	3
E12	0.06251	11
E13	0.06993	9
E14	0.8675	1

Infrastructure development (E7) is fundamental, as access to the internet and communication networks enables the practical implementation of blockchain technologies. Training (E8) is also highlighted as a key element that equips farmers to effectively utilize these technologies. Furthermore, the availability of technology (E9) and supportive government policies (E10) are instrumental in providing the necessary backing and incentives for blockchain adoption.

Research and development (E11) are cited as important for innovating solutions tailored to the unique challenges of African agriculture. Lastly, data quality (E12), transparent communication (E13), and skilled human resources (E14) are underscored as critical for ensuring the accuracy, trustworthiness, and efficient management of agricultural blockchain initiatives.

A sensitivity analysis is conducted to assess the robustness and reliability of the presented results. In this analysis, the weight assigned to each barriers is adjusted, to compute new scores for the enablers. The objective of this analysis is to examine the effectiveness of TOPSIS method by altering its input parameters and observing how these changes affect the ranking of enablers.

The Fig. 4 shows that some enablers, such as skilled human resources (E14), digital literacy (E6) and quality data (E12) exhibit considerable rank stability across all weight scenarios, suggesting they are fundamental elements within the blockchain implementation strategy that retain their significance regardless of weight fluctuations. On the other end, reliable innovation (E2), infrastructure development (E7), and access to technology (E9), among the enablers whose order varies depending on their weight. Which implies that these enablers are less important in the adoption of blockchain.

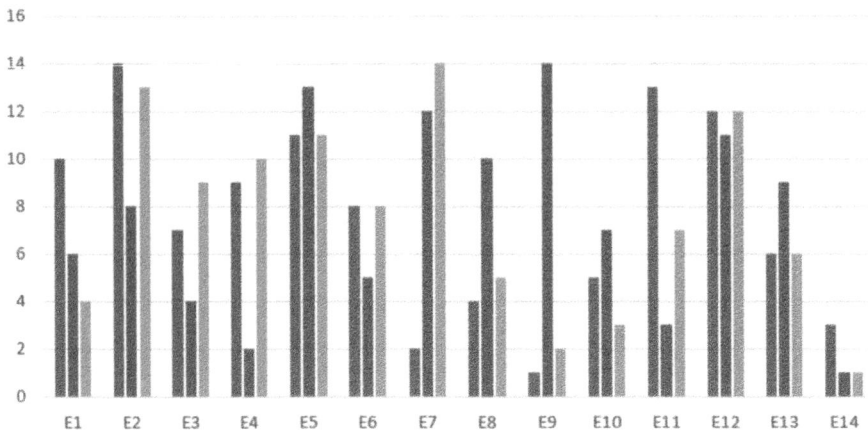

Fig. 4. Impact of criterion weights on TOPSIS final ranking results.

6   Managerial and Researchers Implications

This study provides valuable insights for decision-makers involved in blockchain Technology (BT) adoption in Africa's agricultural sector. It emphasizes the necessity for well-informed and strategic decision-making processes. It delineates the enablers of BT integration, providing a road-map for practical application. Central to the findings is the revelation that the major barriers lie within human resource capabilities and logistical processes. These challenges take precedence over the technological complexity itself. Strengthening communication channels is one aspect to consider. Additionally, ensuring the involvement of marginalized but essential participants in the agricultural supply chain, particularly smallholder farmers, emerges as a significant barrier to be addressed. For BT adoption to flourish, active and committed engagement from supply chain management is imperative. The research also sheds light on the legal gaps within African countries that could impede BT application. It points to a critical need for policy awareness and the judicious selection of technologies. Policymakers are urged to choose technologies that not only fit their national context. They should also select technologies that serve to fill existing voids, ensuring that the agricultural sector can leapfrog to a state of technological empowerment that is inclusive, effective, and sustainable.

7   Conclusion

Blockchain technology is emerging as a promising technology, offering benefits like cryptography and transparency. It holds the potential to transform many sectors, including agriculture. The investigation commenced with a thorough review of existing literature, revealing thirteen significant barriers. Following this, a diverse group of experts was consulted. They not only validated these barriers but also highlighted fourteen key enablers for the integration of BT into these supply chains. However, budgetary constraints require us to prioritize these enablers. To meet this challenge, the TOPSIS method, a rigorous approach, was used to rank these enablers according to their relative importance for the adoption of BT. So that skilled human resources (E14) occupies the first place, access to technology (E9) is in last place. A sensitivity analysis was also conducted to test the robustness of the results. This process enhanced the confidence in the reliability of the obtained rankings. It ensured that the identified key enablers retain their importance even under changing model conditions. This step is essential to ensure that our recommendation for the selective adoption of BT remains valid. It should hold true even in the presence of budget variations or other uncertainties. The careful approach underscores the importance of transparency and integration in successfully implementing blockchain technology (BT). This is essential for advancing towards more transparent and sustainable agricultural supply chains.

References

1. Bai, C., Quayson, M., Sarkis, J.: Analysis of Blockchain's enablers for improving sustainable supply chain transparency in Africa cocoa industry. J. Clean. Prod. (2022)
2. Hastig, G.M., Sodhi, M.S.: Blockchain for supply chain traceability: business requirements and critical success factors. Prod. Oper. Manag. **29**(4), 935–954 (2020)
3. Zkik, K., Belhadi, A., Rehman Khan, S.A., Kamble, S.S., Oudani, M., Touriki, F.E.: Exploration of barriers and enablers of blockchain adoption for sustainable performance: implications for e-enabled agriculture supply chains (2023)
4. Berberi, A., Beaudoin, C., McPhee, C., Guay, J., Bronson, K., Nguyen, V.M.: Enablers, barriers, and future considerations for living lab effectiveness in environmental and agricultural sustainability transitions, a review of studies evaluating living labs. Local Environ. 1–19 (2023)
5. Farooque, M., Jain, V., Zhang, A., Li, Z.: Fuzzy DEMATEL analysis of barriers to Blockchain-based life cycle assessment in China. Comput. Ind. Eng. **147**, 106684 (2020)
6. Leng, J., et al.: ManuChain: combining permissioned blockchain with a holistic optimization model as BiLevel intelligence for smart manufacturing. IEEE Trans. Syst. Man Cybern. Syst. **50**(1), 182–192 (2019)
7. Kamble, S.S., Gunasekaran, A., Sharma, R.: Modeling the blockchain enabled traceability in agriculture supply chain. Int. J. Inf. Manag. **52**, 101967 (2020)
8. Kamble, S.S., Gunasekaran, A., Subramanian, N., Ghadge, A., Belhadi, A., Venkatesh, M.: Blockchain technology's impact on supply chain integration and sustainable supply chain performance: evidence from the automotive industry. Ann. Oper. Res. **327**(1), 575–600 (2023)
9. Oudani, M., Sebbar, A., Zkik, K., El Harraki, I., Belhadi, A.: Green Blockchain based IoT for secured supply chain of hazardous materials. Comput. Ind. Eng. **175**, 108814 (2023)
10. Meafa, A.E., Benabdellah, A.C., Zekhnini, K.: Blockchain technology and smart contracts for smart sourcing process in automotive supply chain. In: International Conference on Artificial Intelligence and Industrial Applications, pp. 225–236. Springer, Cham (2023). https://doi.org/10.1007/978-3-031-43524-9_16
11. Chaouni Benabdellah, A., Zekhnini, K., Cherrafi, A., Garza-Reyes, J.A., Kumar, A., El Baz, J.: Blockchain technology for viable circular digital supplychains: an integrated approach for evaluating the implementation barriers. Benchmarking Int. J. (2023)
12. Dehshiri, S.S.H., Firoozabadi, B.: A novel four-stage integrated GIS based fuzzy SWARA approach for solar site suitability with hydrogen storage system. Energy **278**, 127927 (2023)
13. Sidibé, A., Olabisi, L.S., Touré, H.D.K., Niamba, C.A.: Barriers and enablers of the use of digital technologies for sustainable agricultural development and food security: learning from cases in Mali. Elementa Sci. Anthropocene (2021)

14. Taqui, W., El Hassani, I., Cherrafi, A., Zekhnini, K., Benabdellah, A.C.: Blockchain technology for supply chain resilience. In: 2022 14th International Colloquium of Logistics and Supply Chain Management (LOGISTIQUA), pp. 1–6. IEEE (2022)
15. Yadav, V.S., Singh, A.R.: A systematic literature review of blockchain technology in agriculture. In: Proceedings of the International Conference on Industrial Engineering and Operations Management, pp. 973–981 (2019)

Investigating Critical Success Factors in IoT-Based Marketing Systems

Hamed Nozari[1] , Hamid Reza Irani[2(✉)] , and Maryam Rahmaty[3]

[1] Department of Management, Azad University, Dubai Branch, Dubai, United Arab Emirates
[2] College of Farabi, University of Tehran, Qom, Iran
hamidrezairani@ut.ac.ir
[3] Department of Management, Chalous Branch, Islamic Azad University, Chalous, Iran

Abstract. The Internet of Things connects various types of devices in our daily lives and provides a large amount of data for marketers to improve marketing services. Marketers can use this data to make decisions, identify customer interaction patterns, and predict their behavior and lifestyle. The rise of the Internet of Things is profitable for digital marketers because they can use this huge data set to analyze consumer behavior in their marketing strategy. This allows them to predict consumer behaviors so that they can take better steps to implement their next strategy. Therefore, before implementing any project or program, marketers must correctly identify and evaluate the success factors of these smart technology-based systems. Understanding the correct position of these factors can definitely help improve the performance of the new marketing system. Therefore, the purpose of this research is to extract these critical success factors in the age of digital technologies and evaluate the factors. In this research, first by reviewing the research literature, the criteria and critical success factors (CSFs) in marketing systems based on the Internet of Things have been identified and presented to the experts after screening. In the next step, the research criteria are evaluated and weighted using the fuzzy nonlinear hierarchical analysis method, and then the prioritization of success factors is presented. This research can provide a deep insight into smart marketing systems based on the Internet of Things.

Keywords: Marketing based on Internet of Things · Internet of Things · Critical success factors · fuzzy nonlinear prioritization

1 Introduction

Any business's performance depends on its ability to retain customers over time. The increased competition between companies in finding customers for their products and services and the increase in customer power in today's competitive world demands that companies not only seek to acquire new customers but also nurture existing ones and establish solid connections with them. Today's world is filled with changes and transformations. Changes in technology, changes in information, changes in people's desires, changes in consumers, and changes in global markets are some of the changes that have occurred. Changes in values that can be offered to buyers have been one of the most important changes in the business scene. The leading organizations in every industry owe their success to their ability to supply and provide more value to buyers than their competitors [1].

Gaining a competitive advantage depends on various factors such as optimal use of resources, attracting customers with various policies, increasing productivity, etc. However, what guarantees success in this field is proper strategies that are well structured. In fact, the organization's view in all fields and sectors should be a strategic view. Sometimes, strategies are defined by the position of power, and sometimes, they arise to overcome problems. In any case, it should be studied to gain the ability to take the strategies of organizations and the reason for their importance. In the new era, transformative technologies such as the Internet of Things and artificial intelligence can add many capabilities to businesses' systems. For this reason, it is necessary to understand the critical factors of success in these systems [2].

Businesses use Internet of Things technologies to market their products and services. Connected devices, data, and insights are used to create targeted and personalized marketing strategies that increase customer engagement and improve the overall customer experience. Businesses can use the Internet of Things to create more personalized and relevant marketing strategies by providing them with valuable insights [3].

IoT generates vast amounts of data that marketers can use to gain insight into customer behavior and preferences, enabling more targeted marketing campaigns and better customer segmentation. IoT allows businesses to engage customers in real-time and in context, making marketing more relevant and effective [4]. The Internet of Things can improve the customer experience by providing customized and personalized services, products, and interactions based on data collected from connected devices. Considering these attractive features that the Internet of Things adds to modern marketing systems, it is necessary to understand the critical factors of success in these systems in order to implement them effectively. The system's Critical success factors provide an alert for management and a way to avoid surprises and missed opportunities [5].

Considering the growth of transformative technologies and their tremendous effects on business processes from procurement to marketing, the success of intelligent systems is of great importance. Therefore, the main goal of this article is to identify, extract, and quantitatively evaluate the most important CSFs for intelligent marketing based on the Internet of Things. In this research, according to the research literature, the CSFs of IoT base marketing have been identified and after screening, they have been presented to the experts. In this regard, quantitative techniques (multi-indicator decision-making techniques) have been used in order to weigh and prioritize these factors. The method used in this research is a ranking method based on hierarchical analysis and is nonlinear. On the other hand, due to the fact that the problem space in this research includes qualitative indicators and factors that are presented with expressive words, fuzzy data has been used in the quantitative technique.

2 Literature Review

2.1 IoT Based Marketing

The more we move towards the era of progress, the more professional use of Internet of Things technology we see, and it becomes a part of everyday life. The Internet of Things makes life easier and allows us to focus on the things that are important to us. The development of IoT will dominate the marketing world and allow marketers to conduct efficient marketing activities. This brings consumers and marketers close enough to satisfy each other's needs. The role of the Internet of Things in marketing is one of those very new and practical topics that cannot be ignored if you enter marketing [6].

Marketing intelligence and creative marketing is a modern marketing method that is much wider than traditional marketing. Marketing is the process of collecting information and using it to advertise the company's products or services. In this type of marketing, we creatively use smart tools in the Internet and digital world. Smart marketing combines ingenuity, ingenuity, and knowledge creation, which can be crucial for any business's success [7]. Using this method of promotion makes an organization successful and boosts customer loyalty. Companies use the data obtained from intelligent marketing to determine the market opportunity and formulate marketing campaigns to optimize the presence, brand awareness, and sales numbers of businesses. This is vital to understanding the market situation [8] better.

Business has been affected by the Internet of Things, which has caused change and transformation. A new generation of marketing under the title of smart marketing has been created by the rapid development of communication and mobile computing, providing a better context and opportunity for this change and transformation. Various advantages for marketers have been included in smart interactive marketing based on the Internet of Things. With the existence of social networks, users' experiences regarding the use of a product have become very important, and these experiences are integrated with the information obtained from searches made through Internet search engines. The new inferred information can also be passed on to other users through tools like group media and social networks, influencing their decision to purchase a product [9].

Smart marketing with the Internet of Things approach can have two important benefits. First, it provides the possibility of consulting and using the experiences of customers, and secondly, it provides the possibility of confirming the facilities and features that product express in relation to their supplied goods through the possibility of wide communication that is possible through the Internet of Things. Access to a wide range of information from different points is obtained for marketers with the Internet of Things. From a digital marketing perspective, this is very useful for creating a better understanding of the starting point of the customer journey from start to finish [10].

Marketers can see the customer as they move through the buying process with the help of IoT tech, generating more avenues for uplifting client interactions. With increased access to customer journey information, marketers will be able to find new ways to connect with them to answer questions and build brand relationships. For marketers, access to information about customer lifestyles means they can see how, when, and what certain products or services are used for. Access to information about customer lifestyles means they can see how, when, and what certain products or services are used for [11].

IoT can be integrated with marketing automation tools to streamline marketing processes, manage campaigns, and analyze data more efficiently. Of course, the Internet of Things also raises concerns about privacy and data security, which marketers must address to build trust and maintain regulatory compliance. In order to adapt marketing strategies by the Internet of Things, it is very important to show consumers a high level of security and to ensure that their analytical information is protected. As the way data is collected in digital marketing changes as technology advances and improves, the way data is used by marketers must also change.

In recent years, much research has been conducted in the field of using the Internet of Things in marketing. Nozari et al. [12] in a study investigated the sustainability indicators in marketing systems based on the Internet of Everything. In a study, Najafi et al. [13] evaluated and prioritized sustainability indicators in Internet of Things-based marketing systems using a decision-making method. Liu et al. [14] presented a data mining framework to analyze marketing data with an emphasis on the Internet of Things, artificial intelligence, and cloud computing technologies. This framework provides detailed causal relationships for analyzing marketing data. Singh [15], in applied research, investigated the applications, opportunities, and challenges of using the Internet of Things and machine learning in digital marketing. Zadorozhnyi et al. [16] examined the Application of the Internet of Things and 6G Cellular Communication to Optimize Accounting and International Marketing. Khan [17] examined digital marketing based on the Internet of Things in small and medium-sized companies and outlined the implementation dimensions in detail. They depicted the cause-and-effect relationships of all the effective actors in smart marketing. Bakhshi Movhed et al. [18] examined the opportunities and challenges facing smart marketing based on smart technologies in Industry 5.0. The conceptual model of dimensions and components of marketing 5.0 was presented in this research. Dutt [19] examines the potential benefits of Marketing 5.0, including increased customer engagement, loyalty, and patronage, as well as improved business performance and long-term sustainability. The research also considers challenges and gaps that need to be addressed, such as the need for new skills and capabilities, the difficulty of integrating marketing and sales efforts, and the ethical and privacy concerns associated with data-driven marketing. Overall, this article highlights the potential of Marketing 5.0 to transform marketing practice and create more meaningful and engaging experiences for customers, while highlighting the challenges that must be overcome to realize its full potential.

Therefore, in general, it can be stated that identifying the dimensions, components, and critical success factors of smart interactive marketing systems based on the Internet of Things (and combining this technology with other transformative technologies) can be an effective guide for the powerful implementation of these systems. Figure 1 shows a framework for intelligent marketing systems based on the Internet of Things and Blockchain.

As seen in Fig. 1, smart marketing shapes its interactions in specific domains using smart technologies. In this marketing, the Internet of Things technology and its combination with technologies such as artificial intelligence and blockchain form the core of the framework. The most important feature of the Internet of Things in this framework is that it extracts performance-based big data. Most of these data are semi-structured or

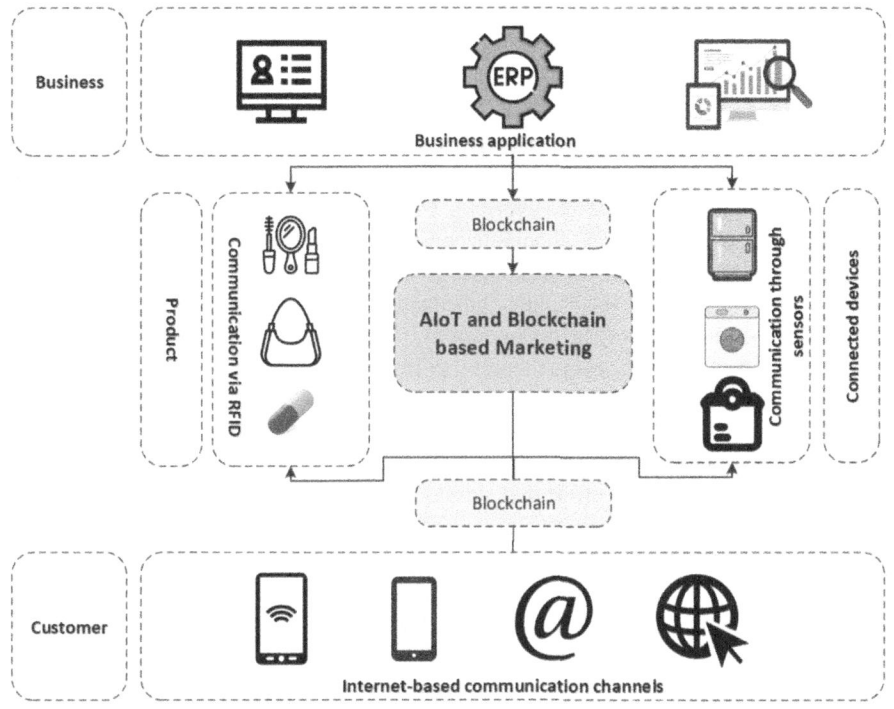

Fig. 1. Smart marketing systems based on the IoT and Blockchain [20]

unstructured and are extracted based on audience performance and using in-house tools. After extracting these data, which are very accurate and transparent, these data are used to analyze the situation. To analyze clear and accurate data, you can use the capabilities of artificial intelligence and various learning systems. In this smart system, the powerful blockchain technology guarantees the accuracy and transparency of the data.

2.2 Critical Indicators of Success in IoT-Based Marketing

The operational goals of a project or organization are achieved by a set of factors called CSFs. The implementation of an organizational project involves many factors and elements. There are some factors that are very important.

Research and studies conducted on successful companies show that certain factors are involved in the success of each project. It is necessary to pay attention to this category that the critical success factors are based on other core competencies. Different researchers declare the critical success factors of an organization. CSFs are a few factors that play a vital role in the success of the organization. To continue functioning, the organization must provide them. Each critical factor is an area that needs to be done in the best possible way for the organization to succeed. On the one hand, these factors are related to the goals of the organization and are necessary to achieve the goals of the organization. On the other hand, they fit the organization's competitive strategy. These factors are the basic requirements that must be achieved as intermediate goals to achieve the main goal [21].

The increased competition between companies in finding customers for their products and services, as well as the increase in customer power in today's competitive world, demands that companies not only seek to acquire new customers but also nurture existing ones and establish solid connections with them. It is also worth considering. The world of today is constantly evolving and evolving. Change in technology, change in information, change in people's desires, change in consumers, and change in global markets are some of the changes that have occurred. However, one of the most important changes in the business scene has been the change in the values that can be offered to buyers, which is known as the main success factor in current organizations. Leading organizations in every industry owe their success to their ability to deliver more value to buyers than their competitors [22].

Identifying the vital success factors for any organization or project is necessary and necessary to achieve its goals. The vital success factors of the guiding rules are defined in order to apply the effort that should be used in the direction of success or an important part of the organization's strategic plan that should be achieved in addition to the organization's goals.

During the last two decades, the use of critical success factors in the fields of management, especially strategic management, operational planning management, resource allocation, and performance evaluation has also expanded. In this research, using the literature review, the most important critical factors of success in marketing systems based on the Internet of Things were investigated and extracted. Then, the concept of refinement was introduced using the opinions of experts. For this purpose, 8 experts with relevant work and research records were used. These critical success factors are shown in Table 1.

Table 1. Critical success factors in the marketing system based on the Internet of Things.

Category	CSF
Strategic factors W1	*Multiple distribution channels* W11
	Research and development W12
	Market share growth W13
Technological factors W2	*Development of technology infrastructure* W21
	Technical skills of employees W22

- **Multiple distribution channels**: The implementation of the Internet of Things brings more integration of distribution channels in marketing systems and can lead to the development and growth of sales, but in this case, it is important that companies seek to ensure appropriate technology and data privacy protection [23].

- **Research and development**: Big data has daily implications for consumers, businesses, and market researchers. Data usage plays an extensive role in market research surveys, as does data processing and analysis. With the digitization of market research in data collection and analysis, traditional methods are no longer sufficient. Therefore, IoT helps market researchers to know consumer habits and behavior [24].
- **Market share growth**: Enterprises and service providers are seeing IoT as a key enabler to drive digital transformation and unlock operational efficiencies. The growing adoption of IoT technology in end consumer industries is positively driving market growth [25].
- **Development of technology infrastructure**: The growth and development of technological infrastructure such as advanced tools and sensors and professional Internet-based communication is considered to be an effective way to absorb and refine large customer data and accurate market information [26].
- **Technical skills of employees**: Specialist employees in the field of marketing who have high ability and expertise in the field of Internet-based information and communication technology can increasingly increase the operational capability of marketing systems [27].

3   Research Methodology

CSFs in Internet of Things-based marketing systems are investigated in this research with a quantitative approach. The background of the subject and expert opinions were used to investigate the factors in this study, and five factors were selected as the most significant ones based on the five-level Likert spectrum. The eight active professionals were surveyed for this purpose. The reason for choosing these people was the access to authors and the work and research records of experts in the field. There were eight experts who had related work histories in marketing and were familiar with concepts of information technology and intelligence. Due to the access of the authors, these specialists were selected from among the manufacturing companies of the food and pharmaceutical industries. There were three experts who had scientific records and professional advice related to the field of study. The most important critical success factors in Internet of Things-based marketing systems were prioritized using a fuzzy hierarchical analysis technique to prioritize the most important critical success factors. The reliability of related questionnaires has been assessed using SPSS software using Cronbach's alpha method. All questionnaires had an alpha value of 0.894, which is considered to be favorable. A separate factor analysis was performed for each of the indicators because five factors were selected as the most significant in this investigation. The quantitative analysis method used in this research is explained in the following.

The most important CSFs in Internet of Things-based marketing systems have been prioritized using a fuzzy nonlinear hierarchical analysis method, known as the Mikhailov method [28] The steps to use this method are listed below.

- Create the decision tree shown in Table 1.
- The matrix of fuzzy pairwise comparisons is formed based on the opinions of experts. The preferences of experts have been expressed using fuzzy numbers.

Prioritizing requires the following actions.

Formulating the model and resolving the issue: This approach assumes that fuzzy pairwise comparisons take the form of triangular fuzzy numbers. The priority rate is almost within the range of the basic fuzzy judgments when the deterministic weight vector is extracted. The following relationship is maintained by the weights being determined in such a way.

$$l_{ij} \leq \frac{w_i}{w_j} \leq u_{ij} \tag{1}$$

The inclusions in the above fuzzy inequalities can be measured through the linear membership function of the following relation, which has a definite weight vector w.

$$\mu_{ij}\left(\frac{w_i}{w_j}\right) = \begin{cases} \frac{(w_i/w_j)-l_{ij}}{m_{ij}-l_{ij}} & \frac{w_i}{w_j} \leq m_{ij} \\ \frac{u_{ij}-(w_i/w_j)}{u_{ij}-m_{ij}} & \frac{w_i}{w_j} \geq m_{ij} \end{cases} \tag{2}$$

The fuzzy prioritization problem becomes a nonlinear optimization problem when we consider the specific form of the membership functions.

$$\begin{aligned}
& \max \lambda \\
& Subject\ to: \\
& (m_{ij} - l_{ij})\lambda w_j - w_i + l_{ij}w_j \leq 0 \\
& (u_{ij} - m_{ij})\lambda w_j + w_j - u_{ij}w_j \leq 0 \\
& i = 1, 2, ..., n - 1, j = 2, 3, ..., n, j > i \\
& \sum_{k=1}^{n} w_k = 1, w_k > 0, k = 1, 2, ..., n
\end{aligned} \tag{3}$$

The non-linearity of the relationship makes it impossible to solve it without using software. The models created in this investigation were solved using the GAMS software. It can be understood that the fuzzy judgments are strongly inconsistent and the weight ratios are almost applied in the initial judgment if the index is positive.

4 Research Findings

In order to emphasize and develop all-round critical success factors in marketing systems based on the Internet of Things, the ranking process is divided into two main parts.

- The matrix of pairwise comparisons is determined based on the integration of expert's opinions.
- Calculating to rank and determine the weight of the variables in the investigation model.

Eight experts and university professors were sent fuzzy questionnaires to prioritize the 5 final factors extracted from this research. Eight questionnaires were answered and sent out. Fuzzy questionnaires are based on linguistic criteria and averaged after receiving from experts. The questionnaires are averaged after receiving from experts. Tables 2, 3 and 4 show pairwise comparison tables of these pairwise comparison tables. The Mikhailov method was used to calculate these tables.

Table 2: Pairwise comparison matrix for overall classification.

	W1			W2		
W1	1.25	2.5	4.1	2.1	3.25	5
W2	2.1	3.1	3.8	1.75	2.75	3.5

Table 3: Pairwise comparison matrix for strategic factors.

	W11			W12			W13		
W11	1.25	2.7	3.8	1.2	2.8	4.1	2.1	3.75	4.5
W12	1.75	2.5	4.1	1.7	2.5	3.5	1.3	2.1	4.2
W13	2.1	3.5	4.8	2.3	3.8	4.7	1.5	2.5	4.2

Table 4: Pairwise comparison matrix for Technological factors.

	W21			W22		
W21	2.1	3.25	4	3.1	3.5	3.8
W22	1.5	2.1	2.8	2	2.5	3

The weight and rank of each CSFs can be obtained by placing the data obtained from Tables 2, 3 and 4 into a non-linear model (3) and solving it using GAMS software. Tables 5, 6 and 7 show the outcomes of calculations involving the nonlinear model for general categories and individual variables.

Table 5. Weight and ranking of the main categories.

Category	code	Weight	rank	objective function (λ)
Strategic factors	W1	0.4100214	2	0.4521
Technological factors	W2	0.5892142	1	

Table 6. Weight and ranking of critical success factors in the strategic category.

Category	code	Weight	rank	objective function (λ)
Multiple distribution channels	W11	0.2612145	2	0.3652
Research and development	W12	0.4821521	1	
Market share growth	W13	0.2521458	3	

Table 7. Weight and ranking of critical success factors in the technological category.

Category	code	Weight	rank	objective function (λ)
Development of technology infrastructure	W21	0.6501251	1	0.4215
Technical skills of employees	W22	0.3395214	2	

As shown in the Tables 5, 6 and 7, a positive value for the compatibility index indicates an acceptable compatibility of the matrices. As can be seen from the Tables 5, 6 and 7, a positive value for the compatibility index indicates an acceptable compatibility of the weights in the general categories and the factors in the specific categories can be normalized to get the total weight. Obtain all the factors regardless of the category and their overall rank. Table 8 shows the results of the normalized calculations (Fig. 2).

Table 8. Ranking of CSFs in marketing systems based on the Internet of Things.

Category	Weight	CSF	Weight	Normalized weight	rank
Strategic factors	0.4100214	Multiple distribution channels	0.2612145	0.107103535	4
		Research and development	0.4821521	0.197692679	3
		Market share growth	0.2521458	0.103385174	5
Technological factors	0.5892142	Development of technology infrastructure	0.6501251	0.383062941	1
		Technical skills of employees	0.3395214	0.20005083	2

As can be seen in Table 8, the development of technological infrastructure is one of the most important CSFs in marketing systems based on the Internet of Things.

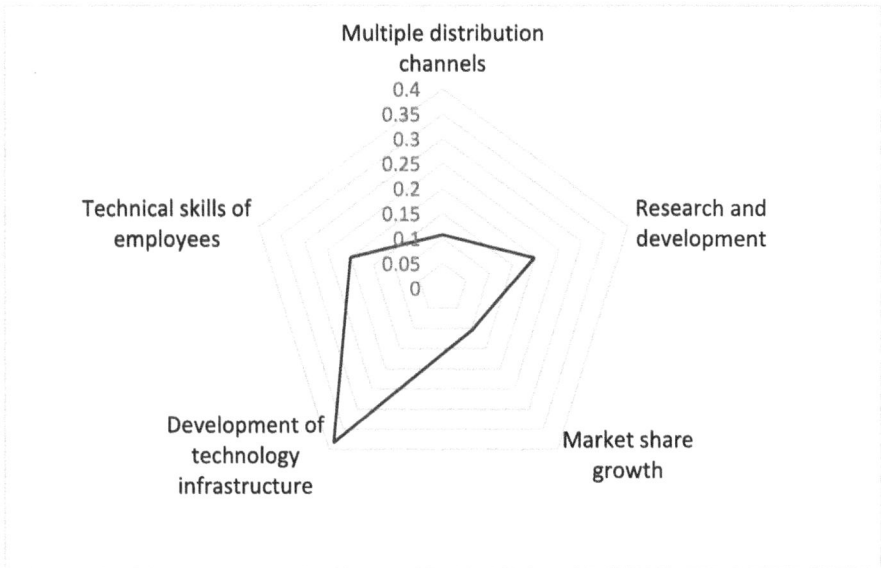

Fig. 2. Normal weight of critical success factors in marketing systems based on the Internet of Things.

5   Conclusion

In today's competitive era, all organizations are affected by challenges such as changing expectations and new demands of customers, improving quality, job accidents, expanding competition, and extensive economic, social, cultural, political, and technological developments. Therefore, they have to choose solutions that they can use to continue their lives. One of the most important components and elements of the process of achieving success is CSFs. Critical success factors are factors that are related to the organization's goals on the one hand and are necessary to achieve the organization's goals, and on the other hand, are proportional to the organization's competitive strategy.

The nature and size of these critical capabilities form the basis of the organization's internal competitive advantages. Based on the research, CSFs refer to limited areas in any project or business that, if they lead to the desired results, will guarantee a suitable competitive advantage and efficiency for the organization. Since sales and marketing is one of the most important organizational units in all production and service organizations. Therefore, it is very important to understand the critical success factors in this sector. Considering the fundamental changes that have been made in the digital age in modern marketing systems, it is necessary to understand the effects of these technologies such as Internet of Things technologies and artificial intelligence and blockchain on marketing systems. Therefore, the critical factors of success in marketing systems based on transformative technologies should be examined in today's era.

For this reason, in this research, in addition to examining the dimensions and components of these interactive intelligent systems, the most important critical success factors in these systems were also examined. In order to evaluate and prioritize factors, a fuzzy nonlinear decision-making method based on hierarchical analysis was used. For the purpose of analysis in this research, pairwise comparison tables with linguistic variables have been used, which were completed and presented by experts. In this research, the opinions of eight experts have been used, five of whom had relevant job records in marketing in the food industry, and three of whom were university experts. Numerical analysis and prioritization showed that the development of technological infrastructure is one of the most CSFs in marketing systems based on the Internet of Things, and special attention should be paid to it.

This research can create an effective insight for the success of intelligent marketing systems. But one of the limitations of this research is the lack of sufficient experts for quantitative analysis. Lack of familiarity with fuzzy concepts and lack of sufficient knowledge in the field of intelligent interactive retrieval has been one of the basic limitations of this research. But due to the ever-increasing growth of transformative technologies, it seems that more and permanent future research in this field is necessary. Some researches can analyze the success of smart marketing with emphasis on security and privacy parameters and criteria, which itself is a prelude to future industrial revolutions.

References

1. Nguyen, B., Simkin, L.: The Internet of Things (IoT) and marketing: the state of play, future trends and the implications for marketing. J. Mark. Manag. **33**(1–2), 1–6 (2017)
2. Lo, F.Y., Campos, N.: Blending Internet-of-Things (IoT) solutions into relationship marketing strategies. Technol. Forecast. Soc. Chang. **137**, 10–18 (2018)
3. Simões, D., Filipe, S., Barbosa, B.: An overview on IoT and its impact on marketing. In: Smart Marketing with the Internet of Things, pp. 1–20 (2019)
4. Nozari, H., Rahmaty, M., Szmelter-Jarosz, A.: A framework for AIoT-based smart sustainable marketing system. In: Artificial Intelligence of Things for Achieving Sustainable Development Goals, pp. 255–271. Springer, Cham (2024)
5. Liu, Y., Alzahrani, I.R., Jaleel, R.A., Al Sulaie, S.: An efficient smart data mining framework based cloud internet of things for developing artificial intelligence of marketing information analysis. Inf. Process. Manage. **60**(1), 103121 (2023)
6. Nozari, H., Ghahremani-Nahr, J., Szmelter-Jarosz, A.: AI and machine learning for real-world problems. Advances in Computers (online first) (2023)
7. Ghahremani-Nahr, J., Nozari, H., Rahmaty, M., Zeraati Foukolaei, P., Sherejsharifi, A.: Development of a novel fuzzy hierarchical location-routing optimization model considering reliability. Logistics **7**(3), 64 (2023)
8. Nozari, H., Tavakkoli-Moghaddam, R., Rohaninejad, M., Hanzalek, Z.: Artificial intelligence of things (AIoT) strategies for a smart sustainable-resilient supply chain. In: IFIP International Conference on Advances in Production Management Systems, pp. 805–816. Springer, Cham (2023)
9. Sturm, M., Weking, J., Böhm, M., Schreieck, M., Krcmar, H.: How two leading partners learn to tango: The case of IoT-based business model co-innovation between a retailer and an electronics supplier. Electron. Mark. **33**(1), 34 (2023)
10. Riyaz, R., Darzi, M.A., Riyaz, R., Bakshi, Y.S.: Internet of things in marketing: applications and concerns. In: Global Applications of the Internet of Things in Digital Marketing, pp. 349–361. IGI Global (2023)
11. Uma, S.: Abridging the digital marketing gap: artificial intelligence (AI) and internet of things (IoT) in boosting global economic growth. In: Global Applications of the Internet of Things in Digital Marketing, pp. 187–212. IGI Global (2023)

12. Nozari, H., Szmelter-Jarosz, A., Ghahremani-Nahr, J.: The ideas of sustainable and green marketing based on the internet of everything—the case of the dairy industry. Future Internet **13**(10), 266 (2021)

13. Najafi, S.E., Nozari, H., Edalatpanah, S.A.: Investigating the key parameters affecting sustainable IoT-based marketing. In: Computational Intelligence Methodologies Applied to Sustainable Development Goals, pp. 51–61. Springer, Cham (2022)

14. Liu, Y., Alzahrani, I.R., Jaleel, R.A., Al Sulaie, S.: An efficient smart data mining framework based cloud internet of things for developing artificial intelligence of marketing information analysis. Inf. Process. Manag. **60**(1), 103121 (2023)

15. Singh, A.K.: Applications of the internet of things and machine learning using python in digital marketing. In: Global Applications of the Internet of Things in Digital Marketing, pp. 213–232. IGI Global (2023)

16. Zadorozhnyi, Z.M., Muravskyi, V., Pochynok, N., Ivasechko, U.: Application of the internet of things and 6G cellular communication to optimize accounting and international marketing. Virtual Economics **6**(1), 38–56 (2023)

17. Khan, S.A.: E-marketing, E-commerce, E-business, and internet of things: an overview of terms in the context of small and medium enterprises (SMEs). In: Global Applications of the Internet of Things in Digital Marketing, pp. 332–348 (2023)

18. Movahed, A.B., Movahed, A.B., Nozari, H.: Opportunities and challenges of marketing 5.0. In: Smart and Sustainable Interactive Marketing, pp. 1–21 (2024)

19. Dutt, V.: Marketing 5.0: the era of technology and the challenges faced by it. Int. J. Adv. Eng. Manag. 1397–1411 (2023)

20. Aliahmadi, A., Nozari, H., Ghahremani-Nahr, J.: A framework for IoT and blockchain based on marketing systems with an emphasis on big data analysis. Int. J. Innov. Mark. Elements **2**(1), 25–34 (2022)

21. Kumar, A., Mani, V., Jain, V., Gupta, H., Venkatesh, V.G.: Managing healthcare supply chain through artificial intelligence (AI): a study of critical success factors. Comput. Ind. Eng. **175**, 108815 (2023)

22. Al-Amin, M., Hossain, T., Islam, J., Biwas, S.K.: History, features, challenges, and critical success factors of enterprise resource planning (ERP) in the era of industry 4.0. Eur. Sci. J. ESJ **19**(6), 31 (2023)

23. Alawag, A.M., et al.: Critical success factors influencing total quality management in industrialised building system: a case of Malaysian construction industry. Ain Shams Eng. J. **14**(2), 101877 (2023)

24. Hussain, M., Rasool, S.F., Xuetong, W., Asghar, M.Z., Alalshiekh, A.S.A.: Investigating the nexus between critical success factors, supportive leadership, and entrepreneurial success: evidence from the renewable energy projects. Environ. Sci. Pollut. Res. **30**(17), 49255–49269 (2023)

25. Naveed, Q.N., Qahmash, A.I., Qureshi, M.R.N., Ahmad, N., Abdul Rasheed, M.A., Akhtaruzzaman, M.: Analyzing critical success factors for sustainable cloud-based mobile learning (CBML) in crisp and fuzzy environment. Sustainability **15**(2), 1017 (2023)

26. Iqbal, M., Ma, J., Ahmad, N., Ullah, Z., Hassan, A.: Energy-Efficient supply chains in construction industry: an analysis of critical success factors using ISM-MICMAC approach. Int. J. Green Energy **20**(3), 265–283 (2023)

27. Su, M., Woo, S.H., Chen, X., Park, K.S.: Identifying critical success factors for the agri-food cold chain's sustainable development: When the strategy system comes into play. Bus. Strateg. Environ. **32**(1), 444–461 (2023)

28. Nozari, H., Najafi, E., Fallah, M., Hosseinzadeh Lotfi, F.: Quantitative analysis of key performance indicators of green supply chain in FMCG industries using non-linear fuzzy method. Mathematics **7**(11), 1020 (2019)

Author Index

A. Mirzazadeh et al. (Eds.): ODSIE 2023, CCIS 2205, pp. 273–274, 2025.
https://doi.org/10.1007/978-3-031-81458-7

The manufacturer's authorised representative in the EU is Springer
Nature Customer Service Centre GmbH, Europaplatz 3, 69115 Heidelberg,
Germany. If you have any concerns regarding our products, please
contact ProductSafety@springernature.com

Printed and bound by CPI Group (UK) Ltd, Croydon, CR0 4YY
24/04/2026
02096375-0007